Sabine Begall
Hynek Burda
Cristian E. Schleich (Eds.)
Subterranean Rodents

AF821597

Sabine Begall
Hynek Burda
Cristian E. Schleich
(Editors)

Subterranean Rodents

News from Underground

With 52 Figures, 2 in Color

Springer

Dr. Sabine Begall
Dept. General Zoology
Institute of Biology, Faculty of Bio- and Geosciences
University of Duisburg-Essen
Universitätsstr. 5
45117 Essen
Germany

E-mail: sabine.begall@uni-due.de

Professor Dr. Hynek Burda
Dept. General Zoology
Institute of Biology, Faculty of Bio- and Geosciences
University of Duisburg-Essen
Universitätsstr. 5
45117 Essen
Germany

E-mail: hynek.burda@uni-due.de

Dr. Cristian E. Schleich
Laboratorio de Ecofisiología
Departamento de Biología
Facultad de Ciencias Exactas y Naturales
Universidad Nacional de Mar del Plata
Casillo de correo 1245, Funes 3250
7600 Mar del Plata
Argentina

E-mail: cschleic@mdp.edu.ar

ISBN 978-3-642-08876-6 e-ISBN 978-3-540-69276-8

This work is subject to copyright. All rights reserved, whether the whole or part of the material is concerned, specifically the rights of translation, reprinting, reuse of illustrations, recitation, broadcasting, reproduction on microfilm or in any other way, and storage in data banks. Duplication of this publication or parts thereof is permitted only under the provisions of the German Copyright Law of September 9, 1965, in its current version, and permission for use must always be obtained from Springer. Violations are liable for prosecution under the German Copyright Law.

Springer is a part of Springer Science + Business Media
springer.com

© Springer-Verlag Berlin Heidelberg 2007
Softcover reprint of the hardcover 1st edition 2007

The use of general descriptive names, registered names, trademarks, etc. in this publication does not imply, even in the absence of a specific statement, that such names are exempt from the relevant protective laws and regulations and therefore free for general use.

Editor: Dr. Dieter Czeschlik, Heidelberg, Germany
Desk Editor: Dr. Jutta Lindenborn, Heidelberg, Germany
Cover design: WMXDesign GmbH, Heidelberg, Germany

We dedicate this book to our friend
and colleague Mathias Kawalika
(*17.07.1962; †19.03.2006)

Main speakers and some participants of the symposium "Light from underground: challenges and insights in the third decade of research on subterranean rodents", held at the International Mammalogical Congress IX in Sapporo, Japan (August 2005)

Last row (from left to right): Cristian Schleich, Ana Paula Cutrera, Simone Lange, Marie-Therese Bappert, Philip Dammann

Middle row: Ema Knotková, Regina Moritz, Sabine Begall, Yanming Zhang, Mathias Kawalika, Hynek Burda, František Sedláček

First row: Sanae Nanbu, Chizuko Suzuki, Shizuka Tokin, Paul Van Daele, Eileen Lacey, Mikhail Moshkin

Front: Eviatar Nevo

Contents

Introduction

Part I Ecophysiology

Part II Sensory Ecology

Part V Molecular Ecology and Evolution

Introduction

1 Subterranean Rodents: News from Underground

Sabine Begall, Hynek Burda, Cristian E. Schleich

Across the globe, in all continents but Australia and Antarctica, at least 250 extant rodent species (38 genera, 6 families – according to the classification applied) spend most of their lives in self-constructed burrows (Table 1.1, Fig. 1.1). Their subterranean ecotope is dark, microclimatically stable, hypoxic and hypercapnic, and deprived of most sensory cues available aboveground. The burrows offer shelter from predators and climatic extremes, but digging is energetically costly, and the yield of foraging is relatively low, because the productivity of the subterranean ecotope is rather low and the food resources (roots and underground plant storage organs like bulbs and tubers) are mostly unpredictably and unevenly scattered. These so-called subterranean rodents are specialized in multiple aspects for their unique way of life in which most events like foraging, mating, and breeding take place underground. Animals that inhabit underground self-made tunnels, but also forage (predominantly) above ground, are called fossorial. Needless to say, a continuum exists between fossorial and subterranean rodents, and in the present volume a categorical differentiation is mostly ignored on purpose. Another mammalian group sharing the same ecotope, but feeding on invertebrates, are subterranean non-rodent mammals like marsupial moles, certain armadillos, as well as "insectivore" moles (e.g. Talpidae) and golden moles (Chrysochloridae). Although the title of the book emphasizes that the focus is on subterranean rodents, we also encouraged the authors to glance at recent findings from studies on other subterranean mammals, and we hope that the reader will profit from this.

Most subterranean rodent taxa have already been scientifically described in the golden ages of alpha taxonomy 100–200 years ago (and, e.g. the blind mole-rat *Spalax* and the mole-vole *Ellobius talpinus* in 1770), and many subterranean dwellers have been familiar to local people for ages (*Spalax*

Sabine Begall: Dept. General Zoology, University of Duisburg-Essen, D-45117 Essen, Germany, E-mail: sabine.begall@uni-due.de

Hynek Burda: Dept. General Zoology, University of Duisburg-Essen, D-45117 Essen, Germany

Cristian E. Schleich: Departamento de Biología, Facultad de Ciencias Exactas y Naturales, CC 1245, Universidad Nacional de Mar del Plata, Funes 3250, Mar del Plata (7600), Argentina

Subterranean Rodents: News from Underground
S. Begall, H. Burda, C.E. Schleich (Eds.)
© Springer-Verlag Berlin Heidelberg 2007

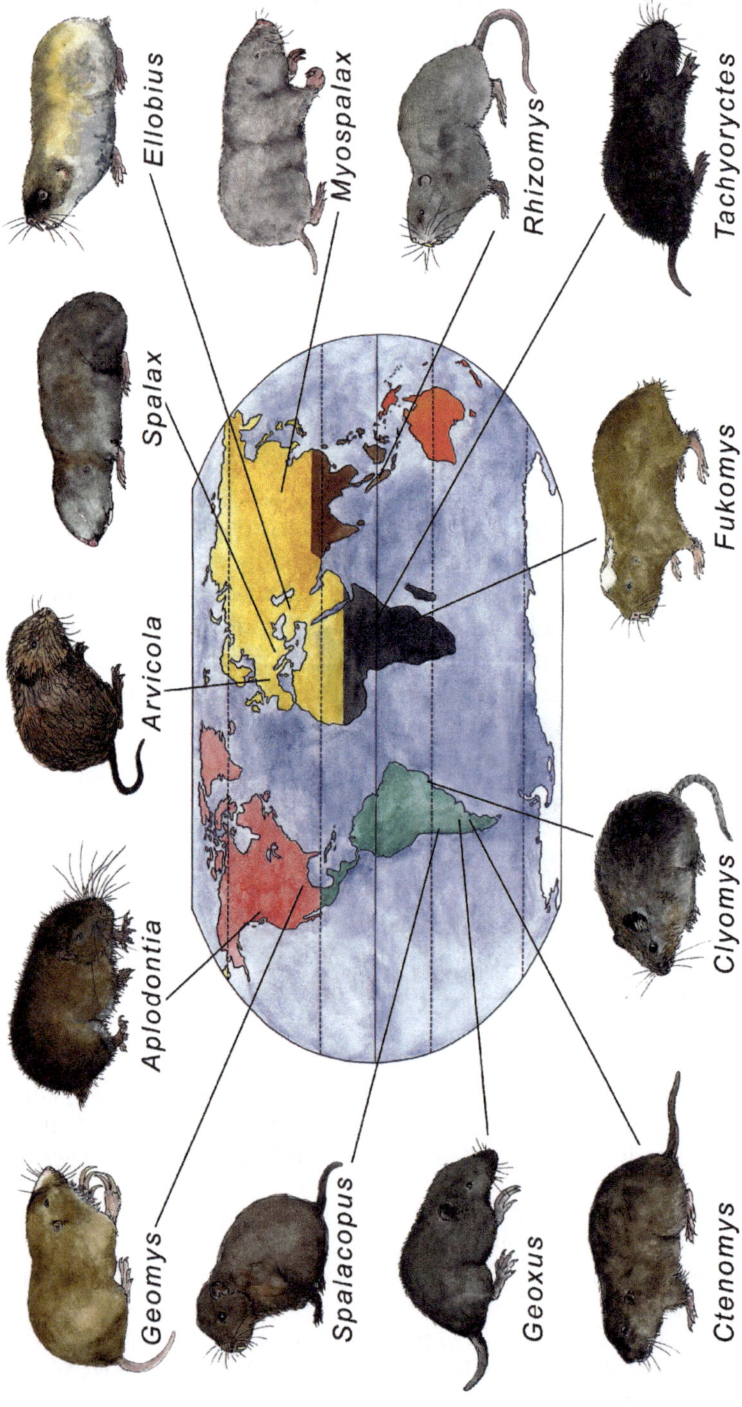

Ellobius
Myospalax
Rhizomys
Tachyoryctes
Spalax
Fukomys
Arvicola
Clyomys
Aplodontia
Geomys
Spalacopus
Geoxus
Ctenomys

Fig. 1.1. Distribution of subterranean rodents across the planet. Different map colours mark the different zoogeographical regions. Only selected genera representing diverse families and subfamilies of subterranean rodents are depicted: *Geomys* (Geomyidae), *Aplodontia* (Aplodontidae), *Arvicola* (Muridae – Arvicolinae), *Spalax* (Muridae – Spalacinae), *Ellobius* (Muridae – Arvicolinae), *Myospalax* (Muridae – Myospalacinae), *Rhizomys* and *Tachyoryctes* (Muridae -Rhizomyinae), *Fukomys* (Bathyergidae), *Clyomys* (Echimyidae), *Ctenomys* (Octodontidae), *Geoxus* (Muridae – Sigmodontinae), *Spalacopus* (Octodontidae). The actual differences in body sizes were not considered. Classification adapted from McKenna and Bell (1997) (Table 1.1). Drawings by Marie-Therese Bappert

having been mentioned in the Bible and by Aristotle). Nevertheless, scientists mostly knew the different species only from museum specimens, and the biology of subterranean mammals has long remained unstudied, which can be partly assigned to their cryptic life-style and technical problems with capturing, keeping, breeding and monitoring underground dwellers. Furthermore, scientists have always been more attracted by animals coping with complicated environments and solving seemingly difficult and complex problems compared to those encountered underground (sensitive vision vs blindness, echolocation in a high frequency range vs hearing in a human auditory range, navigating across hundreds or thousands of kilometers vs maze orientation across tens of meters, thermoregulation in cold environments vs life in a thermally buffered burrow, etc.). However, since the seminal review article by Eviatar Nevo in 1979 (Nevo 1979) and the report of eusociality in the naked mole-rat by Jennifer U.M. Jarvis in 1981 (Jarvis 1981), interest in subterranean mammals in general and in subterranean rodents in particular has been awakened. Admittedly, at the beginning of the 1980s the knowledge was strongly biased, because the majority of papers dealt with Eurasian blind mole-rats (mainly *Spalax ehrenbergi*) and with American pocket gophers (Geomyidae).

In 1989, an international symposium on the evolution of subterranean mammals at the organismal and molecular levels was organized within the International Theriological Congress in Rome by Eviatar Nevo and late Osvaldo A. Reig, and an influential book (actually proceedings of the symposium) edited by both conveners was published in 1990 (Nevo and Reig 1990). Although this book covered such diverse topics as the ecophysiology, sensory biology, comparative morphology, ecology, taxonomy and phylogeny of diverse subterranean and fossorial mammals, it was the naked mole-rat which has become the most popular subterranean mammal, with its popularity also reaching beyond the scientific community. The intensive research resulted in the publication of a highly cited book (Sherman et al. 1991) and about 150 further scientific papers dealing with this species. We dare to say, however, that the impact of the naked mole-rat upon the

Table 1.1. Fossorial and subterranean genera of rodents. Classification adapted from McKenna and Bell (1997)

Suborder	Family Subfamily	Genus
Hystricognatha	Bathyergidae	
	Bathyerginae	*Bathyergus*
		Cryptomys
		Fukomys
		Georychus
		Heliophobius
	Heterocephalinae	*Heterocephalus*
	Octodontidae	
	Octodontinae	*Ctenomys*
		Pithanotomys
		Spalacopus
	Echimyidae	
	Heteropsomyinae	*Carterodon*
		Clyomys
		Euryzygomatomys
Myomorpha	Muridae	
	Arvicolinae	*Arvicola*
		Ellobius
		Hyperacrius
		Microtus
		Prometheomys
	Sigmodontinae	*Aepeomys*
		Akodon
		Blarinomys
		Chelemys
		Euneomys
		Geoxus
		Kunsia
		Notiomys
	Myospalacinae	*Eospalax*
		Myospalax
	Spalacinae	*Nannospalax*
		Spalax
	Rhizomyinae	*Cannomys*
		Rhizomys
		Tachyoryctes
	Geomyidae	
	Geomyinae	*Geomys*
		Orthogeomys
		Pappogeomys
		Thomomys
		Zygogeomys
Sciuromorpha	Aplodontidae	*Aplodontia*

research of subterranean mammals in general was not as high as expected. On the contrary, most general biologists (including text-book authors) and biology students, let alone the interested non-scientific community, still regard the naked mole-rat as a typical, if not exclusive, subterranean rodent. Even many students of the naked mole-rat tend to explain its unique traits as exclusive adaptations to life underground, ignoring the plethora of other rodent taxa which also successfully occupy the underground niche, and display both convergent traits as well as different but not less effective and not less interesting solutions to problems of the underground existence.

Eight years later, in 1997, an international symposium dedicated to subterranean rodents took place within the framework of the International Theriological Congress in Acapulco (note that the topical scope was already narrower and more specific than that in Rome). A book based on the symposium edited by its conveners, Eileen A. Lacey, James L. Patton, and Guy N. Cameron, appeared in 2000 (Lacey et al. 2000). The end of the millennium was generally productive as far as books on subterranean mammals are concerned. In 1999, Nevo published his review of the research on evolution of subterranean mammals (Nevo 1999). In his book he showed how adaptive evolution of subterranean mammals involves structural and functional changes which are both regressive (degenerative) and progressive (compensatory) in nature. The mosaic convergent global evolution of subterranean mammals due to similar constraints and stresses is an example par excellence for an evidence for evolution through natural selection obtained through comparative methods. Subterranean mammals teach us a lot about the nature of adaptive radiation, species diversity in space and time, how phenotypes and genotypes regress, progress, and converge through molecular and organismal tinkering by natural selection. In another book published at that time, Bennett and Faulkes (2000) summarized their research and ideas on mainly South African mole-rats, pointing out that the naked mole-rats have hairy cousins which are not less interesting. In 2001 Nevo and colleagues published their book reviewing the achievements in research of blind mole-rats (Nevo et al. 2001). Finally, a booklet by Yevdokimov published in the same year (written in Russian and therefore, unfortunately, not easily accessible to most scientists) reviewed many studies on the mole-vole, *Ellobius talpinus*, by Russian authors, and particularly, long-term studies on behavioural and population ecology of this species carried out by the author himself (Yevdokimov 2001).

Given the exponential increase in the number of publications dealing with subterranean rodents, growth of the scientific community involved in the research of these subjects, introduction of new taxa, new research aspects, methods, and ideas in the field, we felt that time was more than ripe, after 16 years from the symposium in Rome and 8 years from the symposium in Acapulco, to bring the interested scientists together again. The opportunity was given with the International Mammalogical Congress in Sapporo in 2005. Like the preceding symposia, the symposium in Sapporo was also very successful and we hope that the present book will become a milestone in the research of subterranean dwellers in the same way as its predecessors are. Most chapters of this volume are based on contributions presented at the Sapporo symposium on subterranean rodents. In our symposium the established scholars ("old stagers") working with subterranean rodents for the past decades united with scientists just recently attracted by the subject. Accordingly, very different people and schools contributed to this volume. We were lucky to bring together not only researchers of different generations working in different fields, but also representing different geographical regions – not only by subjects of their interest, but also by their origin and affiliation. All in all, 40 authors from 12 different countries contributed to this volume.

To understand the biology of an organism it is important to look at its adaptations and, of course, scientists are bound to study them. Previous books gave much attention to this topic and are a good basis for students to start; hence, our aim was to expand the knowledge rather than to reproduce the established facts. The present book, which deals with a diverse array of modern topics, is divided into five sections: ecophysiology, sensory ecology, life histories, ecological and economical impacts, and molecular ecology and evolution. Short introductory chapters to each of the five parts give an overview of the respective field, and comment on the chapters. These introductions are followed by reviews on subjects currently studied in the respective field, but naturally, the reviews cannot cover the complete spectrum.

It may appear that a disproportionately large part of the book is dedicated to sensory biology and ecology. However, this fact also reflects the recent and current boom of studies and publications in this field. Thus, along with revisiting classical sensory topics like audition or vision, where recent years marked also significant new findings and ideas changing our view of sensory adaptations in underground dwellers, more exotic or thus far much less studied senses like magnetoreception or somatosensation have also been attracting attention of more and more research.

Besides trying to carry out a complete and current revision of the different areas of investigation on the biology of underground mammals, this book seeks to generate new visions and ideas (through the development of novel research aspects and methods) that constitute the basal stone of future symposiums on subterranean mammals. Judging from earlier developments in the field, we expect that the eight-year periodic cycle between symposia and books on subterranean dwellers will shorten. It is our hope that the impact of this book will reach beyond the ever growing community of researchers in the field. We hope that the case of subterranean rodents will finally attract the attention of authors of text-books in evolutionary biology, animal, sensory and behavioural ecology, biology of behaviour, and general zoology, and that the book will provide ideas and reading texts for seminars in organismal and molecular evolution and ecology.

Acknowledgements. First we wish to thank all contributors for their great work on the chapters, fruitful cooperation, and their patience with the development of the book. We are also grateful to the reviewers who dedicated their time, shared their experience, and gave helpful advice for improving the chapters. Apart from the contributors to the volume who themselves acted as reviewers to other chapters, the external referees were C. Barnard, C. Baudoin, G. Dryden, T. de Freitas, E. Gese, A. Haim, G. Hickmann, M. Kössl, E. Lessa, B. Low, R. Muheim, D. Nolte, R. Pech, J. Randall, B. Rogers, A. Smith, B. Timm, C. White, W. Zielinski, and J. Zook. Last, but not least, we thank our Springer contact - Dr. Jutta Lindenborn - for her interest in the project, and always being ready, patient and constructively helpful with processing the book.

References

Bennett NC, Faulkes CG (2000) African mole-rats: ecology and eusociality. Cambridge Univ Press, Cambridge

Jarvis JUM (1981) Eusociality in a mammal: cooperative breeding in naked mole-rat colonies. Science 212:571–573

Lacey EA, Patton JL, Cameron GN (2000) Life underground: the biology of subterranean rodents. Univ Chicago Press, Chicago

McKenna MC, Bell S (1997) Classification of mammals above the species level. Columbia Univ Press, NY

Nevo E (1979) Adaptive convergence and divergence of subterranean mammals. Ann Rev Ecol Syst 10:269–308

Nevo E (1999) Mosaic evolution of subterranean mammals: regression, progression and global convergence. Oxford Univ Press, Oxford

Nevo E, Reig OA (1990) Evolution of subterranean mammals at the organismal and molecular levels. Wiley-Liss, New York

Nevo E, Ivanitskaya E, Beiles A (2001) Adaptive radiation of blind subterranean mole rats: naming and revisiting the four sibling species of the *Spalax ehrenbergi* superspecies in Israel: *Spalax galili* (2n=52), *S. golani* (2n=54), *S. carmeli* (2n=58), *S. judaei* (2n= 60). Backhuys Publishers b.v., Leiden, The Netherlands

Sherman PW, Jarvis JUM, Alexander RD (1991) The biology of the naked mole-rat. Princeton Univ Press, Princeton, NJ

Yevdokimov NG (2001) Population ecology of the common mole-vole (in Russian). Russ Acad Sci (Ural Sect) Press, Yekaterinburg

Part I
Ecophysiology

2 Adaptive Physiological Mechanisms in the Underground Dwellers

František Sedláček

The underground ecotope is unique and stressful in many aspects (for review see, e.g., Nevo 1999; Burda et al., this volume), and physiological adaptations – although less conspicuous than morphological adaptations – represent the core of underground life. It is assumed that physiology has been molded by convergent evolution to the same extent as morphology has been. However, although physiological adaptations attracted the interest of researchers since a few decades, only certain processes and aspects have been studied in a few species thus far. The problem has been pointed out already in the last comprehensive review of ecophysiology of subterranean mammals by Buffenstein (2000), and this bias has continued in recent years. The diverse subjects can be assigned to two main fields of research: 2.1) Energy and Material Fluxes cover the topics foraging, digestion, metabolic rate, thermal flux, body temperature, thermoneutral zone, blood transport, cardiac responses, ventilatory responses, acid-base balance, water flux, and mineral homeostasis; 2.2) Neuronal and Hormonal Regulation deal with circadian rhythms, sensory ecophysiology, reproductive physiology and endocrinology.

Before we look further into the subjects, we should address data on environmental parameters which are the starting point for ecophysiological studies and might help to create logical insights into adaptations. In the first chapter of this section (Chap. 3), Burda et al. review new data on microclimatic conditions in subterranean burrows. Although the parameters, e.g. temperature, humidity, oxygen concentration etc. are measured in standard ways, the right placement of devices into burrows, and particularly into (occupied) nests is, from a methodological point of view, a demanding operation.

František Sedláček: Faculty of Biological Sciences, University of South Bohemia, Branišovská 31, 37005 České Budějovice, Czech Republic, E-mail: sedlacek@usbe.cas.cz

Subterranean Rodents: News from Underground
S. Begall, H. Burda, C.E. Schleich (Eds.)
© Springer-Verlag Berlin Heidelberg 2007

2.1 Energy and Material Fluxes

Research in this field highlights the potentials as well as the limits of survival and successful existence in the subterranean ecotope. Since foraging is energetically demanding, and the subterranean environment offers generally low plant food supply, digestion of cellulose is expected to be very effective. Following this reasoning, Dyer (1998) suggested that the controlled consumption of symbionts in feces by naked mole-rats (*Heterocephalus glaber*) might have helped to establish and maintain eusociality in these subterranean rodents. Regarding the gastrointestinal system, the animals' age might be an interesting factor (O'Connor et al. 2002; Yang et al. 2002), especially since naked mole-rats as well as Ansell's mole-rats (*Fukomys anselli*) are extraordinary long-lived mammals (Buffenstein 2005; Dammann and Burda 2006). Studies on metabolic rates could help to identify proximate factors of aging. In Chap. 4, Sedláček addresses the energetics of subterranean and fossorial rodents. Recent studies in this field are presented and the author tries to explain how the new data have complemented or changed our understanding of physiology of metabolism in subterranean rodents, or even how they determine new ways and open new horizons for future research in the field.

In the following chapter, Moshkin et al. present a complex study of energy balance and fluxes in the mole-vole (*Ellobius talpinus*) as a model species. Exploring species other than bathyergids or spalacids (the best studied subterranean rodents so far) prove to be very fruitful. Furthermore, the mole-vole originating from moderate zones with changing seasons represents a natural thermal experiment of great importance.

Thermal flux, i.e. heat exchange via conduction or convection, is of particular importance in the underground environment. During periods of high activity, this mechanism represents an effective way to dissipate metabolic heat. Daly and Buffenstein (1998) discussed the skin morphology in *H. glaber* with respect to the poor thermoinsulatory abilities while simultaneously facilitating heat transfer from the environment to the animal by thigmothermy. We may speculate also about another flux that equalizes individually different amounts of energy gained (food digested) by the colony members via inter-individual thermal exchange. Recently, also in some other mole-rats the role of body insulation was evaluated (Cortes et al. 2000, Cutrera and Antinuchi 2004, Zelová et al., in press – see Chap. 4). Naturally, thermal flux is closely bound to body temperature. The extent to which body temperature is regulated varies among subterranean rodents. The reactions to the ambient temperature fluctuations range from strict homeothermy (*Spalax ehrenbergi*) to heterothermy (*F. anselli*) and poikilothermy (*H. glaber*). Further research in this field should focus on questions like: What is the thermoregulatory meaning of the body size in subterranean mole-rats and does Bergmann's rule apply in the subterranean ecotope, or why does the silvery mole-rat have long fur? The third important thermal parameter is the *thermoneutral zone*. In African mole-rats it ranges between 25 and 37°C reflecting probably relatively high and stable burrow temperatures and thermal properties of mole-rats' bodies. However, burrow temperatures measured in the field are often lower (see Chap. 3). The broad thermoneutral zone may be characteristic even within a single species - thus the silvery mole-rat exhibits a very broad thermoneutral zone ranging from 25 to 33°C (Zelová et al., in press). The evolutionary and ecological reasons for these intraspecific differences remain to be determined. It would be also of interest to study thermoregulation in holarctic subterranean voles - particularly those from mountainous regions (see Chap. 5).

In the subterranean environment, where air composition generally differs from above ground conditions, *blood transport* plays an important compensatory role. In subterranean rodents increased oxygen carrying capacity has been observed. This adaptation could be facilitated by elevated hemoglobin concentrations, increased red blood cell counts, and high intrinsic affinity for oxygen. The best studied subterranean rodent in this respect is the blind mole-rat *S. ehrenbergi* living, at least temporarily, under pronounced hypoxic and hypercapnic conditions (Shams et al. 2005a). In this species a series of studies revealed different parts of oxygen supply system on the molecular level (Yang et al. 1998; Avivi et al. 1999, 2005; Shams et al. 2004a, b, 2005b). Attenuated *ventilatory response* to the subterranean atmosphere is considered highly adaptive because it reduces the energetic respiratory costs associated with the deeper and more frequent breathing in a hypercapnic environment. Maina et al. (2001) observed in the naked mole-rat that lungs can functionally adapt to changing environmental conditions without exhibiting any apparent morphological changes. Therefore, studies of respiratory physiology in captive naked mole-rats (and perhaps also in other subterranean rodents) should be regarded cautiously. A similar problem of functional plasticity may apply also to studies of other parameters. Most physiological measurements are made under artificial laboratory conditions, which actually correspond, to surface environment and not to the real conditions faced by the animals underground.

2.2 Neural and Hormonal Regulation

Research in this field follows the course of information from (highly modified) sensory organs through neural and hormonal systems to the effectors. Sensory (and partly also neuronal) adaptations are the topic of a separate section in this volume (Part II: Sensory Ecology) and will therefore not be addressed here. The topics *reproductive physiology* and *endocrinology* are analyzed in detail by Bennett et al. (Chap. 6). Bennett and colleagues show how the reproductive strategies in solitary living as well as in social South African mole-rats are reflected in regulation networks which are, however, sensitive to environmental conditions. Synchronization of breeding with environmental changes and the dilemma between induced versus spontaneous ovulation are discussed. These aspects of reproduction are very important also for successful laboratory breeding of solitary mole-rats, which could be a bottle-neck for further studies.

Another interesting and important aspect of biology of subterranean dwellers, pertinent to their dark monotonous environment, is that of photoperiodicity and *circadian rhythms*. Also, this aspect is briefly reviewed in the sensory ecology section of this volume (chapter by Němec et al.). Circadian (and circannual) rhythmicity involves not only sensory but also more general physiological aspects. Despite the highly pronounced regression of visual organs and pathways, subterranean rodents of several species were shown to perceive changes in photoperiod and to display circadian rhythms of activity. The retina and the infraorbital Harderian glands are likely sources of light detection and activation of the associated neuroendocrine pathways (see Buffenstein 2000). The number of studies on circadian rhythms in subterranean mole-rats has increased drastically within the past years. Riccio and Goldman (2000) have observed in the naked mole-rat (*H. glaber*) an increase in the body temperature and metabolic rate at the end of the light phase in the circadian cycle, probably to prepare the body for following locomotory activity. The ability to entrain to 24-h light-dark cycles is expressed in *H. glaber*, *Georychus capensis*, *Fukomys damarensis*, *Cryptomys hottentotus*, *Fukomys darlingi* (Riccio and Goldman 2000; Oosthuizen et al. 2003; Richter et al. 2003; Gutjahr et al. 2004; Hart et al. 2004; Vasicek et al. 2005a) while in *C. hottentotus*, Negroni et al. (2003) could not demonstrate neither light synchronization nor circadian activity pattern. Large inter- and intraindividual variations in the rate and extent of entrainment, time of activity preference, and activity patterns complicate the studies (Oosthuizen et al. 2003; Vasicek et al. 2005b). Begall et al. (2002) showed that individually housed *Spalacopus cyanus* displayed a clear nocturnal endogenous activity pattern under laboratory conditions. In contrast, Urrejola et al. (2005) demonstrated that free-living *S. cyanus* are diurnal. Physical and social environments in which captive animals are housed may contribute to the observed differences in activity between field and laboratory populations (Urrejola et al. 2005). Obviously, the underground ecotope has not a consistent effect on the temporal patterning of activity. In *S. ehrenbergi* a plausible synchronization mechanism was presented (Tobler et al. 1998, Tobler and Deboer 2001). Further studies have shifted their focus to mechanisms on molecular level (Avivi et al. 2001, 2002, 2004; Oster et al. 2002). Hannibal et al. (2002) revealed that *Spalax* expresses melanopsin in ganglion cells projecting to the circadian clock, thus supporting the role of melanopsin as a circadian photopigment. According to Hough et al. (2002) the expression of a heat shock protein bound to the lens has also undergone adaptive changes corresponding to the subterranean evolution of the blind mole-rat. Surely, more studies in this field will follow in the next few years.

References

Avivi A, Resnick MB, Nevo E, Joel A, Levy AP (1999) Adaptive hypoxic tolerance in the subterranean mole rat *Spalax ehrenbergi*: the role of vascular endothelial growth factor. FEBS Lett 452:133–140

Avivi A, Albrecht U, Oster H, Joel A, Beiles A, Nevo E (2001) Biological clock in total darkness: the Clock/MOP3 circadian system of the blind subterranean mole rat. Proc Natl Acad Sci USA 98:13751–13756

Avivi A, Oster H, Joel A, Beiles A, Albrecht U, Nevo E (2002) Circadian genes in a blind subterranean mammal II: conservation and uniqueness of the three period homologs in the blind subterranean mole rat, *Spalax ehrenbergi* superspecies. Proc Natl Acad Sci USA 99:11718–11723

Avivi A, Oster H, Joel A, Beiles A, Albrecht U, Nevo E (2004) Circadian genes in a blind subterranean mammal III: molecular cloning and circadian regulation of cryptochrome genes in the blind subterranean mole rat, *Spalax ehrenbergi* superspecies. J Biol Rhythms 19:22–34

Avivi A, Shams I, Joel A, Lache O, Levy AP, Nevo E (2005) Increased blood vessel density provides the mole rat physiological tolerance to its hypoxic subterranean habitat. Faseb J 19:1314–1316

Begall S, Daan S, Burda H, Overkamp GJF (2002) Activity patterns in a subterranean social rodent, *Spalacopus cyanus* (Octodontidae). J Mammal 83:153–158

Buffenstein R (2000) Ecophysiological responses of subterranean rodents to underground habitats. In: Lacey EA, Patton JL, Cameron GN (eds) Life underground: the biology of subterranean rodents. Univ of Chicago Press, Chicago, pp 62–110

Buffenstein R (2005) The naked mole-rat: a new long-living model for human aging research. J Gerontol A 60:1369–1377

Cortes A, Miranda E, Rosenmann M, Rau JR (2000) Thermal biology of the fossorial rodent *Ctenomys fulvus* from the Atacama desert, northern Chile. J Therm Biol 25:425–430

Cutrera AP, Antinuchi CD (2004) Fur changes in the subterranean rodent *Ctenomys talarum*: possible thermal compensatory mechanism. Rev Chil Hist Nat 77:235–242

Daly TJM, Buffenstein R (1998) Skin morphology and its role in thermoregulation in mole-rats, *Heterocephalus glaber* and *Cryptomys hottentotus*. J Anat 193:495–502

Dammann P, Burda H (2006) Sexual activity and reproduction delay ageing in a mammal. Curr Biol 16:R117–R118

Dyer BD (1998) A hypothesis about the significance of symbionts as a source of protein in the evolution of eusociality in naked mole rats. Symbiosis 24:369–383

Gutjahr GH, Rensburg LJ, Malpaux B, Richter TA, Bennett NC (2004) The endogenous rhythm of plasma melatonin and its regulation by light in the highveld mole-rat (*Cryptomys hottentotus pretoriae*): a microphthalmic, seasonally breeding rodent. J Pineal Res 37:185–192

Hannibal J, Hindersson P, Nevo E, Fahrenkrug J (2002) The circadian photopigment melanopsin is expressed in the blind subterranean mole rat, *Spalax*. Neuroreport 13:1411–1414

Hart L, Bennett NC, Malpaux B, Chimimba CT, Oosthuizen MK (2004) The chronobiology of the Natal mole-rat, *Cryptomys hottentotus natalensis*. Physiol Behav 82:563–569

Hough RB, Avivi A, Davis J, Joel A, Nevo E, Piatigorsky J (2002) Adaptive evolution of small heat shock protein/alpha B-crystallin promoter activity of the blind subterranean mole rat, *Spalax ehrenbergi*. Proc Natl Acad Sci USA 99:8145–8150

Maina JN, Gebreegziabher Y, Woodley R, Buffenstein R (2001) Effects of change in environmental temperature and natural shifts in carbon dioxide and oxygen concentrations on the lungs of captive naked mole-rats (*Heterocephalus glaber*): a morphological and morphometric study. J Zool Lond 253:371–382

Negroni J, Bennett NC, Cooper HA (2003) Organization of the circadian system in the subterranean mole rat, *Cryptomys hottentotus* (Bathyergidae). Brain Res 967:48–62

Nevo E (1999) Mosaic evolution of subterranean mammals: regression, progression and global convergence. Oxford Univ Press, Oxford

O'Connor TP, Lee A, Jarvis JUM, Buffenstein R (2002) Prolonged longevity in naked mole-rats: age-related changes in metabolism, body composition and gastrointestinal function. Comp Biochem Physiol A 133:835–842

Oosthuizen MK, Cooper HM, Bennett NC (2003) Circadian rhythms of locomotor activity in solitary and social species of African mole-rats (Family: Bathyergidae). J Biol Rhythms 18:481–490

Oster H, Avivi A, Joel A, Albrecht U, Nevo E (2002) A switch from diurnal to nocturnal activity in *S. ehrenbergi* is accompanied by an uncoupling of light input and the circadian clock. Curr Biol 12:1919–1922

Riccio AP, Goldman BD (2000) Circadian rhythms of locomotor activity in naked mole-rats (*Heterocephalus glaber*). Physiol Behav 71:1–13

Richter TA, Malpaux B, Fleming PA, Molteno AJ, Bennett NC (2003) Melatonin secretion in a strictly subterranean mammal, the Damaraland mole-rat (*Cryptomys damarensis*). J Zool Lond 261:313–319

Shams I, Avivi A, Nevo E (2004a) Hypoxic stress tolerance of the blind subterranean mole rat: Expression of erythropoietin and hypoxia-inducible factor 1 alpha. Proc Natl Acad Sci USA 101:9698–9703

Shams I, Nevo E, Avivi A (2004b) Ontogenetic expression of erythropoietin and hypoxia-inducible factor-1 alpha genes in subterranean blind mole rats. Faseb J 18

Shams I, Avivi A, Nevo E (2005a) Oxygen and carbon dioxide fluctuations in burrows of subterranean blind mole rats indicate tolerance to hypoxic-hypercapnic stresses. Comp Biochem Physiol A 142:376–382

Shams I, Nevo E, Avivi A (2005b) Erythropoietin receptor spliced forms differentially expressed in blind subterranean mole rats. Faseb J 19

Tobler I, Deboer T (2001) Sleep in the blind mole rat *Spalax ehrenbergi*. Sleep 24:147–154

Tobler I, Herrmann M, Cooper HM, Negroni J, Nevo E, Achermann P (1998) Rest-activity rhythm of the blind mole rat *Spalax ehrenbergi* under different lighting conditions. Behav Brain Res 96:173–183

Urrejola D, Lacey EA, Wieczorek JR, Ebensperger LA (2005) Daily activity patterns of free-living cururos (*Spalacopus cyanus*). J Mammal 86:302–308

Vasicek CA, Oosthuizen MK, Cooper HM, Bennett NC (2005a) Circadian rhythms of locomotor activity in the subterranean Mashona mole rat, *Cryptomys darlingi*. Physiol Behav 84:181–191

Vasicek CA, Malpaux B, Fleming PA, Bennett NC (2005b) Melatonin secretion in the Mashona mole-rat, *Cryptomys darlingi* – influence of light on rhythmicity. Physiol Behav 83:689–697

Yang H, Nevo E, Tashian RE (1998) Unexpected expression of carbonic anhydrase I and selenium-binding protein as the only major non-heme proteins in erythrocytes of the subterranean mole rat (*Spalax ehrenbergi*). FEBS Lett 430:343–347

Yang T, Buffenstein R, O'Connor TP (2002) Disparate age effects on gastrointestinal enzymes in naked mole rats. Integr Comp Biol 42:1340–1341

Zelová J, Šumbera R, Sedláček F, Burda H (in press) Energetics in the solitary subterranean rodent, the silvery mole-rat, *Heliophobius argenteocinereus*, and the allometry of RMR in African mole-rats (Bathyergidae). Comp Biochem Physiol A

3 Microclimate in Burrows of Subterranean Rodents – Revisited

Hynek Burda, Radim Šumbera, Sabine Begall

3.1 Introduction

The subterranean ecotope provides its inhabitants with several advantages. *Burrowing (fossorial)* mammals construct or visit existing burrow systems regularly or temporarily for shelter but search for their food mainly above ground. *Subterranean* mammals confine their existence and foraging mainly to the underground ecotope, and there is a continuum from fossorial through facultative to strictly subterranean lifestyles (cf., e.g. Nevo 1979, 1999; Nevo and Reig 1990; Kinlaw 1999 for reviews). The shelter function of the underground ecotope has two components: protection from predators and protection from environmental fluctuation or extremes predominating above the ground. Protection is particularly important in periods of enhanced vulnerability: during resting and sleeping, hibernation, aestivation, or breeding. On the other hand, the subterranean niche is highly specialized and challenging. Burrow inhabitants are deprived of most sensory cues available above ground, and face low food supply, high energetic costs of digging and some stressful microenvironmental conditions as high humidity, low gas ventilation, hypoxic and hypercapnic conditions (Kennerly 1964; Arieli 1979; Contreras and McNab 1990; Nevo 1999).

It is assumed that the subterranean niche opened to herbivores in the upper Eocene (45–35 million years ago [mya]) and then extended into the Quaternary (about two mya) when in the course of global cooling and aridisation, open landscapes expanded and geophytes ("staple food" of subterranean rodents) diversified (Nevo 1999; Busch et al. 2000). However, exploiting and colonising the underground niche probably started much earlier. A recently described late Jurassic insectivorous mammal, *Fruitafossor windscheffeli*, representing a basal mammalian lineage, shows

Hynek Burda: Dept. General Zoology, University of Duisburg-Essen, Universitätsstr. 5, D-45117 Essen, Germany, E-mail: hynek.burda@uni-due.de

Radim Šumbera: Faculty of Biological Sciences, University of South Bohemia, Branišovská 31, CZ-370 05 České Budějovice, Czech Republic

Sabine Begall: Dept. General Zoology, University of Duisburg-Essen, Universitätsstr. 5, D-45117 Essen, Germany

Subterranean Rodents: News from Underground
S. Begall, H. Burda, C.E. Schleich (Eds.)
© Springer-Verlag Berlin Heidelberg 2007

morphological traits characteristic of burrowing mammals (Luo and Wible 2005). Its discovery suggests that fossoriality evolved early and probably accompanied evolution of mammals from almost the beginning. Hence, early mammals were already confronted with stressful constraints of the subterranean ecotope and they apparently did so successfully. Nevertheless, contrary to the prevailing belief, recent microenvironmental analyses of burrows, reported below, indicate that the subterranean ecotope is not necessarily as stressful as assumed, so that its colonization might have been in some respect easier.

Detailed studies on microhabitat and microclimate are important to understand the biology and evolution of subterranean mammals at different organisational levels and the adaptive significance of specific morphological, physiological, behavioural and ecological traits. Furthermore, this type of information is needed to simulate (semi-)natural conditions in diverse laboratory studies.

3.2 Microenvironmental Parameters

Microclimate (temperature, humidity) and atmospheric composition (concentration of respiratory gases) in burrows are influenced by corresponding characteristics prevailing above ground, vegetation cover, soil quality (mainly porosity), depth, length, diameter and shape of burrows, as well as the whole architecture of the burrow system and its ventilation (cf. Šumbera et al. 2004). Inhabitants of burrows (including commensals) surely influence the microclimate and atmosphere of burrows through their movement and metabolic activity. Surprisingly, this factor has rarely been considered in the field; studies and published measurements of temperature, humidity etc. in underground nest chambers or burrows refer mainly to abiotic conditions sensu stricto, with dwellers being absent.

3.2.1 Temperature

3.2.1.1 Methodical Considerations

The soil is heated up by direct, short-wave radiation of the sun, which needs to be taken into account when measuring soil and burrow temperatures. Thus standard methods of soil temperature measurements should be applied and any conductance of heat into sealed burrows and bores during the measurements have to be avoided (e.g. Mohr et al. 1972). The soil colour is of importance for the absorption of short-wave radiation and should be noted. These aspects seem to have been underestimated in many existing studies and methods of data acquisition needed to guarantee comparability of results have been seldom described in detail. Most of burrow temperature recordings have been done on short-term basis covering only several days, weeks or even several isolated events, which do not register dynamics of seasonal changes. In spite of these methodical shortcomings and constraints, it is useful to summarize and analyse existing records.

3.2.1.2
Effect of Depth

The microenvironment of subterranean mammals is thermally buffered and in comparison to the surface relatively constant even in open burrows (e.g. Downs and Perrin 1989; Begall and Gallardo 2000; Shenbrot et al. 2002; Moore and Roper 2003). This buffering effect is pronounced in sealed burrows showing stable burrow temperatures as demonstrated for different species of African mole-rats (Jarvis 1979; Bennett et al. 1988; Lovegrove and Knight-Eloff 1988; Brett 1991; Roper et al. 2001; Scharff et al. 2001; Šumbera et al. 2004) (Table 3.1). However, regarding the differences in soil types, humidity, vegetation, and ambient temperatures over the range of distribution of subterranean mammals, the constancy within burrow systems should not be confused or equated with uniformity across burrow systems. Burrow and soil temperatures are affected by fluctuation of ambient temperature in both, open and closed burrow systems (e.g. Kennerly 1964; Hall and Myers 1978; Kay and Whitford 1978; Altuna 1985; Rezende et al. 2003). As expected, the effect of aboveground temperature is more pronounced in superficial tunnels (e.g. about 11 cm deep in *Spalax*, Arieli 1979) and temperature fluctuation decreases with increasing tunnel depth as has been demonstrated for *Dipodomys* (Kay and Whitford 1978). It is well known that soil temperatures in depths >50 cm below ground are diurnally and seasonally more or less constant (Van Wambeke 1992).

Measurements of temperature gradients in burrows of *Fukomys mechowii* in Zambia are illustrative (Scharff 1998; Kawalika, Locker-Grütjen, Scharff, unpublished): the maximum August ambient temperature (recorded 1.5 m above ground in shade) reaches 37°C, while the temperature of the top soil increases up to 58°C. However, *temperatures in burrows* (30 cm deep below ground) are much lower, fluctuating daily from 16.6°C to 18.4°C (17.8±0.07°C SD). Temperatures in unoccupied nest chambers (140 cm deep) are rather low and constant amounting to 19.45°C (±0.02, 19.3–19.5°C). The mean *soil temperatures* changes from 18.3°C (25 cm deep) to 17.4°C (102 cm below ground). Long-term continuous recordings reveal considerable daily and seasonal fluctuations of the temperature in foraging

Table 3.1. Temperature recordings from burrows and (empty) nest chambers in bathyergids (*nests marked by asterisk*)

Species	Site (Country)	Month	Depth (cm)	Mean (°C)	Var. coeff.	SD	Range	Reference
Fukomys damarensis	Dordabis Namibia	VIII	13	18.6	14.5	2.7	15–22.8	Bennett et al. 1988
	Otjiwarango (Namibia)	I	14	34.8	14.1	4.9	26.2–40	
	Kalahari (South Africa)	I	25	33.5	3.9	1.3		Lovegrove and Knight-Eloff 1988
		VII	30	18.8		0.1		
Fukomys kafuensis	Itezhi-Tezhi (Zambia)	VIII	14	20.1			18.0–22.0	Marhold 1989
			70	27.0*			26–28	
Fukomys mechowii	Ndola (Zambia)	VIII	30	17.8	3.9	0.07	16.6–18.4	Kawalika, Locker-Grütjen, Scharff unpubl.
		VIII	50	17.5*				
		VIII	140	19.4*	0.1	0.02	19.3–19.5	
Fukomys whytei	Karonga (Malawi)	VIII	27	26.7		2.1		Burda et al. 2005
	Nyika (Malawi)	IV	55	18.6*				
Cryptomys hottentotus	Wellington (South Africa)	IV	14	25.1	8.0	2	21.0–27.8	Bennett et al. 1988
	Darling (South Africa)	V	20	17.5	5.7	1	15.9–18.8	
Georychus capensis	Cape Town (South Africa)	III	6	21.7	10.1	2.2	16.9–24.2	Roper et al. 2001
	Rondebosch (South Africa)	I	11	26.1	8.4	2.2	22.9–29.7	Bennett et al. 1988
		VII	14	12.2	9.0	1.1	10.2–13.8	
Heliophobius argenteocinereus	Blantyre (Malawi)	VI–X	15	21.3	14.5	3.1	16.5–28.3	Šumbera et al. 2004
		XI–IV	15	22.8	5.3	1.2	20.2–26.9	
	Mulanje (Malawi)	VI–X	30	20.8	4.3	0.9	19.5–23.4	
	Nyika (Malawi)	IV	60	16.8*				Burda and Chitaukali unpubl.
Heterocephalus glaber	Archers (Kenya)	VII	26	31.1	2.0	0.62	30.5–32.0	Bennett et al. 1988
	Lerata (Kenya)	VIII	31	29.6	1.3	0.4	29.0–30.0	
	Kamboyo (Kenya)	I–XII	30	28.0	7.8	2.2		Brett 1991

burrows (less than 15 cm deep) (cf. Table 3.1). In addition to predictable daily fluctuations, an abrupt temperature change within several days can mark a seasonal change, e.g. beginning of the rainy season (cf. Šumbera et al. 2004).

Thermal conductivity of soil is responsible for lapse of the daily burrow temperature cycle. Soil retains heat longer after sunset, when ambient temperature decreases, and remains cold long in the morning when ambient temperature rises faster. The lapse rate is a function of soil thermal capacity (heat storage). Dry soils have lower thermal capacity and cool more rapidly at night. This lapse of temperature cycle has been found in burrows of different subterranean rodents (Kennerly 1964; Bennett et al. 1988; Roper et al. 2001; Šumbera et al. 2004), and may also influence their circadian activity.

3.2.1.3
Burrow Temperatures and Thermoregulation

Burrow systems in most subterranean mammals have a complex three-dimensional arrangement (Hickman 1990; Reichman and Smith 1990). The horizontal level of extensive systems may run across differently vegetated areas and through soils of different types. Vertically, different burrow parts are located at different depths, and accordingly, temperature and its temporal fluctuations vary spatially within a single burrow system. Whereas the horizontal aspect of the burrow system architecture is determined by abundance and distribution of food resources and foraging strategies (Heth 1989; Spinks et al. 2000; Šumbera et al. 2003), the vertical arrangement might be important for safety, thermoregulation, and canalization in the case of flooding.

Burrow temperatures above the upper or below the lower limits of thermoneutrality are physiologically stressful, but animals can easily avoid them by moving up and down to depths with temperatures closer to temperature optimum (Vleck 1979; Lovegrove and Knight-Eloff 1988). We may speculate that as long as burrow temperatures are predictably fluctuating there is no need to evolve special thermoregulatory morphological and physiological adaptations and animals may adapt to different and varying temperatures in shallow burrows behaviourally, i.e., by timing their digging and foraging activity bouts according to prevailing burrow temperatures. The high temperatures near surface may also be the reason why subterranean mammals in hot seasons or during the day dig at deeper levels and transport the excavated soil into abandoned tunnels rather than building mounds (Contreras and McNab 1990 for *Spalacopus*, Nevo 1999 for *Spalax*, Šumbera et al. 2003 for *Heliophobius*).

On the other hand, subterranean mammals spend most of their time in the nest (cf. Dammann and Burda 2006; Šklíba et al., in press; Zhang, this volume; and unpublished observations) and must adapt to the prevailing nest temperature or place their nests in depths/sites with temperatures corresponding to their physiological optimum. Therefore, for eco-physiological comparisons, temperature recordings from nests are of greater interest than those from foraging burrows. Since temperatures at deeper soil layers (>50 cm), where nests in many species are located, are diurnally and seasonally constant, even punctual measurements provide representative data. In spite of that, only few comparative data are available in the literature (Table 3.1). Apparently, temperatures in depths of nest chambers may be rather low (Table 3.1: 16.8°C in the nest of *Heliophobius argenteocinereus*, 17.5°C in *Fukomys mechowii*, and 18.6°C in *F. whytei*), but those of *F. anselli* (26–28°C) and *Heterocephalus glaber* (27–31°C) are markedly higher (Table 3.1). Extremely low temperatures (0–5°C) were reported for nests of Tibetan plateau zokors (*Eospalax fontanierii*) in winter (Zeng et al. 1984; Zhang, this volume). In all studied cases, the temperatures prevailing in the nest areas are below the thermoneutral zone of the respective species (cf. Bennett et al. 1994; Marhold and Nagel 1995; Zelová et al., in press). Subterranean rodents may conserve energy by huddling in the nest in social species and/or by increasing their body size or thickening their fur as exemplified by *H. argenteocinereus* (cf. also Cutrera and Antinuchi 2004 for variation of fur length in *Ctenomys* in dependence on ambient temperatures).

Presence and activity of burrow inhabitants may also influence temperature in some parts of the burrow system and especially in nest chambers. Temperature in the nest of the European badger (*Meles meles*) is about 2.4°C higher than in unoccupied chambers in the same depth (Moore and Roper 2003). A similar "igloo effect" (gain of up to 6°C) has been recorded in occupied beaver lodges in winter (Dyck and MacArthur 1993) but not in burrows of smaller ground squirrels (*Tamias striatus*) (MacLean 1981). Jarvis and Sale (1971) have assessed that decaying nesting material or faecal pellets in nests may provide a heat increase of up to 5°C. The question is whether the temperature benefit of "heating" with decaying wet organic materials could outweigh the potential costs due to enhancing probability of parasite (re)infection and worsening of fur condition. Indeed, at least *Fukomys* and *Heliophobius* mole-rats and coruros (*Spalacopus cyanus*) seem to avoid this kind of "heating" and prefer, in the wild, dry non-decaying materials such as plastic bags and nylon stockings as nest bedding. Also, subterranean rodents frequently abandon old nests with old bedding and build new nest chambers (Scharff and Grütjen 1997; Begall and Gallardo 2000; Scharff et al. 2001; Šumbera et al. 2004).

3.2.2 Humidity

Humidity is probably the most stable factor in burrows with the air of sealed burrows being almost saturated with water even if soil is dry or frozen (Kay and Whitford 1978; Zeng et al. 1984; Moore and Roper 2003). Still, as with other microenvironmental factors, burrow humidity depends on aboveground conditions in open systems as shown in *Dipodomys merriami* (Kay and Whitford 1978) or four species of *Gerbillurus* (Downs and Perrin 1989) originating from arid habitats where humidity in burrows is affected by presence of fog or cloudy weather. On the other hand, humidity in sealed burrows of *Heliophobius* does not differ between the beginning of the dry season following rains and mid of hot dry season (Šumbera et al. 2004). High humidity contributes to economising water balance, because in a highly water-saturated milieu, evaporative cooling from lung surface is minimal and subterranean mammals save water. It should be noted at this point that they do not drink free water even in captivity where evaporative cooling would be possible, but instead derive their water requirements from food.

3.2.3 Ventilation in Subterranean Burrows

Burrows of the European badger, a large carnivore, are open and partly ventilated by air currents created by external winds (Roper and Moore 2003). Ventilation is expected to be restricted or even absent in sealed burrow systems (Roper et al. 2001). Indeed, no significant air currents could be recorded in burrows of *Spalacopus cyanus* or *Fukomys mechowii* even after artificial opening at both ends (Scharff 1998; Begall, unpublished). Nevertheless, a certain degree of ventilation, related to external wind, has been demonstrated also in closed burrows of the European mole (*Talpa europaea*) (Olszewski and Skoczen 1965). Wind blowing above a fresh mound probably creates negative pressure that sucks out burrow air. This kind of ventilation may play a role also in breeding mounds of *Spalax* (Nevo 1999). Presence of ventilating openings in the vicinity of nests or on mounds (Olszewski and Skoczen 1965) or vertical tunnels ("chimneys") ending a few centimetres below the surface, frequently just under the turf, as found in some burrow systems of *Heliophobius* in several localities (Burda, Chitaukali, Šumbera, unpublished) may enable gas exchange between sub- and supraterranean environments. Olszewski and Skoczen (1965) have also measured different ventilation rates in different parts of the European moles' burrow systems. Stronger air currents have been recorded in main tunnels ("runways") than in side tunnels. Hence, burrow architecture, depth of particular components, dead tunnels, openings or mounds, etc. may influence ventilation. In sealed subterranean burrows ventilation of air is thus physically related mainly to two parameters: wind strength and soil porosity. Permeability of soil is generally the main factor influencing diffusion and gas exchange (Wilson and Kilgore 1978). Daily fluctuation of temperature may also cause air circulation in burrows (subtle turnover) as suggested already by Kennerly (1964). Theoretically, air cooled during the night in superficial burrows may displace warmer air in deeper interconnected tunnels, and it is expected that the greater the depth difference, the more significant the air turnover.

Movement of burrow inhabitant(s) is probably the main factor influencing movement of air in burrows. The diameter of tunnels of subterranean mammals is determined by the size of the digger (Šumbera et al. 2004). This relation is particularly apparent in solitary mammals, which are the sole builders and inhabitants of their respective burrows. An animal moving in the tunnel acts as a piston, increasing gas pressure in front of it and decreasing it behind - in a way similar to subway-trains in tunnels. Similarly, pushing excavated soil through tunnels to surface mounds may work in the same way.

3.2.4
Burrow Atmosphere

Concentration of respiratory gases is closely connected to ventilation in burrows. Gas exchange in an animal's burrow is mediated mainly via diffusion (Wilson and Kilgore 1978). This mechanism is particularly effective in highly porous soils. In opened burrows, convection may even play a greater role. Due to the nature of surrounding soil and burrow systems (architecture, length, depth of burrow component), respiratory gas concentration is more variable in space and time in subterranean burrows than above ground (e.g. Roper et al. 2001). It has been assumed that atmosphere in burrows (especially sealed ones) is oxygen poor and carbon dioxide rich, and burrowing animals face hypoxic and hypercapnic respiratory conditions.

Early studies have reported low oxygen concentration (14–18.4%) and high carbon dioxide concentration (0.22–4.8%) in burrows of subterranean rodents (McNab 1966; Darden 1972; Zeng et al. 1984). Similar values have been recorded in burrows of other mammals such as golden hamsters *Mesocricetus auratus* (Kuhnen 1986), ground squirrels *Tamias striatus* (MacLean 1981) or rabbits *Oryctolagus cuniculus* (Hayward 1966). However, no significant differences between burrow and ambient atmospheres have been found for kangaroo rats, *Dipodomys merriami* (Evans and Dill 1969; Soholt 1974), *Dipodomys spectabilis* (Kay and Whitford 1978), or the European badger *Meles meles* (Roper and Kemenes 1997). In recent studies on atmosphere in sealed burrows of African mole-rats, *Fukomys damarensis* and *Georychus capensis* (Roper et al. 2001) as well as in *Heliophobius argenteocinereus* (Šumbera et al. 2004), gas concentrations do not deviate significantly from aboveground conditions. Roper et al. (2001) argued that studies reporting low oxygen and high carbon dioxide concentrations have often been carried out under unnatural conditions or in artificial burrows, where natural gas exchange is limited or even blocked (see literature cited therein). However, large fluctuation in gas composition and high hypoxic (minimal oxygen level 7.2%) and hypercapnic (maximal carbon dioxide level of 6.1%) conditions have been found in *Spalax* burrows in heavy soils during rains (Shams et al. 2005).

Still, soil type and quality influence gas concentration in burrows. Gas ventilation in heavy, compact soils is limited. Thus, Arieli (1979), Shams et al. (2005) and Šumbera et al. (2004) have found lower oxygen concentrations in habitats with heavy compact soil in comparison to light soil, as measured in burrows of *Spalax* and *Heliophobius*, respectively. Similarly, gas concentration in beaver lodges has been related to the building material: Higher carbon dioxide concentrations are found in lodges constructed almost exclusively with mud, clay or compacted vegetation, whereas lodges with walls from twigs interspersed with mud are better ventilated (Dyck and MacArthur 1993). We expect that respiratory gas concentration within a single particular burrow system fluctuates also in spatial and temporal scale. Oxygen concentration in superficial foraging tunnels should be higher than in deeper tunnels, blind tunnels or chambers. Concurrently, carbon dioxide concentration is expected to raise with depth, as has been confirmed in burrows of *Dipodomys spectabilis* (Kay and Whitford 1978). Scharff et al. (2001) have found lower airing expressed as a proportion of coarse pores (diameter > 50 μm)/volume in deeper layers of ferralsol soil.

Concentration of respiratory gases in burrow atmosphere is also influenced by soil water content. Soils comprising more water also contain a higher amount of carbon dioxide (Kennerly 1964; Arieli 1979; Shams et al. 2005). Therefore, rains or flooding, filling pores in the soil, heavily affect gas exchange and concentration in burrows. Indeed, in burrows of *Spalax ehrenbergi* low oxygen- and high carbon dioxide-values have been recorded within a few hours after rainfall (Arieli 1979). Low oxygen concentration probably induces digging activity of *Spalax* to replenish burrow atmosphere (Arieli 1979). Influence of heavy rains could be important and dramatic especially in tropic mesic areas with several months of heavy rains. However, this situation has not been studied so far.

Concentration of respiratory gases may be influenced also by activity and metabolic rate of burrow inhabitants. Oxygen concentration is expected to be lower whereas carbon dioxide concentration should be higher in close vicinity of animals (mainly their noses) especially if they are digging or transporting soil with their heads (as in the case of *Spalax*). The reason is that working animals have higher metabolic rates and breathe into the small space between nose and soil, which is not easily ventilated. Similarly, nest chambers where animals spend much of their time are expected to be hypoxic and hypercapnic – particularly in communal nests. Marked hypoxic and hypercapnic conditions have been found in breeding mounds of *Spalax* (Arieli 1979; Shams et al. 2005). Ventilation of nests may be done by entering animals (pushing air columns in front of them – see above) and by connected ventilation tunnels (cf. Roper and Moore 2003). Commensals like arthropods, toads or other vertebrates may also affect gas composition, but in large burrow systems the effect is probably minimal (Šumbera et al. 2004). The effect of metabolically active plant roots and soil microorganisms on burrow atmosphere is not easy to estimate and has not been studied thus far.

In any case, as with behavioural thermoregulation, animals may also move from unfavourable hypoxic and hypercapnic conditions. Arieli (1979) has suggested that resting in different tunnel parts far from the nest, combined with pronounced physiological capacities of mole-rats to cope with hypoxia and hypercapnia (cf. Nevo 1999), and the extended complex burrow systems, which may be subjected to "behavioural ventilation", enable mole-rats to inhabit heavy soils intolerable to many fossorial rodents.

3.3 Further Perspectives

To understand better the factors affecting the microclimatic conditions of subterranean burrow systems, more comparative data are needed, especially long-term temperature measurements in nests (rather than in burrows) of diverse species from diverse habitats and different geographical regions. More importantly, to get insight about the real conditions, animals are facing in their environment, recordings should be done in occupied nests. Ventilation of burrows and nests is still poorly understood, and effective oxygen and carbon dioxide concentrations faced by working individuals and animals huddling in nests are of great interest. The problem of nitrogen and methane in nests and latrines has not been addressed so far and concentrations of these gases have not been measured. We can conclude that there is still a need for much further investigation to answer such basic open questions, as what are the real physiological limits of subterranean mammals and how do subterranean mammals survive flooding in the area of their occurrence?

References

Altuna CA (1985) Microclima de cuevas de *Ctenomys pearsoni* (Rodentia, Octodontidae) en arroyo carrasco (Montevideo). Actas J Zool Uruguay 1:59–60

Arieli R (1979) The atmospheric environment of fossorial mole rat (*Spalax ehrenbergi*): effect of season, soil texture, rain, temperature and activity. Comp Biochem Physiol A 63:569–575

Begall S, Gallardo M (2000). *Spalacopus cyanus* (Rodentia, Octodontidae): an extremist in tunnel constructing and food storing among subterranean mammals. J Zool Lond 251:53–60

Bennett NC, Jarvis JUM, Davies KC (1988) Daily and seasonal temperatures in the burrows of African rodent moles. S Afr J Zool 23:189–195

Bennett NC, Aguilar GH, Jarvis JUM, Faulkes CG (1994) Thermoregulation in three species of Afrotropical subterranean mole-rats (Rodentia: Bathyergidae) from Zambia and Angola and scaling within the genus *Cryptomys*. Oecologia 97:222–227

Brett RA (1991) The ecology of naked-mole-rat colonies: burrowing, food, and limiting factors. In: Sherman PW, Jarvis JUM, Alexander, RD (eds) The biology of the naked mole-rat. Princeton Univ Press, Princeton, NJ, pp 137–148

Burda H, Šumbera R, Chitaukali WN, Dryden GL (2005) Taxonomic status and remarks on ecology of the Malawian mole-rat, *Cryptomys whytei* (Thomas, 1897) (Rodentia, Bathyergidae). Acta Theriol 50:529–536

Busch C, Antinuchi CD, del Valle JC, Kittlein MJ, Malizia AI, Vassallo AI, Zenuto R (2000) Population ecology of subterranean rodents. In: Lacey EA, Patton JL, Cameron GN (eds) Life underground: the biology of subterranean rodents. Univ Chicago Press, Chicago, pp 183–226

Contreras LC, McNab BK (1990) Thermoregulation and energetics in subterranean mammals. In: Nevo E, Reig OA (eds) Evolution of subterranean mammals at the organismal and molecular levels. Wiley-Liss, New York, pp 231–250

Cutrera AP, Antinuchi CD (2004) Cambios en el pelaje del roedor subterráneo *Ctenomys talarum*: posible mecanismo térmico compensatorio (Fur changes in the subterranean rodent *Ctenomys talarum*: possible thermal compensatory mechanism). Rev Chil Hist Nat 77:235–242

Dammann P, Burda H (2006) Sexual activity and reproduction delays aging in a mammal. Curr Biol 16:R117–R118

Darden TR (1972) Respiratory adaptations of a fossorial mammal, the pocket gopher (*Thomomys bottae*). J Comp Physiol A 78:121–137

Downs MR, Perrin CT (1989) An investigation of the macro- and micro-environments of four *Gerbillurus* species. Cimbebasia 11:41–54

Dyck AP, MacArthur RA (1993) Seasonal variation in the microclimate and gas composition of beaver lodges in boreal environments. J Mammal 74:180–188

Evans DS, Dill DB (1969) Environmental conditions in the burrow of the kangaroo rat *Dipodomys merriami*. Physiologist 12:218

Hall LS, Myers K (1978) Variations in the microclimate in rabbit warrens in semi-arid New South Wales. Austr J Ecol 3:187–194

Hayward JS (1966) Abnormal concentration of respiratory gases in a rabbit burrow. J Mammal 47:723–724
Heth G (1989) Burrow patterns of the mole rat *Spalax ehrenbergi* in two soil types (terra-rossa and rendzina) in Mount Carmel, Israel. J Zool Lond 217:39–56
Hickman GC (1990) Adaptiveness of tunnel system features in subterranean mammal burrows. In: Nevo E, Reig OA (eds) Evolution of subterranean mammals at the organismal and molecular levels. Wiley-Liss, New York, pp 185–210
Jarvis JUM (1979) Ecological studies on *Heterocephalus glaber*, the naked mole-rat, in Kenya. Natl Geogr Soc Res Rep 1979:429–437
Jarvis JUM, Sale JB (1971) Burrowing and burrow patterns of East African mole-rats *Tachyoryctes*, *Heliophobius* and *Heterocephalus*. J Zool Lond 163:451–479
Kay RF, Whitford WG (1978) The burrow environment of the banner-tailed kangaroo rat, *Dipodomys spectabilis*, in South-central New Mexico. Am Mid Nat 99:270–279
Kennerly TE Jr (1964) Microenvironmental conditions of the pocket gopher burrow. Texas J Sci 16:395–441
Kinlaw A (1999) A review of burrowing by semi-fossorial vertebrates in arid environments. J Arid Environ 41:127–145
Kuhnen G (1986) O_2 and CO_2 concentrations in burrows of euthermic and hibernating golden hamsters. Comp Biochem Physiol A 84:517–522
Lovegrove BG, Knight-Eloff A (1988) Soil and burrow temperatures, and the resource characteristics of the social mole-rat *Cryptomys damarensis* (Bathyergidae) in the Kalahari desert. J Zool Lond 216:403–416
Luo ZX, Wible JR (2005) A late Jurassic digging mammal and early mammalian diversification. Science 308:103–107
MacLean GS (1981) Torpor patterns and micro-environment of the eastern chipmunk, *Tamias striatus*. J Mammal 62:64–73
Marhold S (1989) Stoffwechsel, Atem- und Herztätigkeit des afrikanischen Graumulls. MSc Thesis, JW Goethe-Univ, Frankfurt am Main
Marhold S, Nagel A (1995) The energetics of the common mole rat *Cryptomys*, a subterranean eusocial rodent from Zambia. J Comp Physiol B 164:636–645
McNab BK (1966) The metabolism of fossorial rodents: a study of convergence. Ecology 47:712–733
Mohr ECJ, Baren FA, van Schuylenborgh J (1972) Tropical soils – a comprehensive study of their genesis. Mouton Ichtiar Baru Van Hoeve, Den Haag
Moore JAH, Roper TJ (2003) Temperature and humidity in badger *Meles meles* setts. Mammal Rev 33:308–313
Nevo E (1979) Adaptive convergence and divergence of subterranean mammals. Ann Rev Ecol Syst 10:269–308
Nevo E (1999) Mosaic evolution of subterranean mammals: regression, progression and global convergence. Oxford Univ Press, Oxford
Nevo E, Reig OA (1990) Evolution of subterranean mammals at the organismal and molecular levels. Wiley-Liss, New York
Olszewski JL, Skoczen S (1965) The airing of burrows of the mole, *Talpa europaea* Linnaeus, 1758. Acta Theriol 10:181–193
Reichman OJ, Smith SC (1990) Burrows and burrowing behavior by mammals. In: Genoways HH (ed) Curr Mammal, pp 197–243
Rezende EL, Cortes A, Bacigalupe LD, Nespolo RF, Bozinovic F (2003) Ambient temperature limits above-ground activity of the subterranean rodent *Spalacopus cyanus*. J Arid Environ 55:63–74
Roper TJ, Kemenes I (1997) Effect of blocking of entrances on the internal environment of badger *Meles meles* setts. J Appl Ecol 34:1311–1319

Roper TJ, Moore JAH (2003) Ventilation of badger *Meles meles* setts. Mamm Biol 68:277–283
Roper TJ, Bennett NC, Conradt L, Molteno AJ (2001) Environmental conditions in burrows of two species of African mole-rat, *Georhychus capensis* and *Cryptomys damarensis*. J Zool Lond 254:101–107
Scharff A (1998) Systematik und Verhaltensökologie sambischer Sandgräber (Bathyergidae, Rodentia). PhD Thesis, Univ Duisburg-Essen, Essen
Scharff A, Grütjen O (1997) Evidence for aboveground activity of Zambian mole-rats (*Cryptomys*, Bathyergidae, Rodentia). Z Säugetierk 62:253–254
Scharff A, Locker-Grütjen O, Kawalika M, Burda H (2001) Natural history of the giant mole-rat, *Cryptomys mechowi* (Rodentia: Bathyergidae), from Zambia. J Mammal 82:1003–1015
Shams I, Avivi A, Nevo E (2005) Oxygen and carbon dioxide fluctuation in burrow of subterranean blind mole-rats indicate tolerance to hypoxic-hypercapnic stress. Comp Biochem Physiol A 142:376–382
Shenbrot G, Krasnov B, Khokhlova I, Demidova T, Fielden L (2002) Habitat-dependent differences in architecture and microclimate of the burrows of Sundevall's jird (*Meriones crassus*) (Rodentia: Gerbillinae) in the Negev Desert, Israel. J Arid Environ 51:265–279
Soholt LF (1974) Environmental conditions in an artificial burrow occupied by Merriam's kangaroo rat, *Dipodomys merriami*. J Mammal 55:859–864
Spinks AC, Bennett NC, Jarvis JUM (2000) A comparison of ecology of two populations of common mole rat, *Cryptomys hottentotus hottentotus*: the effect of aridity on food, foraging and body mass. Oecologia 125:341–349
Šklíba J, Šumbera R, Chitaukali WN, Burda H (in press) Determinant of daily activity patterns in free-living Afrotropical solitary subterranean rodent. J Mammal
Šumbera R, Burda H, Chitaukali WN, Kubová J (2003) Silvery mole-rat (*Heliophobius argenteocinereus*, Bathyergidae) change their burrow architecture seasonally. Naturwiss 90:370–373
Šumbera R, Chitaukali WN, Elichová M, Kubová J, Burda H (2004) Microclimatic stability in burrows of an Afrotropical solitary bathyergid rodent, the silvery mole-rat (*Heliophobius argenteocinereus*). J Zool Lond 263:409–416
Van Wambeke A (1992) Soils of the tropics – properties and appraisal. McGraw-Hill, New York
Vleck D (1979) The energy costs of burrowing by the pocket gopher *Thommomys bottae*. Physiol Zool 52:122–136
Wilson KJ, Kilgore DL Jr (1978) The effect of location and design on the diffusion of respiratory gases in mammal burrows. J Theor Biol 71:73–101
Zelová J, Šumbera R, Sedláček F, Burda H (in press) The energetics in the solitary subterranean rodent, the silvery mole-rat, *Heliophobius argenteocinereus*, and the allometry of RMR in African mole-rats (Bathyergidae). Comp Biochem Physiol
Zeng J, Wang Z, Shi Z (1984) Metabolic characteristics and some physiological parameters of mole rat (*Myospalax baileyi*) in alpine area. Acta Biol Plat Sin 3:163–171

4 New Data on Metabolic Parameters in Subterranean Rodents

František Sedláček

4.1 Introduction

As a result of similar environmental pressures (cf. Burda et al., this volume), subterranean rodents share a variety of convergent physiological specializations such as low basal metabolic rate, low body temperature and high thermal conductance (reviewed, e.g. by Contreras and McNab 1990; Nevo 1999; Buffenstein 2000).

Three hypotheses attempt to explain low *BMR* in subterranean rodents. 1) The "thermal stress hypothesis" proposes that reduced *BMR* minimizes the risk of overheating in burrows where common cooling mechanisms as evaporative water loss and convective cooling are substantially reduced (McNab 1966, 1979). 2) The "respiratory stress hypothesis" relates low *BMR* to hypoxic and hypercapnic conditions in burrows (Darden 1972; Arieli 1979). 3) The "cost-of-burrowing hypothesis" suggests that reduced *BMR* compensates for the enormous energetic demands of subterranean foraging and represents an energy saving mechanism (Vleck 1979). The latter factor was considered as particularly important in arid regions with widely dispersed and scarce food resources (Jarvis 1978; Lovegrove 1987; Lovegrove and Wissel 1988).

Here, I have revisited the subject and show how recent studies (which could not be reviewed by Buffenstein 2000) have complemented or changed our understanding of physiology of metabolism in subterranean rodents, or even how they determine new ways and open new horizons for future research in the field. I focus mainly on African mole-rats (Bathyergidae), a group of strictly subterranean rodents. To evaluate the situation in mole-rats, data on fossorial rodents are considered too.

František Sedláček: Faculty of Biological Sciences, University of South Bohemia, Branišovská 31, 37005 České Budějovice, Czech Republic, E-mail: sedlacek@usbe.cas.cz

Subterranean Rodents: News from Underground
S. Begall, H. Burda, C.E. Schleich (Eds.)
© Springer-Verlag Berlin Heidelberg 2007

Standard physiological abbreviations used in the text

BMR	= basal metabolic rate (the amount of energy expended while at rest in the thermoneutral zone, in the post-absorptive state – meaning that the stomach is empty and the digestive system is inactive, i.e. the animal is fasting but not starving)
DEE	= daily energetic expenditure
DMR	= digging metabolic rate
NST	= non-shivering thermogenesis
RMR	= resting metabolic rate (corresponds principally to BMR but is measured under less restrictive conditions as far as the requirement of post-absorptive state is concerned)
T_a	= ambient temperature
T_b	= body core temperature
TNZ	= thermoneutral zone (a neutrally temperate environment, range of ambient temperatures in which *BMR* is the lowest)

4.2 Variation of Basal Metabolic Rate on Interspecific Level

4.2.1 Resting Metabolic Rates

From the physiological point of view, the bathyergid mole-rats represent the most profoundly studied family of subterranean rodents. The results of previous studies led to the conclusion that *RMR* was independent of *body mass* and was thus deviating from the general mammalian pattern. The feature was considered a preadaptation and precondition for evolution of eusociality characterizing some bathyergids (cf. Lovegrove and Wissel 1988). However, new data on metabolic rate and thermoregulation in solitary silvery mole-rats, *Heliophobius argenteocinereus*, and meta-analysis of all published data on *RMR* in African mole-rats (see Table 4.1) force us to change the established view (Zelová et al., in press). The presented results combined with data on other mole-rats in a phylogenetically controlled analysis show clearly that *RMR* within the Bathyergidae is mass-dependent. The earlier conclusion of mass independent *RMR* was not sufficiently founded, having been based on a small species sample and confounding effect of an outlying species - the poikilothermic naked mole-rat *Heterocephalus glaber*.

Recently, White (2003) evaluated the influence of foraging mode and aridity on *BMR*s in burrowing mammals. Because the analysis shows that burrowing mammals in arid habitats have a lower *BMR* than those in the mesic ones, subterranean and fossorial species were compared within

Table 4.1. The body mass, body temperature, resting metabolic rate (*RMR*), conductance (*C*) and thermoneutral zone (*TNZ*) in 13 species of African mole-rats (Bathyergidae) from different habitats

Species	Body mass (g)	Body temperature (°C)	RMR (ml $g^{-1}h^{-1}$)	TNZ (°C)	C (ml $g^{-1}h^{-1}$ $°C^{-1}$)	Habitat	References
Bathyergus janetta	406	34.8	0.53	27.5–33.2	0.061	Semi-arid	Lovegrove 1986b
B. suillus	712	35.3	0.49	25–31	0.038	Mesic	Lovegrove 1986b
Cryptomys hottentotus	75	34	0.90	27–30	0.13	Semi-arid	Bennett et al. 1992
C. hottentotus	95	35.8	0.68	30–32	0.14	Semi-arid	Haim & Fairall 1986
C. nimrodi (highveld)	81	34.4	0.83	31–32	0.17	Mesic	Bennett et al. 1996
C. nimrodi (lowveld)	99	34.0	0.95	28–30	0.14	Mesic	Bennett et al. 1996
C. natalensis	102	33.8	0.80	30–31.5	0.13	Mesic	Bennett et al. 1993b
Fukomys anselli	82	36.1	0.76	32.5	0.14	Mesic	Marhold & Nagel 1995
F. anselli	77	33.8	0.63	28–32	0.12	Mesic	Bennett et al. 1994
F. bocagei	94	33.7	0.74	31.5–32.5	0.12	Mesic	Bennett et al. 1994
F. damarensis	131	35	0.66	28–31	0.065	Semi-arid	Bennett et al. 1992
F. damarensis	124.5	35.2	0.57	27–31	0.085	Arid	Lovegrove 1986a
F. darlingi	60	33	0.98	28–31.5	0.19	Mesic	Bennett et al. 1993a
F. mechowii	272	34	0.60	29–30	0.09	Mesic	Bennett et al. 1994
Georychus capensis	193	36.4	0.59	26.3–34	0.046	Mesic	Lovegrove 1987
Heliophobius argenteocinereus	88	35.1	0.85	28–33	0.139	Semi-arid	McNab 1966
H. argenteocinereus	223	33.6	0.68	25–33	0.077	Mesic	Zelová et al. in press
Heterocephalus glaber	39	32.1	0.64	31–37	0.387	Arid	McNab 1966
H. glaber	42	33	1.00	31–34		Arid	Buffenstein & Yahav 1991

these groups. The *BMR* of subterranean and fossorial mammals from mesic habitats could not be distinguished, nor could the *BMR* of large (>77 g) subterranean and fossorial mammals from arid areas. This finding favours the thermal-stress hypothesis, because ecologically different groups appear to have similar *BMR*s despite differences in foraging costs. However, in support of the cost-of-burrowing hypothesis, smaller (<77 g) aridity-adapted subterranean mammals exhibit a significantly lower *BMR* than fossorial aridity-adapted mammals of similar size. Given their high mass-specific metabolic rates, small mammals in arid environments are expected to be under severe energy and water stress. Under such conditions, the greatly reduced *BMR* of small subterranean species may compensate for the enormous energetic demands of subterranean foraging. It is of great importance that selection can differently act in two mass categories. We have to take into account that successful energetic balance is reached by a set of parameters (metabolism, body mass, body insulation, foraging etc.).

4.2.2 Energetic Expenditure During Digging

Digging for food and shelter is an energetically demanding process that can result in energy expenditure more than 300 times higher than that required to move the same distance across the soil surface (Vleck 1979). Therefore this activity determines the total energy balance. Ebensperger and Bozinovic (2000) investigated the energetics and burrowing behaviour in the fossorial degu (*Octodon degus*). Mass-specific *DMR* was higher in animals burrowing in soft soils compared to hard soils. However, this was due to the fact that in soft soils more soil per time unit was removed than in hard soils. As expected, excavating hard soil was energetically more expensive per unit of removed soil compared to soft soil. In the field, increased digging activity coincided with occurrence of rainfall, more humid and relatively softer soils. We could speculate that in hard soils it would be useful to dig slowly or to delay the digging activity (for several days) in order to wait for generally more favourable situations above. Digging costs of degus tend to be similar to (or higher than) those of subterranean geomyids and bathyergids.

Luna et al. (2002) quantified energetic costs of burrowing in the tucotuco *Ctenomys talarum* under natural soil conditions. *DMR* in damp sand was 2.8 times higher than *RMR*. This value is low in comparison with other subterranean rodents: e.g. in *Fukomys damarensis DMR* was 5 times higher than *RMR* (Lovegrove 1989), in *Thomomys bottae* 4.8 times (Vleck 1979). Although body size plays an important role in total energy budget (Vleck 1979), no differences in *DMR* were detected between sexes of *C. talarum* with a sexual dimorphism in body size. Parameters calculated according to the "burrowing model" (Vleck 1979) were comparable to other subterranean rodents. Despite different foraging habits of *C. talarum* compared to other subterranean rodents, the cost of foraging remains energetically comparable, since new tunnel systems are also used to access new plant patches (Luna et al. 2002). In their following study Luna and Antinuchi (2006) tested the effect of soil hardness on digging cost in the same species. Similar to the situation in fossorial degus (Ebensperger and Bozinovic 2000), their results also suggest that in *C. talarum* digging in hard soils is energetically more expensive than in soft soil.

As expected, *DMR* in *C. talarum* is lowest within the thermoneutral zone and increases below and above its limits (Luna and Antinuchi 2007). Interestingly, below thermoneutrality the costs of digging and thermoregulation were additive and heat production for thermoregulation might be compensated by heat produced as a by-product of muscular activity during digging. At higher temperatures conduction between animals and soil provide the primary sources of dissipating metabolic heat. The authors observed that individuals of *C. talarum* pressed the ventral surface of their body against soil in the digging chamber probably to dissipate heat during digging bouts.

The costs of burrowing were also related to the genetic variability among populations. In the pocket gopher *Thomomys bottae* costs of burrowing were significantly higher in populations with lower genetic variability (Hildner and Soule 2004). The authors concluded that these differences are reflected in physiological fitness differences for a trait that is essential to survival.

Bozinovic et al. (2005) tested the thermal-stress and the cost-of-burrowing hypotheses among seven populations of the coruro *Spalacopus cyanus* from different geographic localities with contrasting habitat conditions. Surprisingly, their results support neither the thermal-stress nor the cost-of-burrowing hypothesis. Animals from different habitats exhibited similar *BMR*s and *DMR*s when measured under similar semi-natural conditions. The authors concluded that coruros having originated in the Ands adapted to relatively hard soils. Later, when the populations moved into coastal areas with softer soils, they have retained the original adaptation without further phenotypic changes. Apparently, the phylogeography as well as palaeoecology have to be taken into account when interpreting current features.

4.2.3 Endothermy and Poikilothermy in the Naked Mole-rat

The intensively studied bathyergid, naked mole-rat, *Heterocephalus glaber*, lives in extended families and relies on ectothermic behavioural mechanisms (huddling, basking in warm superficial tunnels or cooling up in deeper burrows) to maintain body temperature. Outside its thermally buffered underground milieu, it is unable to effectively regulate T_b, so that T_b follows T_a (Buffenstein 2000; Woodley and Buffenstein 2002).

Goldman et al. (1999) observed that the metabolic rate decreased with increasing T_a over the range 23–34°C in awake naked mole-rats, whereas in anesthetized animals (suppression of *NST* via prevention of endogenous noradrenaline release) rates of O_2 consumption were very low over the whole range of T_a and tended to increase with increasing T_a. Injections of noradrenaline led to rapid increases in metabolic rate at all T_as in anesthetized subjects and also at *TNZ* in awake mole-rats. However, at lower ambient temperatures, awake subjects given noradrenaline showed little stimulation of O_2 consumption beyond the already elevated baseline rates observed at these T_as. It shows that *H. glaber* has thermoregulatory components like other homeothermic mammals. Relatively high capacity of *NST* has been already reported by Hislop and Buffenstein (1994). From the quantitative point of view, however, the thermoregulatory system below *TNZ* is unable to compensate the heat loss and operates as in a poikilothermic animal. Woodley and Buffenstein (2002) found that prolonged cold exposure did not elicit any increase in *NST* capacity. The authors conclude that naked mole-rats respond to that condition by increasing basal levels of heat production but do not show improved *NST* capacity. The inability to augment their brown adipose tissue thermogenic capacity in response to cold may reflect their evolutionary history in a thermally stable environment. Data on thermogenic capacity of the subterranean octodontid *S. cyanus* also suggest its low physiological plasticity, which is in accordance with a subterranean mode of life (Nespolo et al. 2001).

To get more insight into endothermy of *H. glaber*, Buffenstein et al. (2001) also investigated changes in the thyroid status during chronic cold exposure. Under simulated burrow conditions, free thyroxine (T_4) and thyroid stimulating hormone (*TSH*) levels fell within the reptilian range, one order of magnitude lower than mammalian levels. However, cold induced typical mammalian responses, i.e., during cold exposure, *H. glaber* conforms to typical small mammal profiles and increases thyroid gland activity and basal metabolism. Despite these typical endothermic mechanisms, even after prolonged exposure to a drop in T_a, heat generation is inadequate to maintain body temperature. This inability to regulate T_b may reflect limitations to endothermic metabolism imposed by rate-limiting physiological systems or may reflect inadequate thyroid function. Maintenance of an elevated temperature may be important to this poikilothermic mammal, where physiological function (e.g., gut activity, pregnancy) and locomotion are temperature dependent.

Goldman et al. (1999) found also that during chronic restriction of food to 60–70% of their normal daily consumption, naked mole-rats decrease their metabolic rates, yet metabolic rates were not altered following several hours of acute food deprivation. Food consumption remained somewhat decreased after chronic food restriction, even when animals were returned to ad lib conditions. However, body weights returned to prerestriction values, despite the continued reduction in ad lib food intake. These observations suggest that *H. glaber* may be capable of long-lasting metabolic adaptations as a mean to cope with restricted food supply. Therefore, quantification of starvation should be taken into consideration if energy balance is evaluated during a mole-rat life period.

4.3 Variation of Resting Metabolic Rate on Intraspecific Level

4.3.1 Energetics of Reproduction

As in most other mammals, reproducing subterranean rodent females must meet the increased energetic demands of pregnancy. At the same time, changes in maternal body shape and mass during pregnancy lead to changes in heat transfer properties, and these, too, must be in balance with the physical conditions within the burrows (cf. Buffenstein 2000). For instance, female naked mole-rats are able to maintain higher T_b particularly by increased metabolic heat production as well as by improving insulatory properties of skin whereas pregnant females become less sensitive to changes in T_a (Buffenstein et al. 1996; Urison and Buffenstein 1994). Compared to the extremely precocial guinea-pig (Kunkele and Trillmich 1997), metabolism increase in *H. glaber* during lactation is more pronounced.

Veloso and Bozinovic (2000) examined the effect of food quality on ingestion, digestion, and metabolic rate during pregnancy and lactation in precocial degus and found the highest increase (39%) in *RMR* in lactating females maintained on high-quality food. The study stresses that conversion efficiency of metabolizable energy into tissue growth appears to be linked to quality of food. Similarly to the analysis in *H. glaber*, Zenuto et al. (2002) evaluated the maternal costs of reproduction and pup development in *C. talarum*. Metabolic rates during pregnancy and lactation were 130% and 150% of the resting level observed in non-reproductive females. The total additional energy cost of reproduction above the non-reproductive level was similar for both the gestation and lactation periods. Mass-specific *RMR* revealed an increase of metabolism during lactation but not during gestation. However, no differences were observed in T_b between non-reproductive, pregnant, or lactating females. The differences to *H. glaber* (cf. Urison and Buffenstein 1995) probably reflect the different maternal thermal conditions for embryonic development. Energy and material distribution to embryos as well as their postnatal development show high temperature sensitivity (Hill 1992).

A follow-up study in the tuco-tuco *C. talarum* focused on the influence of mother's contact and huddling with nest mates on *RMR* and T_b of pups (Cutrera et al. 2003). At lower T_a, huddling and contact with the mother reduced pups' body heat loss until they were 15 days old but did not affect their *RMR*. Fifteen-day-old pups showed an increase in their *RMR*, associated with the onset of independent thermoregulation. Pups older than 15 days showed a less variable T_b and their *RMR* decreased. Pups spent 80% of the time in contact with their mother and, when she was absent, they spent less time huddling with their nest mates. However T_a seemed to have no effect upon this behaviour. Huddling had apparently no effect on energy expenditure of young tuco-tucos. This finding was related to stable thermal conditions found in burrows. Furthermore, changing fur density and length in tuco-tucos may function as a compensatory mechanism during seasonal temperature changes in burrows and during pregnancy (Cutrera and Antinuchi 2004). In lactating female *C. talarum* the shorter ventral fur probably indicates heat dissipation from enhanced *RMR* (see Zenuto et al. 2002; Cutrera et al. 2003) and heat flux to pups. However, this energy transfer to pups is not crucial for the postnatal development (Cutrera et al. 2003).

Sexual dimorphism in mammals is frequently connected with high energetic costs of behaviours (e.g. fighting in males). Different reproductive strategies between males and females may lead to different energetic requirements and different physiological performances. Scantlebury et al. (2006a) compared *DEE* and *RMR* in two solitary bathyergids (*Bathyergus janetta* and *Georychus capensis*) with different sexual body size dimorphism during periods of intense digging when males seek females. The authors tested the hypothesis, that larger body size in male could be indicative of greater digging or fighting capacities, and hence greater mass independent *DEE*, in dimorphic *B. janetta*. Their results did not confirm the prediction about mass independent *DEE*. The energy costs of dimorphism are thus simply connected with maintenance of a larger body size rather than with the costs of male-specific behaviours.

4.3.2
Metabolic Rate and Sociality

While in surface-dwelling species aggregation into colony serves often as an anti-predator strategy, underground colonial life may reflect also a kind of energy management. Referring to sociality, several areas of interest and promising topics for future studies in energetics can be identified.

4.3.2.1
Energy Balance per Capita

Spinks and Plaganyi (1999) calculated the energy balance in social common mole-rats *Cryptomys hottentotus* and concluded that increased group size and cooperative foraging lower foraging costs and reduce the risks of unproductive foraging. The authors found that energetic returns per capita decreased with increasing group size suggesting that there was no simple energetic benefit to be gained from increased colony size. Rather, cooperative foraging serves to reduce the risk of starvation, particularly in arid habitats. The model suggests that the energetic rewards gained from foraging in mesic habitats are almost three times greater than those gained in arid regions. The results correspond to the model of Lovegrove and Wissel (1988) based on the "risk-sensitive metabolism hypothesis". Here the mass-independent scaling of mass-specific *RMR* is one of the main presumptions. However, the analysis of *RMR* in African mole rats change the relationship between body mass and *RMR* into a mass-dependent correlation (see Sect. 4.2.1). Therefore, the mentioned hypothesis should be reviewed accordingly.

4.3.2.2
Physiologically Distinct Castes

In social African mole-rats, division of labour between colony members exists. In thoroughly studied eusocial Damaraland mole-rats *Fukomys damarensis* frequent and infrequent workers are distinguished (Bennett and Jarvis 1988). Infrequent workers which perform less than 5% of colony work are probably colony dispersers leaving the colony during relatively short and rare periods when soils are softened by rains. Scantlebury et al. (2006b) found that infrequent workers are larger and fatter than frequent workers and that they also increase their *DEE* compared to frequent workers after rainfall. The authors suggest that infrequent workers constitute a physiologically distinct dispersing caste investing preferably more into their body reserves in preparation to dispersal and reproduction instead of contributing to the work of colony. Increasing *DEE* and sustained metabolic scopes after periods of rainfall may suggest that rainfall triggers energetic demanding activity like, e.g. digging (see Sect. 4.2.2).

4.3.2.3
Metabolic Rate and Aging

O'Connor and Buffenstein (2000) and O'Connor et al. (2002) studied the effect of aging (cf. also Dammann and Burda, this volume) upon some physiological and morphological parameters in the naked mole-rat. The authors found among 5-, 10-, and 20-year-old individuals no age-related changes in body mass, *RMR*, body fat-lean mass ratio, fat-free mass, or bone mineral density. Although *H. glaber* have low metabolic rates, their prolonged longevity (>28 years) results in a lifetime energy expenditure more than 4 times that of mice. The data from these long lasting studies of naked mole rats' ecophysiology have produced an outstanding model for further ageing research (Buffenstein 2005).

Recently, Dammann and Burda (2006, this volume) introduced another potential model for longevity studies – the Ansell's mole-rat *Fukomys anselli*. In this eusocial species, pair bonding and/or sexual activity and/or breeding is apparently the main factor prolonging lifespan. The authors discuss ultimate and proximate explanations of this phenomenon. Although the breeding and non-breeding mole-rats did not differ in time budgets allocated to energy-consuming or -saving or -gaining activities in the laboratory, more detailed analyses are needed regarding the actually gained or consumed energy. As the authors state, differential metabolism in animals of different sexes, ages, sizes, and breeding status has to be measured to recognize the role of energy turnover in determining the lifespan.

Acknowledgements. I thank S. Begall, H. Burda, R. Šumbera, C. Schleich, and J. Zelová for their valuable comments and suggestions on the manuscript. This overview was supported by the grant MSMT6007665801 from the Czech Minitry of Education.

References

Arieli R (1979) The atmospheric environment of fossorial mole rat (*Spalax ehrenbergi*): effect of season, soil texture, rain, temperature and activity. Comp Biochem Physiol A 63:569–575

Bennett NC, Jarvis JUM (1988) The social structure and reproductive biology of colonies of the mole-rat, *Cryptomys damarensis*, (Rodentia, Bathyergidae). J Mammal 69:293–302

Bennett NC, Clarke BC, Jarvis JUM (1992) A comparison of metabolic acclimation in two species of social mole-rats (Rodentia, Bathyergidae) in southern Africa. J Arid Environ 22:189–198

Bennett NC, Jarvis JUM, Cotterill FPD (1993a) Poikilothermic traits and thermoregulation in the Afrotropical social subterranean Mashona mole-rat (*Cryptomys hottentotus darlingi*) (Rodentia: Bathyergidae). J Zool Lond 231:179–186

Bennett NC, Taylor PJ, Aguilar GH (1993b) Thermoregulation and metabolic acclimation in the Natal mole-rat (*Cryptomys hottentotus natalensis*) (Rodentia, Bathyergidae). Z Säugetierk 58:362–368

Bennett NC, Aguilar GH, Jarvis JUM, Faulkes CG (1994) Thermoregulation in three species of Afrotropical subterranean mole-rats (Rodentia: Bathyergidae) from Zambia and Angola and scaling within the genus *Cryptomys*. Oecologia 97:222–227

Bennett NC, Cotterill FPD, Spinks AC (1996) Thermoregulation in two populations of the Matabeleland mole-rat (*Cryptomys hottentotus nimrodi*) and remarks on the general thermoregulatory trends within the genus *Cryptomys* (Rodentia: Bathyergidae) J Zool Lond 239:17–27

Bozinovic F, Carter MJ, Ebensperger LA (2005) A test of the thermal-stress and the cost-of-burrowing hypotheses among populations of the subterranean rodent *Spalacopus cyanus*. J Comp Biochem Physiol A 140:329–336

Buffenstein R (2000) Ecophysiological responses of subterranean rodents to underground habitats. In: Lacey E, Patton JL, Cameron GN (eds) Life underground: the biology of subterranean rodents. Univ Chicago Press, Chicago, pp 62–110

Buffenstein R (2005) The naked mole-rat: a new long-living model for human aging research. J Gerontol A 60:1369–1377

Buffenstein R, Yahav S (1991) Is the naked mole-rat *Heterocephalus glaber* an endothermic yet poikilothermic mammal? J Therm Biol 16:227–232

Buffenstein R, Urison NT, Van der Westhuizen LA, Woodley R, Jarvis JUM (1996) Temperature changes during pregnancy in the subterranean naked mole-rat (*Heterocephalus glaber*): the role of altered body composition and basking behaviour. Mammalia 60:617–626

Buffenstein R, Woodley R, Thomadakis C, Daly TJM, Gray DA (2001) Cold-induced changes in thyroid function in a poikilothermic mammal, the naked mole-rat. Am J Physiol 280:R149–R155

Contreras LC, McNab BK (1990) Thermoregulation and energetics in subterranean mammals. In: Nevo E, Reig OA (eds) Evolution of subterranean mammals at the organismal and molecular levels. Wiley-Liss, New York, pp 231–250

Cutrera AP, Antinuchi CD (2004) Fur changes in the subterranean rodent *Ctenomys talarum*: possible thermal compensatory mechanism. Rev Chil Hist Nat 77:235–242

Cutrera AP, Antinuchi CD, Busch C (2003) Thermoregulatory development in pups of the subterranean rodent *Ctenomys talarum*. Physiol Behav 79:321–330

Dammann P, Burda H (2006) Sexual activity and reproduction delay ageing in a mammal. Curr Biol 16:R117–R118

Darden TR (1972) Respiratory adaptations of a fossorial mammal, the Pocket Gopher (*Thomomys bottae*). J Comp Physiol A 78:121–137

Ebensperger LA, Bozinovic F (2000) Energetics and burrowing behaviour in the semifossorial degu *Octodon degus* (Rodentia: Octodontidae). J Zool Lond 252:179–186

Goldman BD, Goldman SL, Lanz T, Magaurin A, Maurice A (1999) Factors influencing metabolic rate in naked mole-rats (*Heterocephalus glaber*). Physiol Behav 66:447–459

Haim A, Fairall N (1986) Physiological adaptations to the subterranean environment by the mole rat *Cryptomys hottentotus*. Cimbebasia 8A:49–53

Hildner KK, Soule ME (2004) Relationship between the energetic cost of burrowing and genetic variability among populations of the pocket gopher, *T. bottae*: Does physiological fitness correlate with genetic variability? J Exp Biol 207:2221–2227

Hill RW (1992) The altricial/precocial contrast in the thermal relations and energetics of small mammals. In: Tomasi TE, Horton TH (eds) Mammalian energetics. Cornell UP, Ithaca, London, pp 186–212

Hislop MS, Buffenstein R (1994) Noradrenaline induces non-shivering thermogenesis in both the naked mole-rat (*Heterocephalus glaber*) and the Damara mole-rat (*Cryptomys damarensis*) despite very different modes of thermoregulation. J Therm Biol 19:25–32

Jarvis JUM (1978) Energetics of survival in *Heterocephalus glaber* (Rüppell), the naked mole-rat (Rodentia: Bathyergidae). Bull Carnegie Mus Nat Hist 6:81–87

Kunkele J, Trillmich F (1997) Are precocial young cheaper? Lactation energetics in the quinea pig. Physiol Zool 70:589–596
Lovegrove BG (1986a) The metabolism of social subterranean rodents: adaptation to aridity. Oecologia 69:551–555
Lovegrove BG (1986b) Thermoregulation of the subterranean rodent genus *Bathyergus* (Bathyergidae). S Afr J Zool 21:283–288
Lovegrove BG (1987) Thermoregulation in the subterranean rodent *Georychus capensis* (Rodentia: Bathyergidae). Physiol Zool 60:174–180
Lovegrove BG (1989) The cost of burrowing by the social mole rats (Bathyergidae) *Cryptomys damarensis* and *Heterocephalus glaber*: the role of soil moisture. Physiol Zool 62:449–469
Lovegrove BG, Wissel C (1988) Sociality in mole rats - metabolic scaling and the role of risk sensitivity. Oecologia 74:600–606
Luna F, Antinuchi CD (2006) Cost of foraging in the subterranean rodent *Ctenomys talarum*: effect of soil hardness. Can J Zool 84:661–667
Luna F, Antinuchi CD (2007) Energetics and thermoregulation during digging in the rodent tuco-tuco (*Ctenomys talarum*). Comp Biochem Physiol A 146:559–564
Luna F, Antinuchi CD, Busch C (2002) Digging energetics in the South American rodent *Ctenomys talarum* (Rodentia, Ctenomyidae). Can J Zool 80:2144–2149
Marhold S, Nagel A (1995) The energetics of the common mole rat *Cryptomys*, a subterranean eusocial rodent from Zambia. J Comp Physiol B 164:636–645
McNab BK (1966) The metabolism of fossorial rodents: a study of convergence. Ecology 47:712–733
McNab BK (1979) The influence of body size on the energetics and distribution of fossorial and burrowing mammals. Ecology 60:1010–1021
Nespolo RF, Bacigalupe LD, Rezende EL, Bozinovic F (2001) When nonshivering thermogenesis equals maximum metabolic rate: thermal acclimation and phenotypic plasticity of fossorial *Spalacopus cyanus* (Rodentia). Physiol Biochem Zool 74:325–332
Nevo E (1999) Mosaic evolution of subterranean mammals: regression, progression and global convergence. Oxford Univ Press, Oxford
O'Connor TP, Buffenstein R (2000) Resting metabolic rate does not decline with age in long-lived naked mole-rats. Faseb J 14:A440
O'Connor TP, Lee A, Jarvis JUM, Buffenstein R (2002) Prolonged longevity in naked mole-rats: age-related changes in metabolism, body composition and gastrointestinal function. Comp Biochem Physiol A 133:835–842
Scantlebury M, Speakman JR, Bennett NC (2006a) The energy costs of sexual dimorphism in mole-rats are morphological not behavioural. Proc R Soc Lond B 273:57–63
Scantlebury M, Speakman JR, Oosthuizen MK, Roper TJ, Bennett NC (2006b) Energetics reveals physiologically distinct castes in eusocial mammal. Nature 440:795–797
Spinks AC, Plaganyi EE (1999) Reduced starvation risks and habitat constraints promote cooperation in the common mole-rat, *Cryptomys hottentotus hottentotus*: a computer-simulated foraging model. Oikos 85:435–444
Urison NT, Buffenstein R (1994) Shifts in thermoregulatory patterns with pregnancy in the poikilothermic mammal - the naked mole-rat (*Heterocephalus glaber*). J Therm Biol 19:365–371
Urison NT, Buffenstein R (1995) Metabolic and body temperature changes during pregnancy and lactation in the naked mole-rat (*Heterocephalus glaber*). Physiol Zool 68:402–420
Veloso C, Bozinovic F (2000) Effect of food quality on the energetics of reproduction in a precocial rodent, *Octodon degus*. J Mammal 81:971–978
Vleck D (1979) The energy costs of burrowing by the pocket gopher *Thomomys bottae*. Physiol Zool 52:122–136

White CR (2003) The influence of foraging mode and arid adaptation on the basal metabolic rates of burrowing. Physiol Biochem Zool 76:122–134

Woodley R, Buffenstein R (2002) Thermogenic changes with chronic cold exposure in the naked mole-rat (*Heterocephalus glaber*). Comp Biochem Physiol A 133:827–834

Zelová J, Šumbera R, Sedláček F, Burda H (in press) Energetics in the solitary subterranean rodent, the silvery mole-rat, *Heliophobius argenteocinereus,* and the allometry of RMR in African mole-rats (Bathyergidae). Comp Biochem Physiol A

Zenuto RR, Antinuchi CD, Busch C (2002) Bioenergetics of reproduction and pup development in a subterranean rodent (*Ctenomys talarum*). Physiol Biochem Zool 75:469–478

5 Skimping as an Adaptive Strategy in Social Fossorial Rodents: The Mole Vole (*Ellobius talpinus*) as an Example

Mikhail Moshkin, Eugene Novikov, Dmitri Petrovski

5.1 Introduction

Most fossorial or subterranean rodents show similar physiological adaptations to subterranean life. The physical properties of the soil – high density, thermostability, low partial pressure of oxygen and high concentration of carbon dioxide – result in a decrease in basal and maximal metabolic rate, and thermoregulatory ability in fossorial rodents (Ar 1987). Good protection against predators and parasites, together with low dispersion rates (Lovegrove 1991) are assumed to lead to the reduction of reproductive activity and enormously high longevity typical of these species (Jarvis and Bennett 1991; O'Connor et al. 2002). One of the main evolutionary consequences of these adaptations is a social existence with communal nesting and reproductive skew (Jarvis and Bennett 1991; Faulkes and Bennett 2001) that reduces the costs of body maintenance, reproduction and thermoregulation. On the other hand, energetic expenditure on burrowing (Vleck 1979) dramatically increases the metabolic costs of foraging.

Selection pressures exerted by ecological factors lead to the evolution of integrated morphological, physiological and behavioral adaptations (Pianka 1994), so one might expect that all life history and physiological traits of fossorial rodents will, to some extent, reflect trade-offs between burrowing and other activities. Hitherto, however, there have been no studies of overall energy budgeting in fossorial rodents from the perspective of optimal resource allocation theory (Stearns 1992). We have examined species specific adaptive traits in mole-voles (*Ellobius* spp.) and compared these with dwarf hamsters (*Phodopus* spp.) and voles of the genus *Clethrionomys*. These two taxonomically related and sympatric, surface-dwelling rodents have body masses comparable to those of mole-voles. Besides examination of major metabolic demands – cellular maintenance, somatic and generative production, locomotion, thermoregulation and immunity – we include in our analysis an estimation of the adrenocortical response to stress as

Mikhail Moshkin: Institute of Animal Systematics and Ecology, Siberian Branch of Russian Academy of Sciences, Frunze street 11, Novosibirsk, 630091, Russia, E-mail: mmp@eco.nsc.ru

Subterranean Rodents: News from Underground
S. Begall, H. Burda, C.E. Schleich (Eds.)
© Springer-Verlag Berlin Heidelberg 2007

a main mechanism of mobilization and re-allocation of resources in relation to the changing priorities of the organism in the course of adaptation decision-making (Selye 1950). Within- and between-species variation of adrenocortical response to stress in birds and mammals are known to correlate with the severity of ecological conditions (Silverin et al. 1997; O'Reilly and Wingfield 2001; Boonstra 2004). The prevalence of physical or social stressors is also important for intra- and interspecies variation in adrenocortical responsiveness (Ganem and Croset 1990; Moshkin et al. 2001). To take spatial and temporal variability of the physiological parameters into account, we included within-species comparisons in our analysis.

Unlike other fossorial rodents that live mainly in low latitudes, mole-voles occupy a wide range of arid and semiarid landscapes of Eastern Europe, Ural, Middle Asia and West Siberia that coincides in the North with the boundary of the forest-steppe (Bobrinsky et al. 1965). The population ecology of the species has been well described by Evdokimov (1997, 2001) and Evdokimov and Pozmogova (1993). Population structure resembles that of African mole-rats (*Fukomys* and *Cryptomys*), in that family groups of mole-voles consist of one to two dozen individuals with only one reproductive female. Evdokimov and co-workers have also shown that coat color varies with geographic gradients (Bolshakov et al. 1989; Evdokimov 2001) and clarified the mechanism of heritability of this trait (Cheprakov et al. 2005).

5.2 Subjects, Area and Methodological Comments

We studied mole-voles between 1993 and 2005 in Novosibirsk and Altay regions representing the East-Northern part of the species range. For the purpose of comparative analysis we have used red-backed (*Clethrionomys rutilus) and* bank *(C. glareolus)* voles captured between 2001 and 2005 in the mountain taiga of the Altay region, and dwarf hamsters (*Phodopus sungorus*) and bank voles raised in the laboratory. Simultaneous use of both naturally captured and laboratory born individuals allowed us to rule out the effects of natural variability (based on individual experience) and incidental artificial selection. Further details of the methods can be found in Moshkin et al. (1991, 2001, 2002) and Novikov et al. (1996, 2004). In brief, we have estimated standard and maximal metabolic rates by measurement of oxygen consumption in a closed respirometer. Diurnal variations of body temperature in free-living mole-voles have been monitored by means of implanted radiotransmitters. We have used 16 min cold exposure (7–8°C) to helium-oxygen (Heliox) (Rosenmann and Morrison 1974) for measurements of maximal oxygen consumption (MOC) and thermoregulatory ability (measurements of body temperature decrease after cold exposure). Additionally, we have estimated calorigenic and thermoregulatory effects of norepinephrine injection by measuring body temperature and oxygen consumption in restrained individuals. Spontaneous activity, resting metabolism and average daily metabolic rate (ADMR) have been recorded in individuals maintained in metabolic cages equipped with infra-red sensors of activity and an open circuit analyzer of oxygen consumption (Moshkin et al. 2001). Humoral immune response to sheep red blood cells (SRBC) has been estimated according to Cunningham (1965) and expressed as the number of antigen-forming cells per unit body mass. We have defined the adrenocortical response to stress as the ratio of plasma glucocorticoid concentrations after standard cold exposition, 15 min social conflict and 30 min immobilisation to basal hormone level (glucocorticoid index). Glucocorticoids were measured by RIA (Yalow and Berson 1960; see details in Moshkin et al. 2002).

5.3 Standard Metabolic Rate, Cost of Locomotion and Thermoregulation

Among the studied species the mole-vole has the lowest standard metabolic rate (Fig. 5.1). At the same time ADMR and activity of mole-voles does not differ from those in dwarf hamsters, which means that mole-voles spend significantly more energy during activity than hamsters (Table 5.1). During summer, mole-voles demonstrate lower MOC and lower capacities to maintain body temperature when kept in cold in comparison with terrestrial rodents (Fig. 5.1). However, in fall (October), MOC increases significantly up to 9.2±1.2 ml/g·h in comparison with summer (Duncan test, $P<0.01$), and approaches the values of terrestrial species. Like other rodents (Feist and Rosenmann 1976; Heldmaier et al. 1981), mole-voles show significant seasonal variation in the calorigenic effect of norepinephrine (NE) as an indicator of non-shivering thermogenesis, with responses being minimal during summer (see Moshkin et al. 2001). Due to seasonal reduction in metabolic response to the drop in ambient temperature, wild mole-voles show large spontaneous variation in body temperature in July (up to 6°C), but not in September (<3°C). Accordingly, the standard deviation of body temperature is significantly higher in July (0.95±0.05) than in September (0.57±0.03, Fisher test; $P<0.001$). These data taken together suggest that mole-voles demonstrate a passive heterotherm strategy like other social fossorial rodents during summer, whereas they restore homeothermy during spring and autumn. Reduction of non-shivering thermogenesis, which coincides with the low efficiency of muscle performance (Yakimenko et al. 1971), seems to be adaptive for a species with extremely costly locomotion, and prevents overheating while burrowing.

Table 5.1. Locomotor activity, average daily metabolic rate ADMR, resting metabolic rate and relative cost of activity in mole-voles (*Ellobius talpinus*) and dwarf hamsters (*Phodopus sungorus*)

Parameter	*E. talpinus*	*P. sungorus*	Significance
Sample size	8	12	
Activity (sec per hour)	145.0±12.6	139.0±17.9	ns
ADMR (ml $CO_2 \cdot g^{-1} \cdot h^{-1}$)	4.35±0.09	4.18±0.05	ns
Resting metabolic rate (ml $CO_2 \cdot g^{-1} \cdot h^{-1}$)	2.46±0.04	3.53±0.02	t=6.8 ; P<0.001
Cost of activity ($J \cdot g^{-1} \cdot sec^{-1}$)	1.36±0.15	0.87±0.05	t=2.6 ; P<0.01

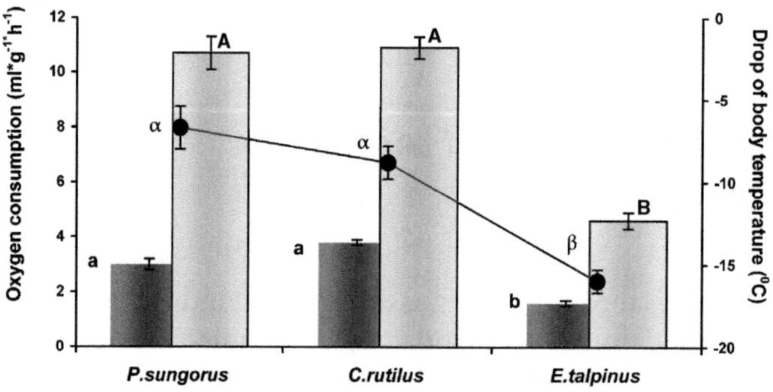

Fig. 5.1. Standard metabolic rate (*closed bars*), MOC (*open bars*) and drop of body temperature after 16 min exposure to Heliox (*line*) in different rodent species. Identical *superscript letters* indicate means that are not significantly different (Duncan test, P<0.05)

5.4 Growth and Reproduction

The sum of costs of successive reproductive events (production of gametes, sexual presentation, courtship behavior, copulation, pregnancy and lactation) constitutes the total cost of reproduction in mammals (Gittleman and Thompson 1988) and shows considerable inter-sexual difference. The most costly reproductive events for males are often territorial, hierarchical and courtship behavior rather than gamete production and copulation. For females, the main costs of reproduction are associated with energy provision for their offspring's somatic growth during pre- and post-weaning periods, i.e. pregnancy and lactation (Gittleman and Thompson 1988; Kaczmarski 1966; Weiner 1987).

In Table 5.2 we present comparative data reflecting the cost of growth and reproduction separately for male and female of mole-voles, red voles and dwarf hamsters. As the major behavioral and morphological features that provide reproductive success of male mammals are androgen-dependent traits, the level of testosterone production seems to be one of the main indicators of reproductive investments of males (Wingfield et al. 1990). In mole-voles the level of serum testosterone, estimated in the reproductive season (April) is considerably lower than in terrestrial rodents. A detectable level of the hormone is found in 65% of males only. During summer, serum testosterone is very low (trace amounts). Testes mass relative to body weight in mole-voles is tenfold lower than that of terrestrial rodents (Table 5.2).

The secondary sexual characteristic most essential for reproductive competition between males is the level of aggression. Tests of freshly captured mole-voles in standard pair-wise tests reveal no aggression between contestants encountered in a neutral arena (Moshkin et al. 1991). However, males tested after six months of laboratory maintenance in family groups show more fights than *Clethrionomys* voles but similar numbers of fights as adult males of the highly aggressive dwarf hamster (Table 5.2). Experimental introduction of mole-voles to inhabited foreign burrows leads to the disappearance of all intruders within two days after introduction (Novikov et al. 2004). Releasing of intruders into empty burrows, however, results in at least half of them being captured two days later. So, the mole-vole's aggression towards strangers seems to be associated with the familiarity of the environment.

Androgen-dependent levels of urinary proteins relative to creatinine concentration, which reflect the intensity and efficiency of scent marking (Cavaggioni and Mucignat-Caretta 2000), are considerably lower in mole-voles (0.37 ± 0.03) than in *Clethrionomys* voles (1.23 ± 0.35; $t=2.5$; $P<0.05$). In contrast to terrestrial rodents, there is no sex difference in protein concentrations in mole-vole urine.

In the northern part of the species range, the reproductive season of mole-voles is shifted to the late winter/early spring period, thus beginning far earlier than in the majority of terrestrial rodents (Evdokimov 2001; our unpublished data). Therefore, dispersion of young individuals coincides in time with the period just before the underground phytomass (bulbs and rhizomes) used for foraging is maximal. Prolonged periods of pregnancy and lactation reduce the number of litters, which do not exceed two per reproductive season. Few litters, in combination

Table 5.2. Androgens, androgen-depent traits, somatic growth and reproduction in mole-vole and terrestrial rodents (sample sizes are given above values). * identical superscript letters indicate means that are not significantly different (Duncan test, P<0.05). ** literature data (Evdokimov, 2001)

Parameter	*Clethrionomys*	*E. talpinus*	*P. sungorus*
	Males		
	22	24	20
Serum testosterone ($ng \cdot ml^{-1}$)	$0.64 \pm 0.12^{a*}$	0.67 ± 0.15^{a}	0.08 ± 0.02^{b}
	24	8	8
Number of fights in 15 min per encounters	3.3 ± 0.9^{a}	14.3 ± 2.9^{b}	13.0 ± 4.1^{b}
	31		38
Protein concentration in urine ($mg \cdot ml^{-1}$)	1.47 ± 0.32^{a}	-	0.37 ± 0.08^{b}
	111	12	64
Mass index of testes	2.02 ± 0.49^{a}	1.70 ± 0.25^{a}	0.16 ± 0.01^{b}
	Females		
	136	17	6
Litter size	5.7 ± 0.1^{a}	$5.5 \pm 0.4^{a,b}$	4.3 ± 0.4^{b}
Number of litters per reproductive season	2–4	3–5	2
Interval between litters (days)	20	20	>30
Duration of reproductive season	May–September	May–September	February–May
Energetic cost of reproduction (kJ)	4300	3000	1500-2000
	Postnatal growth		
		14	10
Body mass of pup at birth	1.3–1.8	1.46 ± 0.03	3.1 ± 0.13
Body mass of pup related to body mass of mother (%)	7.5	5.3–5.6	7.5
Index of mass growth (1–30 days)	0.472	0.588	0.438
Weaning mass related to body mass of mother (%)	60	60	48
Duration of lactation (days)	18–20	15	25
Age of maturation (month)	1–1.5	1.5–2	>12**
Longevity	field: <1 year	lab: max. 3 years	field: max. 6 years**

with small litter sizes, lead to a low total reproductive output per season, which is two- to threefold lower than the output of terrestrial rodents. According to Evdokimov (2001), mole-voles start to breed in their second year after birth, but they live for up to six years, so the maximum expected reproductive output (up to 50 pups) of one female is considerably higher than in terrestrial rodents that usually reproduce in only one season.

Among several hundred mole-vole families in the central part of the species range, Evdokimov (2001) has not found any family with more than one breeding female. In the northeastern periphery of the species range, however, we have captured up to three breeding females simultaneously from a single burrow and have not found any individual older than two years.

5.5 Immune Defense

Despite many apparent trade-offs between immunity and other organismal functions (Sheldon and Verhulst 1996), the direct costs of immune defense are still the subject of debate (Lochmiller and Deerenberg 2000; Raberg et al. 1998; Kurtz et al. 2006). A specific immune response to antigens is considered the most essential defensive reaction in prioritizing resource allocation (Lochmiller and Deerenberg 2000). In this context, the humoral immune response to SRBC in mole-voles is significantly lower than in terrestrial rodents (Fig. 5.2). In mole-voles there was no decrease of humoral immune response in mature individuals in comparison with immature ones that is very evident in *Clethrionomys* voles (Moshkin et al. 1998; Saino et al. 2000). During autumn (i.e., the post-reproductive season), the humoral immune response of mole-voles is even lower than in spring.

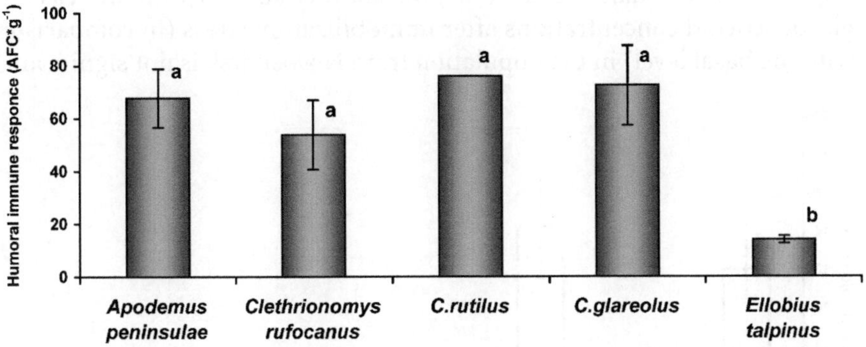

Fig. 5.2. Humoral immune response (AFC per g of body mass) on sheep red bloody cells in mole-voles and terrestrial rodents. Identical *superscript letters* indicate similar means (Duncan test, $P<0.05$)

The humoral immune response in SRBC treated mole-voles varies from 3.7 to 153.8 AFC per g of body mass. One of the reasons of this variation seems to be parasitic load. In autumn, mole-voles parasitized by the nematode *Trichocephalis muris* have a lower immune response than non-infected individuals. In spring, when worm burden is low and some families are free from parasites, there are no differences in humoral immune response between infected and non-infected individuals. On the other hand, humoral immune response of non-infected individuals originating from infected families is significantly higher than those from non-infected families. It seems that helminth infection modifies specific immunity in two ways: first by direct immunosuppression of parasitized individuals; second by positive immunomodulation of non-infected animals which shared nests with infected family members.

5.6 Adrenocortical Response to Stress

Among the species studied, mole-voles have very low values of adrenocortical response (glucocorticoid indices) to acute cooling in Heliox (Fig. 5.3; Moshkin et al. 2001). Mole-voles also had significantly lower adrenocortical responses to immobilization, social conflict and antigenic stimulation of the specific immunity (Fig. 5.3). Comparison of glucocorticoid concentrations measured in animals from the central part of the species range (Bolshakov et al.1989; Moshkin et al. 1991), and Novosibirsk, shows a significant increase in basal level in a north-eastern direction (ranging from 100.2±6.2 ng/ml in northern Kazakhstan to 350.4±38.6 ng/ml in Novosibirsk, Duncan test, $P<0.01$). In contrast to the other studied mole-vole populations (Bolshakov et al. 1989; Moshkin et al. 1991), the increase in glucocorticoid concentrations after immobilization stress (in comparison with the basal level) in the population from Novosibirsk is not significant.

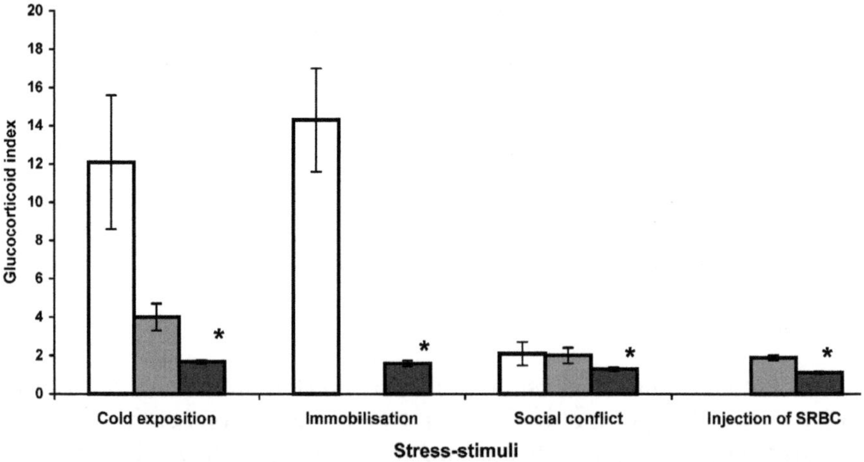

Fig. 5.3. Adrenocortical response (glucocorticoid index) on various stress-stimuli in dwarf hamsters (*white bars*), *Clethrionomys* voles (*gray bars*) and mole-voles (*black bars*). Significant differences in the response to the same stimuli between mole-voles and terrestrial rodents are denoted by *asterisk* (Duncan test, $P<0.05$)

5.7 Conclusion

During summer, mole-voles exhibit a heterothermic thermoregulatory strategy, typical among tropical social fossorial rodents. During spring and autumn, the maintenance of metabolic rate and body temperature in mole-voles does not differ from that in similar-sized terrestrial rodents. Seasonal changes in thermoregulatory strategies appear to be adapted to severe cold during winter in Northern parts of the species range, but the winter biology of mole-voles still has to be examined. On the basis of seasonal dynamics in molar root growth and the amount of winter food stored, Evdokimov (2001) proposed that mole-voles hibernate during winter. This assumption, however, conflicts with the evidence of winter reproduction.

Like other social fossorial rodents, mole-voles demonstrate reproductive skew and relatively low reproductive output. Despite the extreme longevity of fossorial rodents (see Dammann and Burda, this volume) and their ability to reproduce over several seasons, reproductive investment at any given time among both males and females is significantly lower than in terrestrial rodents. This fact, combined with low humoral immunity and decreased ability to mobilize resources under the pressure of stress-stimuli, may be regarded as a manifestation of a frugal adaptive strategy (Szarsky 1983) in its extreme expression. This "skimping" strategy, typical for fossorial rodents, is tied to ecological constraints on all physiological and behavioral demands, the thermostability of the soil environment, and the re-allocation of metabolic resources towards high-cost burrowing activity. Higher physiological costs of surviving and breeding in the periphery of the species range leads to additional resource reallocation that results in a decrease of longevity in those population.

Acknowledgements. The study was supported by the Russian Foundation for Basic Research (RFBR, grants 99-04-49927; 00-04-49210; 02-04-49253; 03-04-48768), by the Siberian Branch of RAS (Integration project No. 56), by funds of the State Contract No. 151, and by the International Science Foundation (grants RAH 300 and RAH 000). Dwarf hamsters and bank voles used in this study were kindly provided by A.V. Surov, A.Y. Telitsina and O.V. Osipova (Institute of Ecology and Evolution Problems, RAS). We thank all colleagues that helped to collect the material and discussed the results: Stephan Steinlechner, Nikolay Evdokimov, Natalya Sineva, Irina Kolosova, Viktoria Mak, Ekaterina Kondratuk, Viktor Panov and Andrey Akulov. Special thanks to S. Abramov for his help with the preparation of this manuscript. We are grateful to Sabine Begall and Christopher Barnard, whose corrections significantly improved the manuscript.

References

Ar A (1987) Physiological adaptation to underground life in mammals. In: Comparative physiology of environmental adaptations, vol. 2. Eighth ESCP Conference, Strasbourg, pp 208–221

Bobrinsky NA, Kuznetsov BA, Kuzyakin AP (1965) Handbook of mammals of USSR. Moscow. Soviet Science (Rus)

Bolshakov VN, Evdokimov NG, Moshkin MP, Pozmogova VP (1989) Coat color polymorphism and it's relation to stress-reactivity in common slepushonka (*Ellobius talpinus* Pallas). Rep Acad Sci USSR (Rus) 308:500–502

Boonstra R (2004) Coping with changing northern environments: the role of the stress axis in birds and mammals. Integr Comp Biol 44:95–108

Cavaggioni A, Mucignat-Caretta C (2000) Major urinary proteins, α2U-globulins and aphrodisin. Biochim Biophys Acta 1482:218–228

Cheprakov MI, Evdokimov NG, Glotov NV (2005) Inheritance of coat color in common slepushonka (*Ellobius talpinus* Pallas). Genetika (Rus) 41:1-6

Cunningham AJ (1965) A method of increased sensitivity for detecting single antibody-forming cells. Nature 207:1106–1107

Evdokimov NG (1997) Dynamics of population structure of common slepushonka (*Ellobius talpinus* Pall.). Ecology (Rus) 2:108–114

Evdokimov NG (2001) Population ecology of common slepushonka (*Ellobius talpinus* Pall.). Ekaterinburg (Rus)

Evdokimov NG, Pozmogova VP (1993) Population structure of common slepushonka in Ural Region. Ecology (Rus) 5:53–60

Faulkes CG, Bennett NC (2001) Family values: group dynamics and social control of reproduction in African mole-rats. Trends Ecol Evol 16:184–190

Feist D, Rosenmann M (1976) Norepinephrine thermogenesis in seasonally acclimatized and cold acclimated red-backed voles in Alaska. Can J Physiol Pharmacol 54:146–153

Ganem G, Croset H (1990) Comparative study of the corticoadrenal response to emotional stress in different kinds of natural populations of the house mouse. Neuroendocrinol 52:62

Gittleman JL, Thompson SD (1988) Energy allocation in mammalian reproduction. Am Zool 28:863–875

Heldmaier G, Steinlechner S, Rafael J, Vsiansky P (1981) Photoperiodic control and effects of melatonin on nonshivering thermogenesis and brown adipose tissue. Science 212:917–919

Jarvis JUM, Bennett NC (1991) Ecology and behavior of the family *Bathyeridae*. In: Sherman PW, Jarvis JUM, Alexander RD (eds) The biology of the naked mole-rat. Princeton Univ Press, Princeton, NJ, pp 66–96

Kaczmarski F (1966) Bioenergetics of pregnancy and lactation in the bank vole. Acta Theriol 11:409–417

Kurtz J, Wegner KM, Kalbe M, Reusch TBH, Schaschl H, Hasselquist D, Milinski M (2006) MHC genes and oxidative stress in sticklebacks: an immuno-ecological approach. Proc R Soc Lond B 273:1407–1414

Lochmiller RL, Deerenberg C (2000) Trade-offs in evolutionary immunology: just what is the cost of immunity. Oikos 88:87–98

Lovegrove BG (1991) The evolution of eusociality in molerats (*Bathyergidae*): a question of risks, numbers and costs. Behav Ecol Sociobiol 28:37–45

Moshkin MP, Evdokimov NG, Miroshnichenko VA, Pozmogova VP, Bolshakov VN (1991) Variability of corticosteroid function in populations of common slepushonka (*Ellobius talpinus* Pallas). Prog Mod Biol (Rus) 111:95–100
Moshkin MP, Dobrotvorsky AK, Mak VV, Panov VV, Dobrotvorskaya EA (1998) Variability of immune response to heterologous erythrocytes during population cycles of red (*Clethrionomys rutilus* Pallas) and bank (*Cl. glareolus* Schreber) voles. Oikos 82:131–138
Moshkin MP, Novikov EA, Petrovski DV (2001) Seasonal changes of thermoregulation in the mole vole *Ellobius talpinus.* Physiol Biochem Zool 74:869–875
Moshkin MĐ, Novikov EA, Kolosova IE, Surov AV, Telitsina AY, Osipova OA (2002) Adrenocortical and bioenergetic responses to cold in five species of murine rodents. J Mammal 83:458–466
Novikov E, Novikova M, Petrovski D, Moshkin M (1996) Geographic and seasonal variations of metabolic activity in Siberian populations of mole vole (*Ellobius talpinus* Pall., Cricetidae). In: Zaim A (ed) Biodiversity and adaptation. Fifth International Conference "Rodens and Spatium", Actes Editions, Rabat, pp 77–88
Novikov E, Petrovski D, Kolosova I, Steinlechner S, Moshkin M (2004) Interactions between intruders and residents in the mole vole *Ellobius talpinus.* Acta Zool Sin 50:19–26
O'Connor TP, Lee A, Jarvis JUM, Buffenstein R (2002) Prolonged longevity in naked mole-rats: age-related changes in metabolism, body composition and gastrointestinal function. Comp Biochem Physiol A 133:835–842
O'Reilly KM, Wingfield JC (2001) Ecological factors underlying the adrenocortical response to capture stress in arctic-breeding shorebirds. Gen Comp Endocrinol 124:1–11
Pianka ER (1994) Evolutionary ecology, 5th edn. HarperCollins, New York, 486 pp
Raberg L, Grahn M, Hasselquist D, Svensson E (1998) On the adaptive significance of stress-induced immunosuppression. Proc R Soc Lond B 265:1637–1641
Rosenmann M, Morrison P (1974) Maximum oxygen consumption and heat loss facilitation in small homeotherms by He-O_2. Am J Physiol 226:490–495
Saino N, Canova L Fasola M, Martinelli R (2000) Reproduction and population density affect humoral immunity in bank voles under field experimental conditions. Oecologia 143:936–938
Selye H (1950) The physiology and pathology of exposure to stress: a treatise based on the concepts of the General-Adaptation-Syndrome and the Diseases of Adaptation. Acta Inc, Medical Publisher, Montreal, Canada
Sheldon BC, Verhulst S (1996) Ecological immunology: costly parasite defenses and trade-offs in evolutionary ecology. Trends Ecol Evol 11:317–321
Silverin B, Arvidsson B, Wingfield J (1997) The adrenocortical responses to stress in breeding willow warblers *Phylloscopus trochilus* in Sweden: effects of latitude and gender. Funct Ecol 11:273–405
Stearns SC (1992) The evolution of life histories. Oxford Univ Press, Oxford
Szarski H (1983) Cell size and the concept of wasteful and frugal evolutionary strategies. J Theor Biol 105:201–209
Vleck D (1979) The energy cost of burrowing by the pocket gopher *Thomomys bottae.* Physiol Zool 52:122–136
Weiner J (1987) Limits to energy budget and tactics in energy investments during reproduction in the Djungarian hamster (*Phodopus sungorus* Pallas 1770). Symp Zool Soc Lond B 57:168–187
Wingfield JC, Hegner RF, Dufty AMJ, Ball GF (1990) The "challenge hypothesis": theoretical implications for patterns of testosterone secretion, mating systems, and breeding strategies. Am Nat 136:829–846

Yakimenko MA, Tkachenko EY, Ivanov KP, Slonim AD (1971) About increase of heat production of muscle contractions under affection of norepinephrine. Rep Acad Sci USSR (Rus) 200:1007–1008

Yalow RS, Berson SA (1960) Immunoassay of endogenous plasma insulin in man. J Clin Invest 39:1157–1175

6 The Reproductive Physiology and Endocrinology of the African Mole-rats: with Special Reference to Southern African Mole-rat Species

Nigel C. Bennett, Gundula H. Gutjahr, Chris G. Faulkes

6.1 Introduction

African mole-rats (family: Bathyergidae) are subterranean hystricomorph rodents that display a wide range of social organisation and reproductive strategies (Jarvis and Bennett 1990). At the one extreme are strictly solitary species that only pair up for short periods of time during the breeding season (Bennett and Jarvis 1988a; Bennett et al. 1991; Šumbera et al. 2003; Herbst et al. 2004). At the opposite extreme are the truly social or eusocial representatives that show monopolization of reproduction by a single female per colony, exhibit extended philopatry and have a work related division of labour (Jarvis 1981; Bennett and Jarvis 1988b; Bennett 1989; Burda and Kawalika 1993; Jarvis and Bennett 1993; Burda et al. 2000; Janse van Rensburg et al. 2002).

The solitary species of southern African mole-rats (*Georychus* and *Bathyergus*) generally inhabit mesic habitats that exhibit a marked seasonality with respect to photoperiod, temperature and rainfall (Jarvis and Bennett 1990, 1991; Bennett and Faulkes 2000). In these habitats the mole-rats respond to the environmental zeitgebers by breeding seasonally. A variety of species that exhibit some degree of social behaviour may also occur in these mesic areas, but also in drier regions (Spinks et al. 1999; Janse van Rensburg et al. 2002). Finally, the species generally regarded as the most social tend to occur in semi-arid to arid habitats or regions where rainfall is unpredictable or sporadic (Jarvis and Bennett 1990, 1991; Jarvis et al.1998). In this chapter we will attempt to synthesize recent advances in the understanding of mole-rat reproductive physiology in the context of seasonal and social control of reproduction, and the interplay between these two major environmental influences. Throughout, we have adopted

Nigel C. Bennett: Mammal Research Institute, Department of Zoology & Entomology, University of Pretoria, Pretoria 0002, South Africa, E-mail: ncbennett@zoology.up.ac.za

Gundula H. Gutjahr: Mammal Research Institute, Department of Zoology & Entomology, University of Pretoria, Pretoria 0002, South Africa

Chris G. Faulkes: School of Biological and Chemical Sciences, Queen Mary, University of London, Mile End Road, London E1 4NS, U.K.

Subterranean Rodents: News from Underground
S. Begall, H. Burda, C.E. Schleich (Eds.)
© Springer-Verlag Berlin Heidelberg 2007

the recent revision of taxonomic nomenclature relating to the genus *Cryptomys*. A clade containing species with a predominantly South African distribution, including *Cryptomys hottentotus*, retains the name *Cryptomys*. A second clade having a more central/west African distribution, containing *Cryptomys mechowi* is now known as *Fukomys* (Kock et al. 2006; Van Daele et al., this volume).

6.2 Reproductive Strategies in Solitary and Social African Mole-rats

6.2.1 Reproductive Strategies in Solitary Mole-rats

Solitary mole-rats are strongly territorial, pairing up briefly during the breeding season (Bennett and Jarvis 1988a). In order to achieve this, southern African mole-rats signal to one another by hind foot drumming, the frequencies of which vary with the sex of the animal. If the initiator receives the correct response signal from the recipient then courtship and mating will ensue. In both *Bathyergus* and *Georychus* the males burrow underground towards the tunnel of the female, mate multiple but brief and then depart by sealing off the excavation to the female's tunnel. The female then remains on her own in the burrow system to give birth and rear the pups. Gestation varies from 40 to 45 days in the Cape mole-rat, *G. capensis*, through to around 50 days for the Namaqua dune mole-rat, *B. janetta*, and Cape dune mole-rat, *B. suillus*, a gestation period in excess of 87 days has been calculated for the silvery mole-rat, *Heliophobius argenteocinereus*, (Jarvis 1969; Bennett et al. 1991; Bennett and Faulkes 2000; Šumbera et al. 2003).

Reproduction in subterranean rodents is ecologically constrained by the burrow environment (Jarvis and Bennett 1990). However, what environmental cues do solitary mole-rats rely on for their seasonal reproduction? The onset of reproduction in seasonally breeding rodents has long been attributed to changes in photoperiod, however, for mammals spending the majority of their life underground other environmental cues such as thermoperiod, changes in soil moisture content or sudden flushes of vegetation associated with good precipitation may be far more important in heralding the onset of reproduction and exhibit classic patterns of hormonal rise and fall associated with a strongly seasonal breeder (Fig. 6.1.a,b).

More recently, Herbst et al. (2004) found an extremely strong correlation of rainfall pattern and hormonal profile in male and female Namaqua dune mole-rats implying that rainfall may be important as an environmental cue

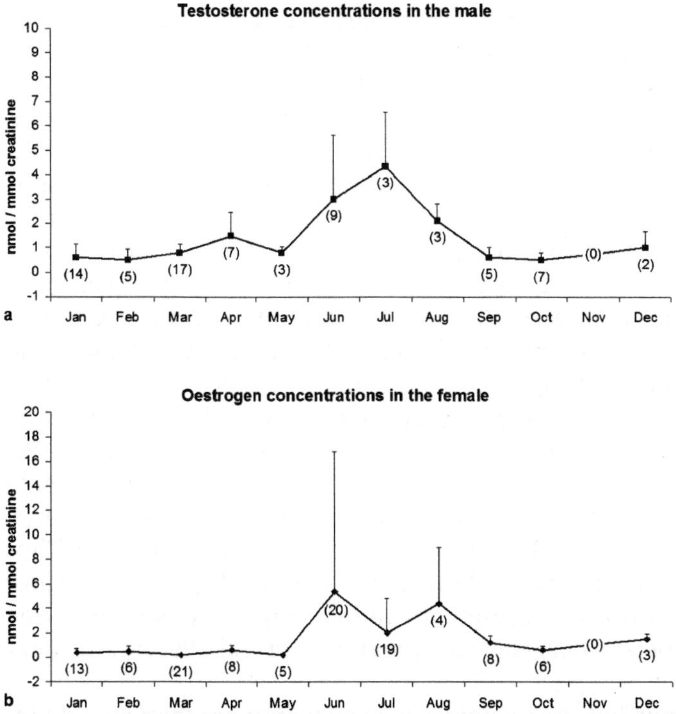

Fig. 6.1. **a** Mean monthly concentrations of urinary testosterone in a captive male *Georychus capensis*. **b** Mean monthly concentrations of urinary oestrogen in a captive female *Georychus capensis*. Sample size shown in parenthesis, *vertical bars* indicate SD about the mean

for timing reproduction. This finding is also supported in the Cape dune mole-rat from the Western Cape where the onset of the rains are correlated with the start of the breeding season (Hart et al. 2006).

6.2.2 Reproductive Strategies in Social Mole-rats

The genera *Cryptomys*, *Heterocephalus* and the newly proposed *Fukomys* contain social and eusocial representatives that are comprised of familial groups of dominant reproductive parents and philopatric non-reproductive offspring (Bennett and Jarvis 1988b; Jarvis and Bennett 1993; Faulkes et al. 1997; Burda et al. 2000; Burland et al. 2002, 2004). Unrelated immigrants of both sexes, but more often males, may also be present or pass through colonies in both *F. damarensis* (Burland et al. 2002, 2004) and *C. hottentotus* (Bishop et al. 2004). DNA fingerprinting studies have shown that immigrant or "floater" males may also sire offspring within the colonies of both of these species (Bishop et al. 2004; Burland et al. 2004). The latter may be also true for *H. glaber*, as disperser males have been observed in the laboratory (O'Riain et al. 1996) and dispersals (of both sexes) have been recorded in the field (Braude 1991, 2000), but detailed genetic studies of parentage in wild colonies remain to be undertaken in this species. A specific male "disperser morph" has been described in naked mole-rat (*H. glaber*) colonies by O'Riain et al. (1996). These individuals are morphologically, behaviourally and hormonally distinct. Although apparently non-breeding within their colony, they had significantly higher concentrations of circulating luteinizing hormone (LH) than a typical non-breeding male and appear to be reproductively primed to breed on dispersal. Colonies of the Damaraland mole-rat (*F. damarensis*) have recently been shown to possess two distinct physiological castes, frequent and infrequent workers (Scantlebury et al. 2006). The infrequent workers constitute a physiologically distinct caste, the members of which, instead of contributing to colony work, rather build up their own body reserves in preparation for dispersal and reproduction when environmental conditions are optimal (Scantlebury et al. 2006). Thus we have parallels in these two mole-rat species in that dispersers are fatter and perform less work when resident within colonies and will, following good rainfall, disperse to start new colonies (O'Riain et al. 1996; Scantlebury et al. 2006).

Within the genus *Fukomys* colonies may contain more than 40 individuals, whereas in *Heterocephalus* some colonies in excess of 290 individuals have been captured (Brett 1991; Burda and Kawalika 1993; Jarvis and Bennett 1993; Bennett and Faulkes 2000; Scharff et al. 2001). Despite the three genera showing a common pattern of colonial living with reproduction monopolized by a single reproductive queen, the mechanism by which reproductive inhibition is orchestrated varies considerably amongst the various species (Bennett et al. 1999). The reproductive females of the social species tend to have long gestation periods, which are generally longer than their solitary counterparts (with the exclusion of *Heliophobius argenteocinereus*) ranging from 66 days in the genus *Cryptomys* (Bennett 1989; Malherbe et al. 2004) through to 111 days in *Fukomys* (Bennett and Aguilar 1995).

Reproductive suppression is taken to the extreme in the naked mole-rat where both males and females are physiologically suppressed. In both captive and wild colonies of the naked mole-rat there are major differences in the size and appearance of the reproductive tracts of breeding and non-breeding males. The breeding males have significantly larger testes relative to body mass compared to their non-breeding counterparts. There is evidence of spermatogenesis and subsequent sperm production in all males. However, non-breeders produce significantly fewer sperm compared to breeders (the estimated numbers in one side of the reproductive tract were 1.8 million compared to 8.6 million respectively). Furthermore the sperm in most non-breeders are non-motile (Faulkes et al. 1994). Endocrine studies have further shown that non-breeders have significantly reduced concentrations of urinary testosterone and low or non-detectable concentrations of basal circulating LH. In addition, the pituitary gland in non-breeders is less responsive to administration of exogenous gonadotropin releasing hormone (GnRH) compared to breeding males. This suggests that there are socially-induced impairments to the hypothalamic-pituitary axis in non-breeding males (Faulkes and Abbott 1991; Faulkes et al. 1991). This reproductive suppression is, however, readily reversible if the social environment is changed: males removed from the inhibitory cues of the natal colony exhibit enhanced secretion of urinary testosterone as little as five days post removal from the colony (Faulkes and Abbott 1991).

In non-breeding female naked mole-rats, anatomical and histological studies suggest that their ovaries are functionally quiescent and prepubescent (Kayanja and Jarvis 1971). This is supported by endocrine studies that reveal that their urinary progesterone and oestrogen concentrations are low, and there is no evidence of progesterone release over time, confirming a lack of ovarian cyclicity and ovulation (Faulkes et al. 1990a; Westlin et al. 1994). The ovulation block appears to be due to inadequate concentrations of circulating LH. As with non-breeding males, a reduced pituitary response to an exogenous GnRH challenge suggests that reproductive suppression may be due to a disruption of GnRH release (Faulkes et al. 1990b). Ovarian activity is normally completely inhibited in non-breeding female naked mole-rats in the confines of the colony. However, removal of a socially suppressed female from their colony and subsequent housing either singly or pairing with a male, may bring about the onset of ovarian activity in as little as eight days (Faulkes et al. 1990a). The re-instatement of ovarian activity can also be initiated following the demise of the breeding female within a colony or presumable following dispersal and outbreeding in the field (Braude 2000). Much laboratory evidence points to the fact that naked mole-rats are facultative inbreeders. In captivity, replacement breeding queens frequently arise within their natal colony and non-breeders may occasionally escape from suppression, kill the existing queen and take over, even if this involves incestuous mating (e.g. Faulkes 1990). While behaviour of this type seems to be restricted to naked mole-rats, laboratory studies have also shown that an outbreeding reproductive strategy is the preferred option (Clarke and Faulkes 1999; Ciszek 2000) and also occurs in the wild (Braude 2000).

The naked mole-rat is thus unique amongst cooperative breeders in that both non-breeding males and females that remain philopatric to the colony are physiologically suppressed from reproducing. The mechanism of reproductive inhibition in all the social species of *Fukomys* and *Cryptomys* studied to date is different from that of *Heterocephalus*, since incest avoidance appears to be the more important component (e.g. *F. anselli*: Burda 1995; *F. darlingi*: Greeff and Bennett 2000; Herbst and Bennett 2001; *F. damarensis*: Bennett et al. 1996; Rickard and Bennett 1997; *F. mechowii*: Bennett et al. 2000a, b; Bappert and Burda 2005). In male Damaraland mole-rats there is little or no anatomical difference in the reproductive tracts between breeders and non-breeders, and no suppression of sperm production or associated deficiencies in sperm motility in non-breeding males (Faulkes et al. 1994). Despite this, breeding male Damaraland mole-rats like the naked mole-rat have heavier testis relative to their body mass. However, larger testes do not translate into increased production of spermatozoa as no significant differences were found in sperm numbers between breeding and non-breeding males (Faulkes et al. 1994). Endocrine studies support these anatomical findings in that both reproductive and non-reproductive males have similar concentrations of urinary testosterone concentrations, basal LH concentrations and pituitary responses to an exogenous GnRH challenge. The male Damaraland mole-rat thus conforms more to the usual pattern in social mammals where there is little or no physiological suppression, but rather it is incest avoidance that leads to a lack of opportunity to mate and reproduce.

In the female Damaraland mole-rat, both incest avoidance and physiological suppression appears to be involved in maintaining reproductive skew. Interestingly, the ovaries of non-breeding Damaraland mole-rats are much more functionally developed than those of the non-breeding naked mole-rats; they exhibit a wide range of follicular development through to Graafian follicles. Similar observations have also been noted in *F. anselli* (Willingstorfer et al. 1998). However, in Damaraland mole-rats corpora lutea are only found in the ovaries of non-breeding females when they are removed from the colony and housed alone, supporting the notion of a spontaneous method of ovulation. While they remain in the confines of the colony non-reproductive females are anovulatory (Molteno and Bennett 2000). As with the naked mole-rats, the pituitary of non-breeding female Damaraland mole-rats exhibits a reduced sensitivity to exogenous GnRH, compared to breeding females (Bennett et al. 1993; Fig. 6.2.a). The absolute concentrations of urinary progesterone in non-breeding female Damaraland mole-rats are higher than those in non-breeding female naked mole-rats, but they do not reach the concentrations of reproductive females suggesting a block to ovarian cyclicity and ovulation. This slight elevation in progesterone concentration observed in the non-reproductive females arises from the luteinization of unruptured secondary and tertiary follicles (Bennett 1994; Bennett et al. 1994a).

Unlike the naked mole-rat which may facultatively inbreed, in the social species of *Fukomys* and *Cryptomys* if the reproductive female dies, succession of a new reproductive female is not from within, but rather the colony

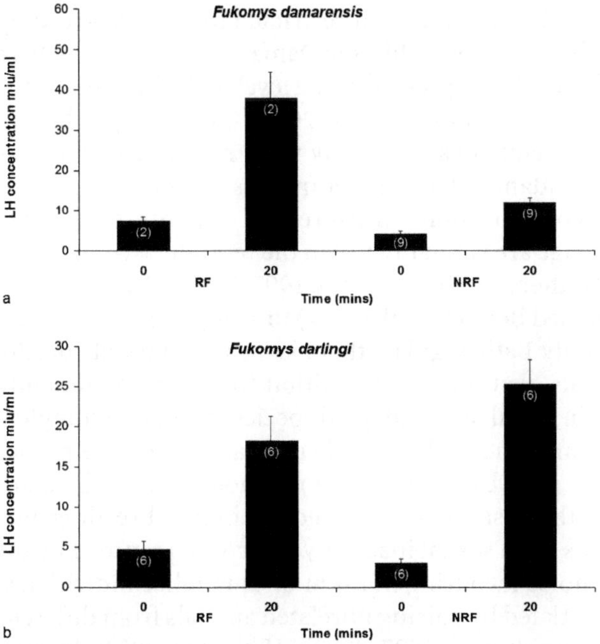

Fig. 6.2. a Concentrations of plasma luteinizing hormone (LH) (means±SD) in reproductive (RF) and non-reproductive (NRF) female *F. damarensis* before (0) and 20 min after (20) a single s.c. injection of GnRH.
b Concentrations of plasma LH (means±SD) in RF and NRF female *F. darlingi* before (0) and 20 min after (20) a single s.c. injection of GnRH. Sample size shown in parenthesis, *vertical bars* indicate SD about the mean

fragments during periods of good rainfall. Unrelated opposite sexed conspecifics could then pair up to form new colonies. Both field and laboratory studies of diverse *Fukomys* species have shown that following the death or experimental removal of the reproductive female, breeding ceases (Burda 1995; Bennett et al. 1996; Rickard and Bennett 1997; Bappert and Burda 2005). In the case of Damaraland mole-rats, basal concentrations of LH rise in non-reproductive females that are still maintained in the natal colony but in which the reproductive female has been removed. Likewise, the pituitary response to an exogenous GnRH challenge is greater than that in the same non-reproductive females housed in the presence of the reproductive female (Bennett et al. 1996). This clearly illustrates that the presence of the reproductive female is important for maintaining reproductive inhibition (Molteno et al. 2004), but that without an unrelated mate breeding does not take place.

Moving down the social scale, the Mashona mole-rat, *F. darlingi*, is a species that occurs in mesic habitats in small to medium sized colonies of two to nine individuals, with reproduction restricted to a single breeding pair (Bennett et al. 1994b). However, unlike the Damaraland mole-rat where the females are physiologically suppressed at the level of the hypothalamo-pituitary axis (Bennett et al. 1996; Molteno et al. 2004), the non-reproductive male and female colony members appear only to be restrained from sexual activity by incest avoidance. The circulating basal concentrations of LH as well as the LH concentrations measured in response to a single exogenous GnRH challenge are similar between the reproductive and non-reproductive groups of either sex (Bennett et al. 1997; Fig. 6.2.b). Comparative studies such as these led Bennett et al. (1997) to suggest that, within the social species of the family Bathyergidae, there is a continuum of socially induced infertility. In this continuum a transition from a predominantly behavioural repression in social mesic adapted species through to complete physiological suppression in the arid adapted eusocial naked mole-rat occurs (Bennett et al. 1997, 2000b). In the Mashona mole-rat and the giant mole-rat (*F. mechowii*), the absence of reproduction in non-breeding animals of both sexes results from sexual inactivity as a consequence of incest avoidance, with mate choice favouring non-kin or unfamiliar individuals. Sexual activity is only initiated by pairing unrelated animals from different colonies (Burda 1995; Bennett et al. 1997, 2000b; Heth et al. 2004; Bappert and Burda 2005).

6.2.3 How are Non-breeding Females Physiologically Suppressed?

In both naked (*Heterocephalus glaber*) and Damaraland (*Fukomys damarensis*) mole-rats, the block to reproduction appears to be due to a disruption of the GnRH secreting system. However, there are no distinct differences in either the number of cells, the morphology or the size of the cell bodies of the GnRH neurosecretory cells between reproductive and non-reproductive female Damaraland mole-rats (Molteno et al. 2004). However, there is a significant difference in the amount of GnRH in the hypothalamic neurosecretory cells of these two groups.

The GnRH concentrations in the median eminence and proximal pituitary stalk are significantly higher in non-reproductive females compared with reproductive females. As expected, in males, the concentrations of GnRH measured by radioimmunoassay in the hypothalami of reproductive and non-reproductive males did not differ. These findings suggest that the release of GnRH must be inhibited in non-reproductive females causing accumulation in the cells, which leads to the observed increase in GnRH concentrations. Endogenous opioid peptides are known to modulate GnRH secretion in many mammals, and their role in the suppression of GnRH/LH secretion in mole-rats has been investigated. Neither single nor multiple administration of the opioid antagonist naloxone had any significant effect on LH secretion, in intact or hystero-ovariectomized females, suggesting that opioids are not involved in socially induced reproductive suppression (Molteno and Bennett 2002).

6.3 Seasonal Breeding in Social Mole-rats

Many of the social species of bathyergids do not show a seasonal component to reproduction and reproduce throughout the year (Bennett et al. 1991, 1994a, b; Bennett and Aguilar 1995). The common mole-rat, *Cryptomys hottentotus*, and the highveld mole-rat, *C. pretoriae*, are the only two social species of bathyergid studied so far that exhibit a marked seasonality of reproduction (Spinks et al. 1997; Janse van Rensburg et al. 2002, 2003). The common mole-rat occurs in a winter rainfall region and rears young during the southern hemisphere summer (late November through to January; Spinks et al. 1997, 1999). In contrast, the highveld mole-rat occurs in a summer rainfall zone and rears the young in the winter (early June to August; Janse van Rensburg et al. 2002, 2003). While the females of both species show a marked seasonality in reproduction, male (both reproductive and non-reproductive) common and highveld mole-rats produce motile sperm throughout the year. The maintenance of reproductive activity outside of the breeding period is uncommon amongst seasonally breeding mammals, but may possibly be retained in these species because they disperse during the wet, non-breeding time of the year. By maintaining functional testes and motile sperm, pair bond formation could be facilitated at this time of the year, and it may also open the possibility of attaining reproductive success throughout the year. Frequent dispersal of non-reproductive males would favour the production of highly motile sperm, hence promoting colony genesis (Janse van Rensburg et al. 2003). These two species also inhabit environments which are relatively mesic and have a predictable season of rainfall. As a consequence, the number of animals remaining philopatric for extended periods of time is low. This equates to a higher lifetime reproductive success for individuals and the opportunity to eventually become a breeding animal is increased dramatically.

Despite being strictly subterranean, southern African mole-rats have a functional pineal gland. The day and night time levels of plasma melatonin have been measured in both the Damaraland (*Fukomys damarensis*) and the highveld mole-rat (*Cryptomys pretoriae*) (Richter et al. 2003; Gutjahr et al. 2004). A day–night rhythm of plasma melatonin was found in all animals housed on a 12L: 12D schedule (12 h of light followed by 12 h of darkness) with significantly higher concentrations of melatonin recorded in the dark phase. Both species of mole-rats when transferred to constant light or dark maintained an endogenous rhythm of melatonin secretion similar to that on a 12L: 12D cycle. The magnitude of this response was reduced under constant light. However, African mole-rats in general rarely if ever come onto the surface or are exposed to light/dark transitions. The ability to respond to light suggests that the surface dwelling ancestors of bathyergid mole-rats relied on chronobiological rhythms for their survival. Over evolutionary time, the adoption of a subterranean lifestyle may have reduced the dependence on light changes in the regulation of activity. Seasonal timing appears to be of significant importance in these seasonally breeding species. However, it is unlikely that photoperiod plays any major role in determining the onset of breeding.

6.4 Induced vs Spontaneous Ovulation and the Role of Penile Ornamentation

In addition to social and seasonal environmental cues, the mode by which ovulation occurs is also an important controlling factor. Ovulation is achieved essentially by two means, either spontaneously (seasonally or continuously) or by induction following mating (Milligan 1980). Female mammals that exhibit induced ovulation normally demonstrate a spontaneous growth of follicles, but without copulation and the mechanical stimulation of the vagina and cervix with the penis, they will not ovulate. Spontaneous ovulation on the other hand is characterized by a continuous cycling of reproductive hormones and subsequent ovulation in the absence of mechanical stimulation by the penis (Knobil and Neill 1988).

The solitary *Georychus capensis* is an induced ovulator (van Sandwyk and Bennett 2005) and has conspicuous epidermal spines on the shaft and glans of the penis. The loosely social mole-rats, *Cryptomys pretoriae* and *C. natalensis* (8–12 individuals per colony with a single breeding pair) also exhibit induced ovulation (Malherbe et al. 2004; Jackson and Bennett 2005) and possess penile protrusions (rounded, raised structures) rather than spines over the entire surface of the penis. The lack of penile spines in these two species suggests that, although being induced ovulators, the females can receive multiple copulations over a greater time period since they are colonial living and have more opportunities to be with a male within the colony versus that of the solitary living species. The two eusocial species, *Fukomys damarensis* and *Heterocephalus glaber* lack penile spines and protrusions altogether, and possess smooth ridges over the entire surface and run the length of the penis. Both these eusocial species exhibit spontaneous ovulation (Faulkes et al. 1990a; Snyman et al. 2006). Spontaneous ovulation may have arisen or persisted (it is likely that there have been independent gains and losses within the family) in these two species because colony turnover is less frequent. As a consequence of extreme natal philopatry and more prolonged mating opportunities, there may be no selective advantage to induced ovulation, and hence the requirement for elaborate ornamentation of the penis (Parag et al. 2006).

It is particularly noteworthy that within the genera *Cryptomys* and *Fukomys* there are representative species that show both spontaneous and induced ovulation (Malherbe et al. 2004; Jackson and Bennett 2005; Snyman et al. 2006). Induced ovulation is a trait that would be adaptive in situations where mating opportunities are unpredictable, so that ovulations are not wasted when there are no suitable mates available, and females can rapidly come into breeding condition. Hence in a family of social mole-rats where incest avoidance alone is operational, the ovaries of non-reproductive females remain quiescent but primed for action. During times of the year when dispersal opportunities become available, for example, following good rains, these previously non-reproductive females can leave the natal group and rapidly establish their own colony. Since males in these species are not physiologically suppressed, mating between dispersing animals and induction of ovulation may provide immediate opportunities for conception (Faulkes and Bennett 2001).

6.5 Habitat Shapes the Reproductive Physiology of African Mole-rats

In outbred social *Cryptomys* and *Fukomys* mole-rats inhabiting mesic environments the opportunities for dispersal and establishing independent colonies are high. For example, in *C. hottentotus* dispersal from the natal colony is seasonal and coincides with the rains when the soil is more workable (Spinks et al. 2000). Since mesic adapted species usually rely solely on incest avoidance to maintain reproductive suppression in colonies, the acquisition of an unrelated male to the colony may allow the reproductive female extra pair copulations. However, an unrelated female entering the colony would pose a threat to the queen and it is possible in these circumstances that fighting would ensue. This may explain why dispersal in some mole-rat species is male biased (Hazell et al. 2000).

In marked contrast, naked mole-rats (*Heterocephalus glaber*) inhabit xeric conditions during which relatively few opportunities to disperse present themselves (rainfall is sporadic). The naked mole-rat exhibits one of the most extreme forms of social suppression of reproduction where physiological blocks to fertility are apparent in both sexes. In captive colonies, unlike in other social mole-rats, daughters may succeed mothers as breeders. The Damaraland mole-rat (*F. damarensis*) is an enigma in that it has both components of incest avoidance and physiological suppression of reproduction. Incest avoidance alone would only be sufficient to maintain reproductive skew in the Damaraland mole-rat if colonies are founded by unrelated opposite sexed conspecifics and contain exclusively their offspring. Damaraland mole-rats show a high degree of natal philopatry, and usually colony fragmentation and subsequent dispersal arises only in colonies in which the reproductive female or male has died. However, during periods of good rain, unrelated dispersing individuals may enter functionally complete colonies. In such instances an unrelated male entering the colony could potentially mate with non-reproductive females, were they not physiologically suppressed by the breeding female (Bennett et al. 1996, 1999; Burland et al. 2002, 2004). An alternative explanation is that incest avoidance based on individual recognition cannot work in larger groups and/or groups with a more frequent turnover because rodents cannot memorise all family members (Burda 1995, 1999). However, O'Riain and Jarvis (1997) show that individual recognition is present in *Heterocephalus*, which occur in much larger colonies than *Cryptomys* and *Fukomys*.

6.6 Monopolization of Reproduction: Plural Breeding in Males and Females

All social African mole-rats show extreme female reproductive skew with in most cases one female per colony responsible for procreation. Numbers of breeding males have until recently been unclear. Concession theory predicts that reproductive females may allow some subordinates to breed in order to provide staying or peace incentives. These incentives are normally directed at older and less related colony members (Keller and Reeve 1994). A critical assumption of concession theory is that the dominant female controls reproduction in the subordinates. However an alternative point of view, the theory of incomplete control, posits that there is a struggle for reproduction status between the dominant breeding female and subordinate females (Clutton-Brock 1998). Recent studies using microsatellite genotyping to investigate relatedness and parentage in a number of species have revealed that the colonies are far from being derived solely from a single breeding male and female. Colonies of *Fukomys damarensis* and *Cryptomys hottentotus* have traditionally been thought to comprise a single breeding female, one or two breeding males and their offspring. Parentage analysis has found that multiple and unidentified paternity was widespread within wild colonies and immigrants of both sexes were regularly present in colonies (Burland et al. 2004). These findings suggest that in the wild, non-reproductive female Damaraland mole-rats (*F. damarensis*) can come into contact with unrelated males, even when they do not disperse from the natal colony. Similarly, in the common mole-rat (*C. hottentotus*), analysis of parentage of entire colonies showed that paternity was not assigned to a single male, but rather a number of males. This suggests that a "sneak" mating strategy may be adopted by some males. Established breeding males may also sire extra colony young in the same season as siring young within their own colonies (Bishop et al. 2004). Colonies of the common mole-rat are also known to have a higher turnover compared with Damaraland mole-rats; thus the skew in lifetime reproductive success is less in the former than the latter (Spinks et al. 2000).

The most plausible scenario to explain the degree of skew exhibited in *Heterocephalus glaber* is that of incomplete control by the dominant female (although "peace incentives" cannot be ruled out). Female-female competition occasionally gives rise to fighting and plural breeding (more than one queen per colony). Laboratory studies have shown that queen succession and reproductive take-overs frequently occur from within the colony. Plural breeding among queens has been recorded in 11% of 53 colonies in the laboratory, and 7% of 26 colonies in the wild (Braude 1991; Sherman et al. 1992). In species that are obligate outbreeders there is apparent incest avoidance also between father and daughters, and between brothers and sisters in families of Zambian *Fukomys*, but the incest avoidance between mother and sons seems to be weaker. Experiments and observations to date (Burda 1995, 1999) suggest that behavioural control (mate guarding) by the dominant male (the father) may also play an important role in inhibition of male reproduction. This fact may have represented a selective pressure leading to significant increase in the body mass (and sexual dimorphism) of breeding males.

6.7 Perspective

One of the main unanswered questions in mole-rat sociobiology is to determine the factor(s) that bring about the disrupted release of GnRH from the neurosecretory cells in non-reproductive females. The role of primer pheromones (chemical signals released by conspecifics and bringing about a physiological response in the recipient) in the modulation of reproductive function in rodents is well known (Vandenbergh 1988). It was originally posited that pheromones in the urine of the reproductive or breeding female could suppress reproduction in naked mole-rats, *Heterocephalus glaber* (Jarvis 1981). However, Faulkes and Abbott (1993) and Smith et al. (1997) have shown that primer pheromones from urine or other body fluids contained in soiled bedding or litter fail to prevent activation in females removed from the inhibitory effects of the social environment or housed with a male. Another possibility is that overt aggression is responsible for social suppression – breeding female naked mole-rats characteristically shove colony mates to impose their dominance. This behaviour develops as females become reproductive active (Clarke and Faulkes 2001), and larger animals are shoved more frequently since they may pose a greater threat to the breeding female's reign (Reeve and Sherman 1991). In Damaraland mole-rats, *Fukomys damarensis*, there is no obvious aggression directed towards kin of either sex in family groups (where skew may be maintained by incest avoidance; Bennett et al. 1999), and agonistic behaviour has never been fully investigated in mixed-kin groups. We still know relatively little about the ways in which social mole-rats maintain the reproductive skew. Further investigation might enlighten us into the means by which reproductive inhibition is brought about in this interesting admixture of social species occupying diverse and distinct habitats.

Acknowledgements. We would like to extend our thanks to Sabine Begall and Hynek Burda for the invitation to write this review for the book. NCB acknowledges funds from the National Research Foundation, South Africa and University of Pretoria that enabled much of this work to take place. Hynek Burda and Abraham Haim are gratefully thanked for critically reviewing earlier drafts of the chapter.

References

Bappert MT, Burda H (2005) Individual recognition as means to maintain incest avoidance in the giant mole-rat, *Cryptomys mechowi* (Rodentia, Bathyergidae). Abstracts of the 9th International Mammalogical Congress (IMC9), Sapporo (Japan)

Bennett NC (1989) The social structure and reproductive biology of the common mole-rat, *Cryptomys h. hottentotus* and remarks on the trends in reproduction and sociality in the family Bathyergidae. J Zool Lond 219:45–59

Bennett NC (1994) Reproductive suppression in social *Cryptomys damarensis* colonies:- a lifetime of socially-induced sterility in males and females (Rodentia: Bathyergidae). J Zool Lond 234:25–39

Bennett NC, Aguilar GH (1995) The reproductive biology of the giant Zambian mole-rat, *Cryptomys mechowi* (Rodentia: Bathyergidae). S Afr J Zool 30:1–4

Bennett NC, Faulkes CG (2000) African mole-rats: ecology and eusociality. Cambridge Univ Press, Cambridge

Bennett NC, Jarvis JUM (1988a) The reproductive biology of the Cape mole-rat, *Georychus capensis* (Rodentia, Bathyergidae). J Zool Lond 214:95–106

Bennett NC, Jarvis JUM (1988b) The social structure and reproductive biology of colonies of the mole-rat *Cryptomys damarensis* (Rodentia, Bathyergidae). J Mammal 69:293–302

Bennett NC, Jarvis JUM, Aguilar GH, McDaid EJ (1991) Growth and development in six species of African mole-rats (Rodentia: Bathyergidae). J Zool Lond 225:13–26

Bennett NC, Jarvis JUM, Faulkes CG, Millar RP (1993) LH responses to single doses of exogenous GnRH by freshly captured Damaraland mole-rats, *Cryptomys damarensis*. J Reprod Fert 99:81–86

Bennett NC, Jarvis JUM, Millar RP, Sasano H, Ntshinga KV (1994a) Reproductive suppression in eusocial *Cryptomys damarensis* colonies: socially-induced infertility in females. J Zool Lond 233:617–630

Bennett NC, Jarvis JUM, Cotterill FPD (1994b) The colony structure and reproductive biology of the afrotropical Mashona mole-rat, *Cryptomys darlingi*. J Zool Lond 234:477–487

Bennett NC, Faulkes CG, Molteno AJ (1996) Reproductive suppression in subordinate, non-breeding female Damaraland mole-rats: two components to a lifetime of socially induced infertility. Proc R Soc Lond B 263:1599–1603

Bennett NC, Faulkes CG, Spinks AC (1997) LH responses to single doses of exogenous GnRH by social Mashona mole-rats: a continuum of socially induced infertility in the family Bathyergidae. Proc R Soc Lond B 264:1001–1006

Bennett NC, Faulkes CG, Jarvis JUM (1999) Socially induced infertility, incest avoidance and the monopoly of reproduction in cooperatively breeding African mole-rats, family: Bathyergidae. Adv Study Behav 28:75–114

Bennett NC, Faulkes CG, Molteno AJ (2000a) Reproduction in subterranean rodents. In: Lacey EA, Patton JL, Cameron GN (eds) Life underground: the biology of subterranean rodents. Univ Chicago Press, Chicago, pp 145–177

Bennett NC, Molteno AJ, Spinks AC (2000b) Pituitary sensitivity to exogenous GnRH in giant Zambian mole-rats, *Cryptomys mechowi* (Rodentia: Bathyergidae): support for the "socially induced infertility continuum". J Zool Lond 252:447–452

Bishop JM, Jarvis JUM, Spinks AC, Bennett NC, O'Ryan C (2004) Molecular insight into patterns of colony composition and paternity in the common mole-rat, *Cryptomys hottentotus hottentotus*. Mol Ecol 13:1217–1229

Braude SH (1991) The behavior and demographics of the naked mole-rat, *Heterocephalus glaber*. Unpubl Doctoral dissertation. Univ Michigan, Ann Arbor, USA

Braude SH (2000) Dispersal and new colony formation in wild naked mole-rats: evidence against inbreeding as the system of mating. Behav Ecol 11:7–12

Brett RA (1991) The ecology of naked mole-rat colonies: burrowing, food and limiting factors. In: Sherman PW, Jarvis JUM, Alexander RD (eds) The biology of the naked mole-rat. Princeton Univ Press, Princeton, NJ, pp 137–184

Burda H (1995) Individual recognition and incest avoidance in eusocial common mole-rats rather than reproductive suppression by parents. Experientia 51:411–413

Burda H (1999) Syndrome of eusociality in African subterranean mole-rats (Bathyergidae, Rodentia), its diagnosis and aetiology. In: Wasser SP (ed) Evolutionary theory and processes: modern perspectives. Kluwer Academic Publ, NL-Dordrecht, pp 385–418

Burda H, Kawalika M (1993) Evolution of eusociality in the Bathyergidae: the case of the giant mole rats (*Cryptomys mechowi*). Naturwiss 80:235–237

Burda H, Honeycutt RL, Begall S, Locker-Grütjen O, Scharff A (2000) Are naked and common mole-rats eusocial and if so, why? Behav Ecol Sociobiol 47:293–303

Burland TM, Bennett NC, Jarvis JUM, Faulkes CG (2002) Eusociality in African mole-rats: new insights from patterns of genetic relatedness in the Damaraland mole-rat (*Cryptomys damarensis*). Proc R Soc Lond B 269:1025–1030

Burland TM, Bennett NC, Jarvis JUM, Faulkes CG (2004) Colony structure and parentage in wild colonies of cooperatively breeding Damaraland mole-rats suggest incest avoidance alone may not maintain reproductive skew. Mol Ecol 13:2371–2379

Ciszek D (2000) New colony formation in the "highly inbred" eusocial naked mole-rat: outbreeding is preferred. Behav Ecol 11:1–6

Clarke FM, Faulkes CG (1999) Kin discrimination and female mate choice in the naked mole-rat, *Heterocephalus glaber*. Proc R Soc Lond B 266:1995–2002

Clarke FM, Faulkes CG (2001) Intracolony aggression in the eusocial naked mole-rat, *Heterocephalus glaber*. Anim Behav 61:311–324

Clutton-Brock TH (1998) Reproductive skew, concessions and limited control. Trends Ecol Evol 13:288–292

Faulkes CG (1990) Social suppression of reproduction in the naked mole-rat, *Heterocephalus glaber*. Unpubl doctoral dissertation. Univ London, UK

Faulkes CG, Abbott DH (1991) Social control of reproduction in breeding and non-breeding male naked mole-rats (*Heterocephalus glaber*). J Reprod Fert 93:427–435

Faulkes CG, Abbott DH (1993) Evidence that primer pheromones do not cause social suppression of reproduction in male and female naked mole-rats (*Heterocephalus glaber*). J Reprod Fert 99:225–230

Faulkes CG, Bennett NC (2001) Family values: group dynamics and social control of reproduction in African mole-rats. Trends Ecol Evol 16:184–190

Faulkes CG, Abbott DH, Jarvis JUM (1990a) Social suppression of ovarian cyclicity in captive and wild colonies of naked mole-rats, *Heterocephalus glaber*. J Reprod Fert 88:559–568

Faulkes CG, Abbott DH, Jarvis JUM, Sherriff FE (1990b) LH responses of female naked mole-rats, *Heterocephalus glaber*, to single and multiple doses of exogenous GnRH. J Reprod Fert 89:317–323

Faulkes CG, Abbott DH, Liddell CE, George LM, Jarvis JUM (1991) Hormonal and behavioral aspects of reproductive suppression in female naked mole-rats. In: Sherman PW, Jarvis JUM, Alexander RD (eds) The biology of the naked mole-rat. Princeton Univ Press, Princeton, NJ, pp 426–445

Faulkes CG, Trowell SN, Jarvis JUM, Bennett NC (1994) Investigation of numbers and motility of spermatozoa in reproductively active and socially suppressed males of two eusocial African mole-rats, the naked mole-rat (*Heterocephalus glaber*) and the Damaraland mole-rat (*Cryptomys damarensis*). J Reprod Fert 100:411–416

Faulkes CG, Bennett NC, Bruford MW, O'Brien HP, Aguilar GH, Jarvis JUM (1997) Ecological constraints drive social evolution in the African mole-rats. Proc R Soc Lond B 264:1619–1627

Greeff JM, Bennett NC (2000) Causes and consequences of incest avoidance in the cooperatively breeding mole-rat, *Cryptomys darlingi* (Bathyergidae). Ecol Lett 3:318–328

Gutjahr GH, Janse van Rensburg L, Malpaux B, Richter TA, Bennett NC (2004) The endogenous rhythm of plasma melatonin and its regulation by light in the highveld mole-rat (*Cryptomys hottentotus pretoriae*): a microphthalmic seasonally breeding rodent. J Pineal Res 37:185–192

Hart L, O'Riain MJ, Jarvis JUM, Bennett NC (2006) Is the Cape dune mole-rat, *Bathyergus suillus* a seasonal or aseasonal breeder? J Mammal 87:1078–1085

Hazell RWA, Bennett NC, Jarvis JUM, Griffin M (2000) Adult dispersal in the co-operatively breeding Damaraland mole-rat (*Cryptomys damarensis*): a case study from the Waterberg region of Namibia. J Zool Lond 252:19–25

Herbst M, Bennett NC (2001) Recrudescence of sexual activity in a colony of the Mashona mole-rat (*Cryptomys darlingi*): an apparent case of incest avoidance. J Zool Lond 254:163–175

Herbst M, Jarvis JUM, Bennett NC (2004) A field assessment of reproductive seasonality in the threatened wild Namaqua dune mole-rat (*Bathyergus janetta*). J Zool Lond 263:259–268

Heth G, Todrank J, Begall S, Wegner R, Burda H (2004) Genetic relatedness discrimination in eusocial, *Cryptomys anselli* mole-rats, Bathyergidae, Rodentia. Folia Zool 53:269–278

Jackson CR, Bennett NC (2005) Is the Natal mole-rat (*Cryptomys hottentotus natalensis*) a spontaneous or induced ovulator? J Mammal 86:1–6

Janse van Rensburg L, Bennett NC, Van der Merwe M, Schoeman AS (2002) Seasonally reproduction in the highveld mole-rat, *Cryptomys hottentotus pretoriae* (Rodentia: Bathyergidae). Canad J Zool 80:810–820

Janse van Rensburg L, Bennett NC, Van der Merwe M, Schoeman AS, Brinders J (2003) Are non-reproductive male highveld mole-rats, *Cryptomys hottentotus pretoriae* physiologically suppressed while in the confines of the natal colony? J Zool Lond 260:73–78

Jarvis JUM (1969) Aspects of the biology of East African mole-rats. Unpubl doctoral dissertation. Univ East Africa, Nairobi, Kenya

Jarvis JUM (1981) Eusociality in a mammal: cooperative breeding in naked mole-rat colonies. Science 212:571–573

Jarvis JUM, Bennett NC (1990) The evolutionary history, population biology and social structure of African mole-rats: family Bathyergidae. In: Nevo E, Reig OA (eds) Evolution of subterranean mammals at the organismal and molecular levels. Wiley-Liss, New York, pp 97–128

Jarvis JUM, Bennett NC (1991) Ecology and behavior of the family Bathyergidae. In: Sherman PW, Jarvis JUM, Alexander RD (eds) The biology of the naked mole-rat. Princeton Univ Press, Princeton, NJ, pp 66–96

Jarvis JUM, Bennett NC (1993) Eusociality has evolved independently in two genera of bathyergid mole-rats – but occurs in no other subterranean mammal. Behav Ecol Sociobiol 33:253–260

Jarvis JUM, Bennett NC, Spinks AC (1998) Food availability and foraging by wild colonies of Damaraland mole-rats (*Cryptomys damarensis*): implications for sociality. Oecologia 113:290–298

Kayanja FIB, Jarvis JUM (1971) Histological observations on the ovary, oviduct and uterus of the naked mole-rat. Z Säugetierk 36:114–121

Keller L, Reeve HK (1994) Partitioning of reproduction in animal societies. Trends Ecol Evol 9:98–102

Knobil E, Neill JD (1988) Physiology of reproduction. Raven Press, New York

Kock D, Ingram CM, Frabotta LJ, Honeycutt RL, Burda H (2006) On the nomenclature of Bathyergidae and *Fukomys* n. gen. (Mammalia: Rodentia). Zootaxa 1142:51–55

Malherbe GP, Schoeman AS, Bennett NC (2004) Is the highveld mole-rat *Cryptomys hottentotus pretoriae* (Rodentia: Bathyergidae) an induced or spontaneous ovulator? J Zool Lond 263:159–165

Milligan SR (1980) Pheromones and rodent reproductive physiology. Symp Zool Soc Lond 45:251–275

Molteno AJ, Bennett NC (2000) Anovulation in non-reproductive female Damaraland mole-rats (*Cryptomys damarensis*) J Reprod Fert 119:35–41

Molteno AJ, Bennett NC (2002) Social suppression in non-reproductive female Damaraland mole-rats, *Cryptomys damarensis*: no apparent role for endogenous opioid peptides. Horm Behav 41:115–125

Molteno AJ, Kallo I, Bennett NC, King JA, Coen CW (2004) A neuroanatomical and neuroendocrinological study into the relation between social status and the GnRH system in cooperatively breeding female Damaraland mole-rats, *Cryptomys damarensis*. Reproduction 127:13–21

O'Riain MJ, Jarvis JUM (1997) Colony member recognition and xenophobia in the naked mole-rat. Anim Behav 53:487–498
O'Riain MJ, Jarvis JUM, Faulkes CG (1996) A dispersive morph in the naked mole-rat. Nature 380:619–621
Parag A, Bennett NC, Faulkes CG, Bateman PW (2006) Penile morphology of African mole-rats (Bathyergidae): structural modification in relation to mode of ovulation and degree of sociality. J Zool Lond 270:323–329
Reeve HK, Sherman PW (1991) Intracolonial aggression and nepotism by the breeding female naked mole-rat. In: Sherman PW, Jarvis JUM, Alexander RD (eds) The biology of the naked mole-rat. Princeton Univ Press, Princeton, NJ, pp 337–357
Richter TA, Malpaux B, Fleming PA, Molteno AJ, Bennett NC (2003) Melatonin secretion in a strictly subterranean mammal, the Damaraland mole-rat (*Cryptomys damarensis*). J Zool Lond 261:313–319
Rickard CA, Bennett NC (1997) Recrudescense of sexual activity in a reproductively quiescent colony of the Damaraland mole-rat (*Cryptomys damarensis*), by the introduction of an unfamiliar and genetically unrelated male – a case of incest avoidance in "queenless" colonies. J Zool Lond 241:185–202
Scantlebury M, Speakman JR, Oosthuizen MK, Roper TJ, Bennett NC (2006) Energetics reveals physiologically distinct castes in a eusocial mammal. Nature 440:795–797
Scharff A, Locker-Grütjen O, Kawalika M, Burda H (2001) Natural history of the giant mole-rat, *Cryptomys mechowi* (Rodentia: Bathyergidae) from Zambia. J Mammal 82:1003–1015
Sherman PW, Jarvis JUM, Braude SH (1992) Naked mole-rats. Sci Am 267:72–78
Smith TE, Faulkes CG, Abbott DH (1997) Combined olfactory contact with the parent colony and direct contact with nonbreeding animals does not maintain suppression of ovulation in female naked mole-rats (*Heterocephalus glaber*). Horm Behav 31:277–288
Snyman PC, Jackson CR, Bennett NC (2006) Do dispersing non-reproductive female Damaraland mole-rats, *Cryptomys damarensis* (Rodentia: Bathyergidae) exhibit spontaneous or induced ovulation? Physiol Behav 87:88–94
Spinks AC, Van der Horst G, Bennett NC (1997) Influence of breeding season and reproductive status on male reproductive characteristics in the common mole-rat, *Cryptomys hottentotus hottentotus*. J Reprod Fert 109:79–86
Spinks AC, Bennett NC, Jarvis JUM (1999) Regulation of reproduction in female common mole-rats (*Cryptomys hottentotus hottentotus*), the effects of breeding season and reproductive status. J Zool Lond 248:161–168
Spinks AC, Jarvis JUM, Bennett NC (2000) Comparative patterns of philopatry and dispersal in two common mole-ratpopulation: implications for the evolution of mole-rat sociality. J Anim Ecol 69:224–234
Šumbera R, Burda H, Chitaukali WN (2003) Reproductive biology of a solitary subterranean bathyergid rodent, the silvery mole-rat (*Heliophobius argenteocinereus*). J Mammal 84:278–287
Vandenbergh JG (1988) Pheromones and mammalian reproduction. In: Knobil E, Neill JD (eds) Physiology of reproduction. Raven Press, New York, pp 1679–1689
Van Sandwyk JHdT, Bennett NC (2005) Do solitary, seismic signaling Cape mole-rats (*Georychus capensis*) demonstrate spontaneous or induced ovulation? J Zool Lond 267:75–80
Westlin LM, Bennett NC, Jarvis JUM (1994) Relaxation of reproductive suppression in non-breeding female naked mole-rats, *Heterocephalus glaber*. J Zool Lond 234:177–188
Willingstorfer WJ, Burda H, Winckler J (1998) Ovarian growth and folliculogenesis in breeding and nonbreeding females of a social rodent, the Zambian common mole-rat, *Cryptomys* sp. J Morphol 237:33–41

Part II
Sensory Ecology

7 Sensory Ecology of Subterranean Rodents

Sabine Begall, Cristian E. Schleich

Ever since Dusenbery's seminal monograph published in 1992, sensory ecology has been defined as the study of how organisms acquire and respond to information. Hence, in contrast to classical ecology with its focus on *energy flows*, sensory ecology deals with *information flows* (Dusenbery 1992). Since this information is closely linked to the animal's environment, sensory ecology tries to find answers to the question of *how* the animal interacts with its environment. On the one hand, the animal's environment consists of the physical properties of its habitat, and on the other hand it also implies the interactions with other animals. In the case of subterranean mammals, the ecotope is dark, rather monotonous and harsh in comparison to the situation found above the ground (Burda et al., this volume). These differences between the underground and aboveground ecotopes led early scientists working with subterranean mammals to speculate about the sensory equipment of their study objects (Eloff 1951, 1958; Crawford 1966; Quilliam 1966; Poduschka 1978). A first review of studies on sensory adaptations in subterranean rodents conducted until the late 1980s is given by Burda et al. (1990a).

Within the past decade, the number of papers dealing with sensory ecology in subterranean rodents, including a relevant review by Francescoli (2000), has increased drastically, covering "basic senses" like olfaction, vision, and hearing, but also "exotic senses" (i.e. rarely found in mammals) like magnetoreception. Interestingly, many studies on sensory ecology in subterranean rodents revealed mammalian peculiarities or extremes: unusual photoreceptor properties in blind (David-Gray et al. 2002) and Ansell's mole-rats (Peichl et al. 2004), magnetoreception in small Zambian mole-rats (Burda et al. 1990b; Marhold et al. 1997), or reduced cutaneous nociception (Park et al. 2003) in naked mole-rats to mention just a few studies, making subterranean mammals a fruitful field for investigations. This "publication boom" is also reflected in this volume, as the sensory ecology part amounts to approximately a third of this book. The senses

Sabine Begall: Dept. General Zoology, University of Duisburg-Essen, D-45117 Essen, Germany, E-mail: sabine.begall@uni-due.de

Cristian E. Schleich: Departamento de Biología, Facultad de Ciencias Exactas y Naturales, CC 1245, Universidad Nacional de Mar del Plata, Funes 3250, Mar del Plata (7600), Argentina

Subterranean Rodents: News from Underground
S. Begall, H. Burda, C.E. Schleich (Eds.)
© Springer-Verlag Berlin Heidelberg 2007

addressed in this section are olfaction, audition, vision, magnetoreception, and somatosensation, and although we tried to give a broad overview, some senses like equilibrioception or proprioception are not outlined here due to space restrictions. Most chapters in this part follow different approaches, because sensory ecology – as a true multi-discipline – bases on behavioral, morphological, neurobiological, and physiological studies, which are interpreted in evolutionary terms taking ecological variables into account. In the following, we are going to shortly present the chapters belonging to this section of the book.

The first chapter of this section deals with presumably the oldest, yet least understood of all basic senses, i.e. olfaction. Heth and Todrank review in Chap. 8 the role of odors and olfaction in subterranean rodents. Their main focus lies on odor-guided foraging by means of kairomones, and the importance of odors in social interactions. For the latter, two odor-based discriminative processes have been recognized: G-Ratios (i.e. genetic relatedness assessment through individual odor similarities) and individual recognition.

In the following (Chap. 9), Begall and colleagues describe in short the acoustical properties of the subterranean habitat and how these contribute to what the animals probably hear and also to the convergent evolution of the auditory system. Sounds of relatively low frequencies propagate best in subterranean tunnels where certain frequencies might even be enhanced. Accordingly, audition is shifted to the low-frequency range, and also the hearing apparatus is tuned to these sounds. In light of these findings, it is not surprising that subterranean rodents use preferably low-frequency vocalizations to communicate. Supplementing the information provided in this chapter, Schleich et al. (Chap. 10) give details on vocalizations and show in a short excursus that juvenile vocalizations of subterranean rodents might not be considered as honest signals.

The largest chapter in this part is the one by Němec and colleagues (Chap. 11) going along with the attention that has been paid to the subterranean mammals' eyes for the past years. They describe the astonishing diversity found in the visual system of subterranean rodents. These intriguing findings contradict the previous view of convergent evolution towards degenerate eyes in the majority of subterranean mammals. The authors review the morphological studies on the eye and on related brain structures and account possible functions of the more or less well-preserved structures. Nevertheless, in many subterranean species, vision is restricted, and may not serve for spatial orientation purposes.

Magnetoreception might fill the gap and help the subterranean animal to orient itself. In Chap. 12, Moritz and colleagues review the newest findings on magnetoreception and undergo a virtual journey from the incoming magnetic stimulus to the brain.

Tied up to the orientation task, an extensive review on the somatosensory system by Park and colleagues finishes the sensory ecology part (Chap. 13). So far, the best studied subterranean rodent in this respect is the naked mole-rat showing some remarkable specifications and becoming a model species. This final chapter shows why subterranean rodents can be regarded as touch specialists, showing the importance of this sense for life in the complex tunnel systems.

Despite the great diversity of the topics presented in this part, some subjects still need to be addressed. The perception of vibrations, for instance, so typical for subterranean rodents (cf., Narins et al. 1997; Rado et al. 1998; Mason and Narins 2001), first described 20 years ago (Heth et al. 1987; Rado et al. 1987), is not covered by this volume; but then, only little progress has been made in this field within the last five years. New inspiration - not only for the study of the seismic aspect - could be expected if the sensory environment of subterranean rodents was thoroughly examined both physically and chemically applying new methods. Doing so, Crawford's approach (Crawford 1966) could be extended and deepened, so that we could better estimate what the animals actually do perceive: e.g. temperature gradients, electromagnetic radiation, odor gradients, or which previously unforeseen sensory strategies may be applied (cf. "infrasound echolocation" for spatial orientation - Kimchi et al. 2005). All in all, plenty of work is still waiting out there for subterraneologists and sensory ecologists.

References

Burda H, Bruns V, Müller M (1990a) Sensory adaptations in subterranen mammals. In: Nevo E, Reig OA (eds) Evolution of subterranean mammals at the organismal and molecular levels. Wiley-Liss, New York, pp 269–294

Burda H, Marhold S, Westenberger T, Wiltschko R, Wiltschko W (1990b) Magnetic compass orientation in the subterranean rodent *Cryptomys hottentotus* Bathyergidae. Experientia 46:528–530

Crawford BH (1966) Perception underground: review of physical aspects and measurements. J Zool Lond 149:103–106

David-Gray ZK, Bellingham J, Munoz M, Avivi A, Nevo E, Foster RG (2002) Adaptive loss of ultraviolet-sensitive/violet-sensitive (UVS/VS) cone opsin in the blind mole rat (*Spalax ehrenbergi*). Europ J Neurosci 16:1186–1194

Dusenbery DB (1992) Sensory ecology: how organisms acquire and respond to information. WH Freeman, New York

Eloff G (1951) Orientation in the mole-rat *Cryptomys*. Brit J Psychol 1/2:134–145

Eloff G (1958) Functional and structural degeneration of the eye of the South African rodent mole, *Cryptomys bigalkei* and *Bathyergus maritimus*. S Afr J Sci 54:292–302

Francescoli G (2000) Sensory capabilities and communication in subterranean mammals. In: Lacey EA, Patton JL, Cameron GN (eds) Life underground: the biology of subterranean rodents. Univ Chicago Press, Chicago, pp 111–144

Heth G, Frankenberg E, Raz A, Nevo E (1987) Vibrational communication in subterranean mole rats (*Spalax ehrenbergi*). Behav Ecol Sociobiol 21:31–33

Kimchi T, Reshef M, Terkel J (2005) Evidence for the use of reflected self-generated seismic waves for spatial orientation in a blind subterranean mammal J Exp Biol 208:647–659

Marhold S, Wiltschko W, Burda H (1997) A magnetic polarity compass for direction finding in a subterranean mammal. Naturwiss 84:421–423

Mason MJ, Narins PM (2001) Seismic signal by fossorial mammals. Am Zool 41:1171–1184

Narins PM, Lewis ER, Jarvis JUM, O'Riain J (1997) The use of seismi signals by fossorial southern African mammals: a neuroethological gold mine. Brain Res Bull 44:641–646

Park TJ, Comer C, Carol A, Lu Y, Hong HS, Rice FL (2003) Somatosensory organization and behavior in naked mole-rats: II. Peripheral structures, innervation, and selective lack of neuropeptides associated with thermoregulation and pain. J Comp Neuro 465:104–120

Peichl L, Němec P, Burda H (2004) Unusual cone and rod properties in subterranean African mole-rats (Rodentia, Bathyergidae). Europ J Neurosci 19:1545–1558

Poduschka W (1978) Zur Frage der Wahrnehmung von Lichtreizen durch die Mullratte, *Cryptomys hottentotus* (Lesson, 1826). Säugetierk Mitt 1978:269–274

Quilliam TA (1966) The mole's sensory apparatus. J Zool Lond 149:76–88

Rado R, Levi N, Hauser H, Witcher J, Adler N, Intrator N, Wollberg Z, Terkel J (1987) Seismic signalling as a means of communication in a subterranean mammal. Anim Behav 35:1249–1251

Rado R, Terkel J, Wollberg Z (1998) Seismic communication signals in the blind mole-rat (*Spalax ehrenbergi*): electrophysiological and behavioural evidence for their processing by the auditory system. J Comp Physiol A 183:503–511

8 Using Odors Underground

Giora Heth, Josephine Todrank

8.1 Introduction

Subterranean rodents that have adapted to living primarily underground have conquered numerous sensory and behavioral challenges that the harsh, though relatively secure, subterranean habitat presents (see reviews of Nevo 1999; Bennett and Faulkes 2000; Lacey et al. 2000). The advantages of colonizing the subterranean niche, primarily environmental stability and safety from predation, may seem trivial, however, in light of the disadvantages. Life-sustaining food can only be acquired by digging through solid soil. In the darkness of the subterranean tunnel territories, surrounded by earth, the sensory information necessary for locating food and mates is highly constrained. Edible plants may be widely scattered and prospective partners are dispersed somewhere in neighboring tunnel territories. In contrast to surface dwellers, subterranean rodents cannot see plants or other animals either close-up or far away nor can they smell plants or other animals from a distance. In contrast to the sensory adaptations made in the visual, auditory, and somatosensory systems of subterranean rodents (Nevo 1999), the olfactory system, such as in blind mole rats (Zuri et al. 1998), functions similarly to that of other rodents despite the different context. This chapter summarizes recent studies that shed light on and extend the limited knowledge about olfaction in subterranean rodents. These studies demonstrate that olfaction plays an important role not only in finding food but also in social interaction, especially recognizing familiar individuals and choosing mates. There is also evidence that urine and body odors of subterranean rodents provide individually distinctive chemosensory cues that are also indicative of the individual's sex, population, and species, and that subterranean rodents respond to the similarities and differences between odors of conspecifics and heterospecifics. Although the availability of suitable mates may be restricted underground, subterranean rodents, given

Giora Heth: Institute of Evolution, University of Haifa, Haifa 31905, Israel, E-mail: heth@research.haifa.ac.il

Josephine Todrank: Institute of Evolution, University of Haifa, Haifa 31905, Israel

Subterranean Rodents: News from Underground
S. Begall, H. Burda, C.E. Schleich (Eds.)
© Springer-Verlag Berlin Heidelberg 2007

the opportunity, prefer the odors of (and to mate with) genetically dissimilar individuals, i.e., non-kin as opposed to kin. The odor-based processes enabling individual recognition and differential responses based on genetic relatedness in subterranean rodents are a particular focus of this chapter.

8.2 Foraging Underground

Because the underground parts of plants cannot be seen, heard, or smelled from a distance, it was typically assumed that finding food underground unavoidably depended on chance encounters with food items (e.g. Heth et al. 1989; Jarvis et al. 1998; Spinks et al. 1999), in other words that subterranean rodents found food items unpredictably by running into them while digging rather than by determining their location in advance. Foraging was not considered random, that is without a method or plan, however, despite its being considered "blind" to sensory cues, because subterranean rodents observably used strategies, such as digging in straight lines (rather than circles) and branching in areas where a food item was encountered, to enhance their foraging success (Heth 1989; Benedix 1993; Jarvis et al. 1998). Thus, subterranean rodents were thought to use logistical strategies or methods to improve their odds of encountering food items in the absence of sensory cues to predict their location in advance. To determine whether subterranean rodents could also use strategies based on sensory (odor) cues to improve their chances of finding food efficiently (at least from short distances), several laboratories participated in a series of experiments designed to assess subterranean rodents' abilities to use odor-guided foraging (Heth et al. 2002a).

In the process of growing, plants take in water and nutrients from the soil through their roots and then release waste products of their metabolism back into the soil (Kolek and Holobrada 1992). If these plant waste products (known as "exudates" because they are "exuded" by the roots) contained odorous substances, perhaps even some that were distinctive for a particular plant, and if, in addition, subterranean rodents were able to detect and perhaps even identify these exudates in the soil, subterranean rodents may be able to optimize their foraging efforts by orienting toward the exudates, thereby locating the plant itself more efficiently. In the laboratory, blind mole rats, *Spalax ehrenbergi*, and naked mole rats, *Heterocephalus glaber*, can use odors to discriminate between the roots of edible and poisonous plants out of the soil (Brett 1991; Heth et al. 2000); thus it is plausible that they would make similar distinctions underground. To evaluate the hypothesis that subterranean rodents could use odors underground to orient their digging while foraging, three species of mole rats (Israeli blind mole rats; Zambian Ansell's mole rats, *Cryptomys anselli* (recently reclassified as *Fukomys anselli* by Kock et al. 2006); Kenyan naked mole rats) and Chilean coruros, *Spalacopus cyanus*, were tested in a soil-filled Perspex tunnel T-maze (Heth et al. 2002a). One arm of the crossbar contained "exudates soil", namely soil in which edible plants had been growing but from which the plants and their roots had been removed; the other arm was filled with soil maintained under the same conditions but in which no plants had been growing. Animals entered the open runway from a start box and proceeded to the soil "choice point". The trial lasted until the animal dug through all the soil and touched the end of one arm. A significant proportion of animals from all the species that were tested dug through the soil with the exudates, indicating that they were able to discriminate between the soil that did or did not contain exudates and that they preferred to dig in the soil in which the plants had been growing. Blind mole rats were tested with soil in which different types of edible plants had been growing, and they chose the exudates soil even when the plants that had been growing there were a less preferred food. When they were tested with soil in which poisonous plants had been growing, however, they showed no preference between the soil with and without exudates. In follow-up tests, Ansell's mole-rats dug preferentially through soil moistened with water in which carrots had been growing for three days and through soil toward a box in which a carrot was still growing (Lange et al. 2005), further confirming the discriminability of and preference for soil containing edible exudates. Thus, although it has yet to be demonstrated in nature, subterranean rodents presumably could use odors underground to save energy and increase the efficiency of their foraging. In light of this evidence, a re-evaluation of models on energetic costs of underground foraging that presume food search to be non-random but "blind" is clearly warranted.

8.3 Recognizing and Discriminating Between Conspecifics Underground

Odor-based mechanisms for recognizing and discriminating among kin and conspecifics remains a controversial topic (see, e.g. Tang-Martinez 2001; Todrank and Heth 2003) because recent evidence contradicts the traditional theories (Holmes 2004). Evidence from our studies, conducted in collaboration with various colleagues and using aboveground and fossorial as well as subterranean rodent species (reviewed in Todrank and Heth 2003), suggested that there are two types of odor-based discriminative processes functioning in rodents: individual recognition and "genetic relatedness assessment through individual odor similarities" (abbreviated

"G-ratios"). Individual recognition refers to a process in which, through interacting together, rodents learn associations between particular individuals and their individual odors that enable them to recognize and identify those individuals and their odors in subsequent encounters. Their relatedness to these familiar individuals is irrelevant in the functioning of this process. In contrast, "G-ratios" refers to a process in which rodents respond differentially to other rodents of differential genetic relatedness to themselves by comparing the degree of similarity between another individual's odor and their own odor (Heth et al. 2003). G-ratios is possible because individuals that share proportions of genes in common share proportional similarities in the qualities of their individual odors ("odor-genes covariance"; Heth and Todrank 2000); thus odor similarity can be used to approximate genetic similarity. Prior association and learning are irrelevant in the functioning of this process. In the following subsections we summarize the evidence for these mechanisms in subterranean rodents.

8.3.1 Individual Odors

Each individual produces an odor in its bodily secretions and excretions that is a unique proportion of compounds and is perceived as a composite or gestalt (see references in Todrank and Heth 2003). This composite provides information about individual identity and group membership, which depend on unchanging genetic traits, and about the individual's varying biological state(s). Although the mechanisms by which individual genotypes are expressed in individual odors and how environmental interactions affect the odors are not known, evidence from experiments assessing odor-genes covariance (in which the animals' biological states are carefully controlled) indicates that, along a continuum from siblings to across species, degrees of genetic similarity are evident in the degree of overlap in the qualities of the individual odors (see Todrank and Heth 2003). The similarities and differences between individual odors may seem like two sides of the same coin, but note that it is the similarities between individual odors that enable grouping those individuals together and the differences between individual odors that enable telling individuals apart.

8.3.2 Odor-genes Covariance in Subterranean Rodents

In our early studies of chemical communication in subterranean rodents, we assessed responses of blind mole rats to odorous stimuli, such as urine of conspecifics and heterospecifics, to ascertain the most appropriate paradigms and types of apparatus for testing the odor-mediated behavior of subterranean rodents, particularly those that are functionally blind, and to determine whether their odor discrimination behavior was comparable to that of aboveground rodents (Heth and Todrank 1995, 1997; Heth et al. 1996b; Todrank and Heth 1996). As with aboveground rodents, individual odors of subterranean rodents are individually distinctive and also provide information about reproductive status as well as the individual's sex, population, and species (Heth et al. 1996a, b; Todrank and Heth 1996; Zenuto and Fanjul 2002; Fanjul et al. 2003; Schwanz and Lacey 2003; Zenuto et al. 2004).

When tested in a standard habituation-discrimination paradigm (Halpin 1986), blind mole rats spontaneously discriminate between individual odors of both conspecifics and heterospecifics (Todrank and Heth 1996) and tuco-tucos, *Ctenomys talarum*, spontaneously discriminate between individual odors of conspecifics (Zenuto and Fanjul 2002), but it is not possible to determine from the standard habituation methodology whether rodents respond to the odors of more closely related individuals as if they are perceived as having similar odor qualities compared with the odor of a less closely related individual. To assess the perceptual similarities of individual odors of rodents from the same kin group, we developed the habituation-generalization technique (Todrank et al. 1998). As in the standard habituation paradigm, subjects are presented with the odor of one individual (the habituation odor donor) for repeated trials, and the time spent investigating the odor is measured during each trial. A significant reduction in investigation time during the last habituation trial compared with the first indicates that the subjects recognized the stimulus as the same odor in each trial and "habituated" to it. In the standard habituation paradigm, the odor of a second individual is presented during a test trial, and a significant increase in investigation of the odor compared with the last habituation trial indicates that the subjects discriminated between the two odors. In the habituation-generalization paradigm, two test trials are conducted with two different odors: the first test odor is from a donor that is a close relative, such as a same-sex sibling, of the habituation odor donor; the second test odor is from a same-sex individual that is not related to the other two odor donors. As in the standard habituation paradigm, a significant increase in investigation during the first test trial compared with the last habituation trial indicates that subjects discriminated between the odors of the different individuals, but in the habituation-generalization paradigm, a significant increase in investigation during the second test trial compared with the first test trial indicates that subjects treated the odors of the relatives as similar compared with the odor of the non-relative (Todrank et al. 1998). Giant mole rats (*Cryptomys mechowii* (recently reclassified as *Fukomys mechowii* by Kock et al. 2006)) tested in this paradigm with odors from ano-genital secretions demonstrated the odor-genes covariance phenomenon within families because odors of same-sex siblings were treated as having similar qualities (Heth et al. 2002b).

In pilot tests using this habituation-generalization technique to investigate odor-genes covariance among individuals of the same species (i.e. that the odors of two unrelated conspecifics were treated as similar compared with the odor of a heterospecific), the increase in investigation time during the first test trial was so large compared with the last habituation trial, presumably because the individual odors of conspecifics from different families are readily discriminable, that it was impossible to show a statistically significant increase in investigation when the heterospecific odor was presented in the second test trial. Thus we developed a second type of habituation-generalization methodology (Heth and Todrank 2000) that was also a variation on a standard habituation-discrimination paradigm in which the habituation odor is presented for one long habituation trial and then the habituation odor and the novel odor are presented together during the test trial (Halpin 1986). In this new variation, after one long habituation trial, the two test odors were presented together in a single test trial. Although it was not possible with this second habituation-generalization technique to demonstrate, for example, that the odors of two conspecifics are discriminable, it was possible to demonstrate that conspecific odors were perceptually similar compared with heterospecific odors when the odor of the heterospecific was investigated significantly longer during the test trial than the odor of the second conspecific. We used this paradigm to investigate odor similarities across species within four species of blind mole rats from the *Spalax ehrenbergi* complex (Nevo 1999) in which the phylogeny was known, and in this study, the odors of the donors from the phylogenetically closer species were treated as perceptually similar (demonstrated by differential investigation) compared with more distant heterospecifics (Heth and Todrank 2000).

The series of studies of odor-genes covariance in subterranean and aboveground rodents (reviewed in Todrank and Heth 2003) contributed to understanding odor-based mechanisms for recognizing and discriminating between conspecifics in two important ways: one had to do with individual recognition and the other with discriminative responses based on differential genetic relatedness.

8.3.3 Individual Recognition

When we developed the initial habituation-generalization technique to explore perceptual similarities between odors of siblings, we expected to find a small but significant increase in investigation of the second sibling's odor during the first test trail and a larger significant increase in response to the odor of the non-sibling during the second test trial. During extensive studies using this technique – first with aboveground but then also with subterranean rodents – we discovered that there is a second equally valid response pattern in this paradigm: the odor of the second sibling can be perceived as sufficiently similar to the habituation odor that subjects do not increase their investigation during the first test trial, and, in fact, the investigation time can even be shorter on average than during the last habituation trial (Todrank et al. 1998, 1999; Heth et al. 1999, 2002b, 2004). Interestingly, a clear differential response pattern emerged: when subjects were familiar with the sibling odor donors through rearing together, because they were either from the same biological family (Todrank et al. 1998; Heth et al. 1999, 2002b) or foster siblings (Todrank et al. 1999), they discriminated between the sibling donors' odors. In contrast, when the subjects were not familiar with the sibling odor donors because they were either from a different biological family (Todrank et al. 1998; Heth et al. 1999, 2002b) or from the same biological family but reared apart (Todrank et al. 1999), subjects did not show spontaneous discrimination between the odors. We concluded from the combined results that (1) subjects discriminated between the odors of familiar donors (even when they were so similar because the donors were siblings) because they recognized the odors as belonging to a specific donor but (2) subjects failed to discriminate between the similar odors of unfamiliar sibling donors because they had not had an opportunity to make an association between the particular individuals and their odors (Todrank and Heth 2003).

Social African mole rats (such as those from the *Cryptomys* and *Fukomys* complex) live and reproduce in colonies of a single breeding pair and their non-breeding offspring (Burda et al. 2000); thus it is possible to test their responses to odors of familiar biological siblings or unfamiliar siblings from a different colony; the reproduction within each colony is sufficiently sporadic and the development of the pups sufficiently slow (Begall and Burda 1998), however, that it is not possible to conduct studies that require cross-fostering of newborns, and thus it is impossible to arrange for familiar siblings from another colony or biological siblings that are unfamiliar because they were reared apart. Thanks to the clever procedure developed in Hynek Burda's laboratory to study incest avoidance in colonies of *Fukomys* mole rats (Burda 1995), it is possible to arrange for biological siblings to be separated for a sufficient period of time to have forgotten one another. Two brothers are removed from their colony for one or two days, then just before the brothers are returned to the colony, two sisters are removed for one or two days, and then the brothers are removed just before the sisters are returned, etc. By maintaining this switching practice over a period of three weeks with Ansell's mole rats, the brothers and sisters that are removed periodically retain their familiarity with other members of their home colony but lose their familiarity with one another. This "forgetting" or "estrangement" can be confirmed by assessing their willingness to mate with their forgotten sibling in contrast with their continued unwillingness to mate with other familiar opposite-sex colony members (Burda 1995). Ansell's mole rats from several colonies did not spontaneously discriminate between the ano-genital odors of their estranged opposite-sex siblings when tested in the habituation-generalization paradigm, but they did discriminate between the ano-genital odors of their familiar opposite-sex siblings (Heth et al. 2004). This differential response to the odors of familiar vs unfamiliar siblings is another indication that subterranean rodents learn enough about the association between familiar individuals and their odors to recognize the odors of their familiar siblings individually. They can fail, however, to discriminate between the odors of unfamiliar siblings when the odors have perceptually similar qualities. Surprisingly, in a later study with Chilean coruros testing responses to odors from pairs of familiar biological siblings and pairs of unfamiliar, unrelated siblings from another colony in the habituation-generalization paradigm (Hagemeyer and Begall, 2006), subjects spontaneously discriminated between the odors of the unfamiliar siblings as well as familiar siblings, but it was not clear whether the larger differences in the qualities of the odors were due to slight differences in diet or to greater genetic variability in the coruros, and thus in their odors, than in the mole rats.

8.3.4 Genetic Relatedness Distinctions

The cumulative evidence indicates that the odor-genes covariance principle is comparably demonstrable along a continuum of genetic relatedness, from siblings to populations to species and across species, in subterranean mole rats as well as in aboveground rodents (see Todrank and Heth 2003), thus raising the intriguing possibility that subterranean rodents may also use the predictable relationship between individual odor similarities and genotypic similarities in their responses to odors. That is, they may respond indirectly to varying degrees of genetic relatedness to other individuals by responding directly to varying degrees of similarity between another individual's odor and their own odor. This "genetic relatedness assessment through individual odor similarities" ("G-ratios") process was first discovered in two species of wild mice (*Mus spicilegus* and *M. musculus*) that were cross-fostered and reared in mixed litters of both species (Heth et al. 2003). Their odor preferences across a series of experiments (which included two additional species of wild mice (*M. macedonicus* and *M. domesticus*) as odor donors) clearly demonstrated not only that they used their own odor as a referent but also that their responses were graded along a continuum of genetic relatedness that extended from within their own population of conspecifics to closer as opposed to more distant heterospecifics.

Cross-fostering experiments are logistically impossible with slowly developing non-seasonally breeding molerats, but the "forgotten siblings paradigm" (see Sect. 8.3.3) was helpful for assessing preferential responses to other individuals and their odors depending on the degree of genetic relatedness between the subject and the other individuals/odor donors. Ansell's mole rats of both sexes were first given a two-choice ano-genital odor preference test between an unfamiliar, unrelated conspecific and a "forgotten" sibling and then given a two-choice individual preference test when the conspecific and the sibling that had been odor donors for the first test were presented to the subject separated by a wire screen, which permitted full olfactory and limited tactile contact. Subjects spent significantly more time investigating the unrelated conspecifics and their odors than the forgotten siblings and their odors, and they initiated courtship behavior with the unrelated conspecific (Heth et al. 2004). To be sure that the siblings had truly been forgotten, pairs of estranged siblings were observed when placed together in an aquarium, and their initiation of courtship behavior confirmed that these siblings treated each other as suitable mates. Thus the longer investigation of the unrelated individuals and their odors in the preference tests could be considered indicative of a mate preference (Heth et al. 2004). Nothing is known about the mode of breeding dispersal or about how partners meet to establish new mole rat colonies in nature (Burda 1999), but there is some evidence that *Cryptomys* and *Fukomys* mole rat colonies are founded by genetically unrelated pairs (Bennett and Faulkes 2000; Burland et al. 2002). Contrary to previous assumptions, there is evidence suggesting that outbreeding may also be the rule in the foundation of new colonies of naked mole rats (Braude 2000; Ciszek 2000). Although additional experiments at differing degrees of genetic relatedness will be necessary to confirm that subterranean rodents use G-ratios, the findings certainly are suggestive that a G-ratios process could be involved in mate choice and thus in the founding of new colonies.

8.3.5 Two Separate Mechanisms – Two Separate Functions

Subterranean rodents clearly recognize familiar individuals and discriminate between unfamiliar individuals on the basis of differential genetic relatedness. To recognize familiar individuals (or their odors), subterranean rodents must, during interactions with one another, form sufficiently strong memories of the particular individual's distinctive phenotypic characteristics (which would include the individual's odor) that they can match the phenotypic characteristics observed on a subsequent encounter accurately with the remembered characteristics. The extent of the genetic relatedness between the individuals would not affect the process of remembering and later matching their phenotypic characteristics to the memory. In contrast, to assess the degree of genetic relatedness between themselves and another individual, subterranean rodents must, upon encountering one another, assess each other's phenotypic characteristics and compare them with their own and respond differentially depending on the degree of similarity between them. Whether the individuals are familiar with one another or not does not affect the process of assessing the degree of similarity between their phenotypic characteristics. Thus individual recognition requires comparisons with memories of other individuals' characteristics whereas G-ratios requires comparisons of other individuals' characteristics (whether familiar or not) with one's own. Although both processes involve evaluation of other individuals' characteristics, what is done during the comparison process for individual recognition as opposed to G-ratios is sufficiently different to warrant treating these processes as separate.

8.4 Conclusion

The evidence from an ever-growing number of studies on using odors underground conducted with a variety of subterranean rodent species suggests that olfaction plays an important role in the lives of subterranean rodents that have made a successful adaptation to life underground. Subterranean rodents can respond to the odors of plant exudates in the soil and thus orient their digging to enhance their foraging success. Subterranean rodents can recognize familiar conspecifics and their odors individually. They can also respond differentially to other individuals and their odors, familiar or not, based on differential degrees of genetic relatedness between them. Future studies will, no doubt, further extend the understanding of the myriad functions that odors serve in living successfully underground.

References

Begall S, Burda H (1998) Reproductive characteristics and growth rate in the eusocial Zambian common mole-rat (*Cryptomys* sp., Bathyergidae). Z Säugetierk 63:297–306

Benedix JH Jr (1993) Area-restricted search by plains pocket gopher (*Geomys bursarius*) in tall grass prairie habitat. Behav Ecol 4:318–324

Bennett NC, Faulkes CG (2000) African mole-rats: ecology and eusociality. Cambridge Univ Press, Cambridge

Braude S (2000) Dispersal and new colony formation in wild naked mole-rats: evidence against inbreeding as the system of mating. Behav Ecol 11:7–12
Brett RA (1991) The population structure of naked mole-rat colonies. In: Sherman PW, Jarvis JUM, Alexander RD (eds) The biology of the naked mole-rat. Princeton Univ Press, Princeton, NJ, pp 137–184
Burda H (1995) Individual recognition and incest avoidance in eusocial common mole-rats rather than reproductive suppression by parents. Experientia 51:411–413
Burda H (1999) Syndrome of eusociality in African subterranean mole-rats (Bathyergidae, Rodentia), its diagnosis and aetiology. In: Wasser SP (ed) Evolutionary theory and processes: modern perspectives. Kluwer Academic Publishers, Dordrecht, pp 385–418
Burda H, Honeycutt RL, Begall S, Locker-Grütjen O, Scharff A (2000) Are naked and common mole-rats eusocial, and if so why? Behav Ecol Sociobiol 47:293–303
Burland TM, Bennett NC, Jarvis JUM, Faulkes CG (2002) Eusociality in African mole-rats: new insights from patterns of genetic relatedness in the Damaraland mole-rat (*Cryptomys damarensis*). Proc R Soc Lond B 269:1025–1030
Ciszek D (2000) New colony formation in the "highly inbred" eusocial naked mole-rat: outbreeding is preferred. Behav Ecol 11:1–6
Fanjul MS, Zenuto RR, Busch C (2003) Use of olfaction for sexual recognition in the subterranean rodent *Ctenomys talarum*. Acta Theriol 48:35–46
Hagemeyer P, Begall S (2006) Individual odour similarity and discrimination in the coruro (*Spalacopus cyanus*, Octodontidae). Ethology 112:529–536
Halpin ZT (1986) Individual odors among mammals: origins and functions. Adv Study Behav 16:40–70
Heth G (1989) Burrow patterns of the mole rats *Spalax ehrenbergi* in two soil types (Terra-rossa and Rendzina) in Mount Carmel, Israel. J Zool Lond 217:39–56
Heth G, Todrank J (1995) Assessing chemosensory perception in subterranean mole rats: different responses to smelling versus touching odorous stimuli. Anim Behav 49:1009–1015
Heth G, Todrank J (1997) Patterns of urination of a blind subterranean rodent, *Spalax ehrenbergi*. Ethology 103:138–148
Heth G, Todrank J (2000) Individual odour similarities across species parallel phylogenetic relationships in the *S. ehrenbergi* superspecies of mole rats. Anim Behav 60:789–795
Heth G, Golenberg EM, Nevo E (1989) Foraging strategy of a subterranean rodent, *Spalax ehrenbergi*. Oecologia 79:496–505
Heth G, Beauchamp G, Nevo E, Yamazaki K (1996a) Species, population and individual specific odors in urine of mole rats (*Spalax ehrenbergi*) detected by laboratory rats. Chemoecology 7:107–111
Heth G, Nevo E, Todrank J (1996b) Seasonal changes in urinary odors and in responses to them by blind subterranean mole rats. Physiol Behav 60:963–968
Heth G, Todrank J, Johnston RE (1999) Similarity in the qualities of individual odors among kin and species in Turkish (*Mesocricetus brandti*) and golden (*Mesocricetus auratus*) hamsters. J Comp Psych 113:321–326
Heth G, Todrank J, Nevo E (2000) Do *Spalax ehrenbergi* blind mole rats use food odors to forage for and select food? Ethol Ecol Evol 12:75–82
Heth G, Todrank J, Begall S, Koch R, Zilbiger Y, Nevo E, Braude SH, Burda H (2002a) Odours underground: subterranean rodents may not forage "blindly". Behav Ecol Sociobiol 52:53–58
Heth G, Todrank J, Burda H (2002b) Individual odor similarities within colonies and across species of eusocial *Cryptomys* mole rats. J Mammal 83:569–575
Heth G, Todrank J, Busquet N, Baudoin C (2003) Genetic relatedness assessment through individual odour similarities (G-ratios) in mice. Biol J Linn Soc 78:595–603

Heth G, Todrank J, Begall S, Wegner R, Burda H (2004) Genetic relatedness discrimination in a eusocial mole-rat, *Cryptomys anselli* (Bathyergidae, Rodentia). Folia Zool 53:269–278

Holmes WG (2004) The early history of Hamiltonian-based kin recognition research theory/past and future. Ann Zool Fenn 41:691–711

Jarvis JUM, Bennett NC, Spinks AC (1998) Food availability and foraging by wild colonies of Damaraland mole-rats (*Cryptomys damarensis*): implications for sociality. Oecologia 113:290–298

Kock, D, Ingram CM, Frabotta LJ, Honeycutt RL, Burda H (2006) On the nomenclature of Bathyergidae and *Fukomys* n. gen. (Mammalia: Rodentia). Zootaxa 1142:51–55

Kolek J, Holobrada M (1992) Ion uptake and transport. In: Kolek J, Kozinka V (eds) Physiology of the plant root system. Kluwer Academic Publishers, Dordrecht, pp 204–285

Lacey EA, Patton JL, Cameron GN (eds) (2000) Life underground: the biology of subterranean rodents. Univ Chicago Press, Chicago

Lange S, Neumann B, Hagemeyer P, Burda H (2005) Kairomone-guided food location in subterranean Zambian mole-rats (*Cryptomys* spp., Bathyergidae). Folia Zool 54:263–268

Nevo E (1999) Mosiac evolution of subterranean mammals: regression, progression, and global convergence. Oxford Univ Press, Oxford

Schwanz LE, Lacey EA (2003) Olfactory discrimination of gender by colonial tuco-tucos (*Ctenomys sociabilis*). Mamm Biol 68:53–60

Spinks AC, Branch TA, Croeser S, Bennett NC, Jarvis JUM (1999) Foraging in wild and captive colonies of the cooperatively breeding common mole-rat, *Cryptomys hottentotus hottentotus* (Rodentia: Bathyergidae). J Zool Lond 249:143–152

Tang-Martinez Z (2001) The mechanisms of kin discrimination and the evolution of kin recognition in vertebrates: a critical re-evaluation. Behav Proc 53:21–40

Todrank J, Heth G (1996) Individual odours in two chromosomal species of blind subterranean mole rat (*Spalax ehrenbergi*): Conspecific and cross-species discrimination. Ethology 102:806–811

Todrank J, Heth G (2003) Odor-genes covariance and genetic relatedness assessments: Rethinking odor-based "recognition" mechanisms in rodents. Adv Study Behav 32:77–130

Todrank J, Heth G, Johnston RE (1998) Kin recognition in golden hamsters: evidence for kinship odours. Anim Behav 55:377–386

Todrank J, Heth G, Johnston RE (1999) Social interaction is necessary for discrimination and memory for odours of close relatives in golden hamsters. Ethology 105:771–782

Zenuto RR, Fanjul SM (2002) Olfactory discrimination of individual scents in the subterranean rodent *Ctenomys talarum* (tuco-tuco). Ethology 108:629–641

Zenuto RR, Fanjul MS, Busch C (2004) Use of chemical communication by the subterranean rodent *Ctenomys talarum* (tuco-tuco) during the breeding season. J Chem Ecol 30:2111–2126

Zuri I, Fishelson L, Terkel J (1998) Morphology and cytology of the nasal cavity and vomeronasal organ in juvenile and adult blind mole rats (*Spalax ehrenbergi*). Anat Rec 251:460–471

9 Acoustics, Audition and Auditory System

Sabine Begall, Simone Lange, Cristian E. Schleich, Hynek Burda

9.1 Introduction

In the darkness of the subterranean ecotope, hearing seems to be predestined to counterbalance the visual restrictions similar to the situation familiar in cave dwellers like bats. It is therefore generally assumed that audition and acoustic signals play a major role in communication and alertness of subterranean mammals, and it is not surprising that studies on sensory ecology of subterranean species are biased toward hearing and vocalisations.

The auditory system represents the receiver for acoustic (and also seismic) signals. What the animals can really hear in their natural surroundings depends, however, not only on the capabilities of their auditory systems (ears and neural processing), but also on the tunnels' acoustic characteristics. Acoustics of the environment, auditory systems, and audition are inevitably interwoven.

Acoustic signals propagate either directly along the burrows (air-borne) or through the soil (substrate-borne), and in both cases lower frequencies are transmitted best. Thus, it is expected that subterranean mammals have tuned their hearing towards lower frequencies. As demonstrated below (Sect. 9.3), the audiograms of subterranean rodents of diverse species show the same characteristics; nevertheless, in ultimate terms the findings have been interpreted differently (vestigial hearing vs specialized hearing). The detailed studies on ear morphology presented in Sect. 9.4 shed light on this controversy.

Sabine Begall: Dept. General Zoology, University of Duisburg-Essen, D-45117 Essen, Germany, E-mail: sabine.begall@uni-due.de

Simone Lange: Dept. General Zoology, University of Duisburg-Essen, D-45117 Essen, Germany

Cristian E. Schleich: Departamento de Biología, Facultad de Ciencias Exactas y Naturales, CC 1245, Universidad Nacional de Mar del Plata, Funes 3250, Mar del Plata (7600), Argentina

Hynek Burda: Dept. General Zoology, University of Duisburg-Essen, D-45117 Essen, Germany

Subterranean Rodents: News from Underground
S. Begall, H. Burda, C.E. Schleich (Eds.)
© Springer-Verlag Berlin Heidelberg 2007

In this chapter we review the current findings on the acoustics in burrows, and summarise available data on hearing in subterranean mammals. Last but not least, we demonstrate that the middle and inner ears of subterranean rodents are tuned to low frequencies – the main component of sounds transmitted best underground.

9.2 Acoustics in Burrows

Until recently, only a single publication addressed propagation of air-borne sounds in burrow systems of subterranean rodents (Heth et al. 1986). This study showed that in natural burrows of the blind mole rat (*Spalax ehrenbergi*) from Israel, air-borne sounds of 440 Hz propagated best, whereas lower and higher frequencies were strongly attenuated, and transmission was effective over short distances (up to 5 m) only. Since vocalizations, auditory and oto-morphological characteristics are very similar across a variety of species of subterranean mammals (see below and Chap. 10), a rather uniform acoustic environment (corresponding to that described by Heth et al. 1986 for *Spalax*) is assumed for diverse subterranean rodents.

In a recent study, Lange and colleagues (Lange 2006; Lange et al., 2007) refined the results by Heth et al. (1986) when measuring the sound transmission in burrow systems of Zambian mole-rats of the genus *Fukomys* (formerly known as *Cryptomys*; see Kock et al. 2006). They studied propagation of air-borne sounds of diverse frequencies (pure tone signals between 100 Hz and 3.2 kHz) in natural burrows of giant mole-rats (*F. mechowii*) and common mole-rats (*F. kafuensis*) at several sites in Zambia. In the field, straight burrows were opened, leaving an undisturbed tunnel in situ in the length of 1–3 m between both openings. The signals were played back via a loudspeaker inserted into one tunnel opening, which was tightly sealed with a cotton cloth. Each frequency signal was played back with 75 dB SPL and the transmitted signals were recorded via the microphone, inserted into the tunnel at the opposite opening, also sealed with cotton cloth, and connected to a DAT recorder. Attenuation was calculated (using Avisoft software) as the SPL-difference between 75 dB and the value recorded at the tunnel end. It was found that sounds of low frequencies (200–800 Hz) consistently showed the lowest attenuation, i.e. they propagated better than the tested signals of lower and higher frequencies (Fig. 9.1). The combined data of both studies show that burrows of different diameters (4.5–9 cm), at different depths (15–55 cm), in different habitats and different soils exhibit very similar acoustical properties. Thus, the assumption that acoustics is similar in tunnels of different species of subterranean rodents is strongly supported.

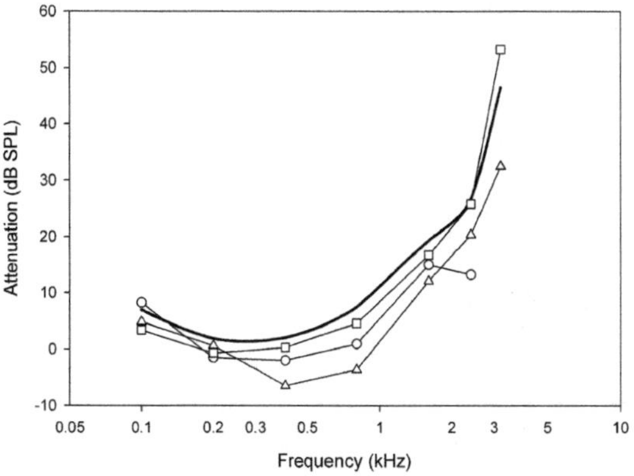

Fig. 9.1. Mean attenuation (*bold solid line*) of seven recordings over a distance of 1 m measured in mole-rat burrows at different locations in Zambia. Additionally the results of the single recordings of tunnels, which showed an amplification effect are depicted (*square* – locality: Kaleya, *Fukomys kafuensis*, diameter 6 cm; *circle* – locality: Mansansa, *F. mechowii*, diameter 8 cm; *triangle* – locality: Kaleya, *F. kafuensis*, diameter 5.5 cm). Frequency = tested pure tone signal in kHz, attenuation in dB SPL

Quilliam (1966) suggested that certain sound frequencies in burrows might be amplified like in an ear trumpet and this so-called "stethoscope effect" has been cited quite often since then. Lange and colleagues were able to measure this proposed amplification effect for the first time. At three different burrows a selective amplification at certain frequencies (200 Hz, 400 Hz, and 800 Hz) was measured. The maximum recorded amplification amounted to more than two times in amplitude (6.5 dB) at a distance of 1 m. Although in this study, the sample size was small and the equipment used was simple, the results are important, since the same "stethoscope effect" has been demonstrated in burrows of different diameters. There is, however, no doubt that a more thorough approach involving a higher sample size covering different well-defined tunnel parameters is desirable using sophisticated technical apparatuses to exclude possible artefacts and reveal the source and nature of the amplification.

9.3 Hearing in Subterranean Rodents

In general, hearing in mammals may be studied by different approaches. Electrophysiological audiograms (recordings from brain auditory centres) are available for *Spalax ehrenbergi* (Bruns et al. 1988) and *Fukomys anselli* (Müller and Burda 1989). The so-called cochlear audiogram based on the study of acoustic distortion products from the cochlea was described in

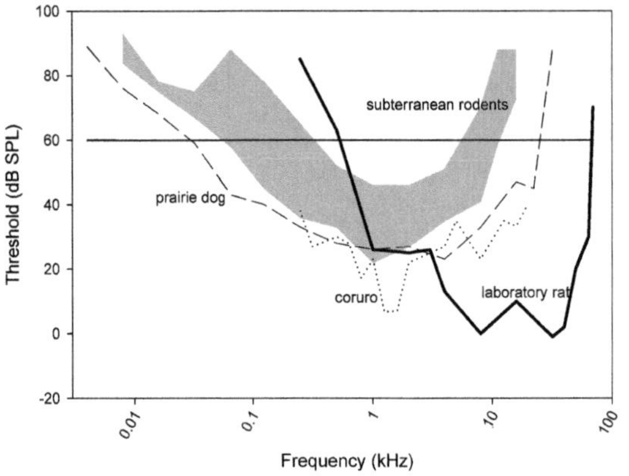

Fig. 9.2. Combined behavioural audiograms of subterranean rodents in comparison to more fossorial rodents like the coruro (*Spalacopus cyanus*, *dotted line*, Begall et al. 2004) and the prairie dog (*Cynomys sp.*, *dashed line*, Heffner RS et al. 1994). The *bold solid line* illustrates the audiogram of the laboratory rat (Heffner HE et al. 1994) as an example of a hearing generalist. The *shaded area* indicates the minimum and maximum threshold values (dB SPL) at different frequencies for strictly subterranean rodents (Müller and Burda 1989; Heffner RS and Heffner 1990, 1992, 1993; Kössl et al. 1996; Brückmann and Burda 1997; Begall et al. 2004). The *simple horizontal line* shows the 60 dB SPL level of hearing. Redrawn from Begall et al. (2004)

F. anselli (Kössl et al. 1996). Behavioural audiograms are available for several species of subterranean mammals: *Geomys bursarius* (Heffner RS and Heffner 1990), *Spalax ehrenbergi* (Heffner RS and Heffner 1992), *Heterocephalus glaber* (Heffner RS and Heffner 1993), *Fukomys anselli* (Brückmann and Burda 1997), *Spalacopus cyanus* (Begall et al. 2004), and *Heliophobius argenteocinereus* (Knotková et al. 2005) (cf. Table 9.1, Fig. 9.2). These audiograms were obtained using operant conditioning and are basically comparable. They illustrate that subterranean rodents have restricted hearing ranges with low high-frequency cut-offs at 60 dB SPL with the lowest value found for *S. ehrenbergi* (5.9 kHz). Frequencies of best hearing were at 4 kHz (*H. glaber*) or in the range of 0.8–2 kHz in other species (Table 9.1). Although the overall sensitivity is relatively low, subterranean rodents are more sensitive in the low-frequency range (<0.8 kHz) compared to aboveground living generalists like rats (Fig. 9.2), and resemble in this respect aboveground living rodents with good low-frequency hearing like kangaroo rats (Heffner H and Masterton 1980) or gerbils (Ryan 1976).

Hearing in mammals classified as intermediate between strictly subterranean and fossorial (i.e. those that are not obligatorily bound to the subterranean habitat, e.g. in search of food) has been studied only in coruros (Begall et al. 2004) and prairie dogs (Heffner RS et al. 1994). These rodents apparently possess wider hearing ranges and greater sensitivity

Table 9.1. Audiological parameters of hearing in subterranean and fossorial mammals (obtained by behavioural audigrams). [a]Frequency-limits could not be obtained due to technical limitations [b]Values obtained by means of sound-evoked potentials of the inferior colliculus

Subterranean/fossorial mammals	60 dB-lowest frequency	60 dB-highest frequency	Best frequency (kHz)	Best sensitivity (dB SPL)	Source
Talpa europaea[b]			0.5	34	Aitkin et al. 1982
Spalacopus cyanus	< 0.25[a]	> 20[a]	1.25–1.6	7	Begall et al. 2004
Heliophobius argenteocinereus	< 0.25[a]	11.7	1	22	Knotková et al. 2005
Heterocephalus glaber	0.065	11.5	4	35	Heffner RS and Heffner 1993
Fukomys anselli	0.225	18.5	0.8	24	Brückmann and Burda 1997
Spalax ehrenbergi	0.054	5.9	1	32	Heffner RS and Heffner 1992
Geomys bursarius	0.27	8.7	2	24	Heffner RS and Heffner 1990
Cynomys leucurus	0.044	26			Heffner RS et al. 1994
C. ludovicianus	0.029	26	4	20.3	Heffner RS et al. 1994
Mean	0.06	15.47	1.84	24.79	

than strictly subterranean mammals (Fig. 9.2). Frequencies characterizing highest hearing sensitivity are generally lower than those for aboveground rodents (Table 9.1).

The only strictly subterranean non-rodent mammal for which audiological data are available is the European mole (*Talpa europaea*). Rough data on the hearing range (0.2–15 kHz) and the frequency of best hearing (0.5 kHz – 34 dB SPL) obtained by behavioural tests and sound-evoked responses in the inferior collicle (Aitkin et al. 1982) comply with the general pattern of hearing in subterranean rodents (Table 9.1).

Since the subterranean mammals studied thus far belong to different families and originate from different geographical regions, their hearing abilities appear to be adapted to the subterranean habitat, and are thus the result of convergent evolution. Generally, the convergence is adaptive but may be the result of two different processes: regressive degeneration due to lack of stimulation or progressive specialization upon perception of restricted stimuli. Accordingly, Heffner RS and Heffner (1990, 1992, 1993) point out low overall sensitivity and poor high-frequency hearing of subterranean rodents and regard their hearing as vestigial or degenerated. However, it should be noted that the hearing range in subterranean rodents compared to their aboveground counterparts is not generally restricted (it still covers about eight octaves), but is only shifted towards lower frequencies. Furthermore, hearing sensitivity in the low frequency range is conserved or even improved.

The question, why (in ultimate terms) sensitivity at frequencies of the "best hearing" is much lower than that in other rodents, can be explained by the existence of the "stethoscope effect" (Lange et al., 2007). It may have been advantageous for subterranean mammals to compensate the increase in sound intensity in the environmental channel by reducing hearing sensitivity and thus to avoid over-stimulation. Hearing in subterranean rodents was studied in laboratory in epigeic open field conditions. We may expect that the best hearing sensitivity of subterranean rodents in their natural environment, i.e. in burrows, is comparable to that of aboveground living rodents.

Hence, both frequency tuning and low sensitivity of hearing can be considered convergent adaptive specializations to the unique acoustic environment.

9.4 Morphological Adaptations of the Ear

9.4.1 Outer and Middle Ear

The ear pinnae of most subterranean mammals are reduced or completely lacking (Burda et al. 1990), and it has been commonly assumed that this reduction is necessary to prevent the ear pinnae to shovel loosened soil, which might get stuck in the ear canal. However, many burrowing rodents, possessing normally-sized pinnae, demonstrate that pinnae and digging do not exclude each other. Hence, the reason for the reduction is supposedly a different one. Generally, pinnae are important for binaural sound localisation, which is, however, of little use in underground burrows (Burda et al. 1989). Obviously, there is no selective pressure to maintain this anatomical structure in strictly subterranean mammals.

The outer ear canals of subterranean rodents tend to be tight (diameter 0.6 mm in pocket gophers, Heffner RS and Heffner 1990). Furthermore, ear canals of subterranean rodents are often filled with cerumen (Burda et al. 1992). Both aspects, small lumen and ceruminous plugs are expected to reduce hearing sensitivity.

Bullar volumes of subterranean rodents are relatively (taking body weight into account) large in comparison to aboveground living hearing generalists (Schleich and Vassallo 2003; Lange 2006), but they are still smaller than those of fossorial low-frequency hearing specialists like kangaroo rats or gerbils (Lay 1972; Webster and Webster 1975). However, different bulla shapes (spherical - conical), presence or absence of pneumatised bulla walls, mode of measurement (volume of the bulla in toto vs volume of the middle ear cavity), and calculation formulas (cuboid - conical) complicate comparisons between different species and studies. The tympanic membranes are relatively large, rather round and the pars flaccida is small or absent (Burda et al. 1992).

In all subterranean mammal species examined to date, the middle ears are of the freely mobile type according to Fleischer's classification (Fleischer 1978), i.e. the gonial is reduced or absent, the fused malleus-incus complex is only weakly attached to the bulla wall by means of ligaments allowing the malleus-incus complex to swing relatively freely around the rotatory axis, and the mallear manubrium runs parallel to the incudal crus longum (Burda et al. 1989, 1992; Mason 2001, 2003a, 2004; Schleich and Busch 2004; Lange and Burda 2005; Begall and Burda 2006; Lange 2006). The middle ear muscles, m. tensor tympani and m. stapedius, are strongly reduced or absent (with *Spalacopus cyanus*, possessing weakly developed stapedius muscles, being an exception among subterranean mammals,

Begall and Burda 2006). The freely mobile middle ear type and the reduced middle ear muscles result in reduced stiffness, which makes the system less sensitive towards high frequencies. The incus in *Spalax ehrenbergi* possesses a peculiar bony cap, a structure not seen in other mammals studied to date. This feature has been correlated with the ability of *Spalax* to perceive (and communicate by means of) seismic signals (Burda et al. 1989; Rado et al. 1989). The stapedes are usually of normal stirrup-like shape typical of mammals; however, there is a remarkable intraspecific variability in the shape with a tendency for reduction of the intercrural foramen resulting in a columella-shape (Burda et al. 1989). Stapedes in pocket gophers (*Geomys*, *Cratogeomys*) exhibit large, concave stapedial footplates (Burda et al. 1992). The stapedial footplates are significantly larger with respect to body size in fossorial species compared to non-fossorial species (Burda et al. 1992; Mason 2001) resulting in relatively low area ratios between eardrum and footplate area. Similarly, the lever ratios (length of the mallear lever divided by length of incudal lever) are smaller in subterranean rodents than in non-fossorial species (Burda et al. 1992; Mason 2001; Lange 2006). The resulting transformer ratio, i.e. the product of the area and the lever ratio, as a measure of the efficiency of sound transmission (pressure and force amplification respectively) is markedly lower in mole-rats compared to their epigeic counterparts and reflects the low middle ear sensitivity.

Among fossorial mammals, some (but not all) chrysochlorids possess the most-developed middle ears, often with hypertrophied mallei and unusually large stapedial footplates, the latter contributing to extremely low area ratios (Fleischer 1973; Burda et al. 1992; Mayer et al. 1995; Mason 2001, 2003a). The extraordinary morphological structures found in chrysochlorids probably mediate detection of low-frequency seismic sounds transmitted through the soil (Mason 2003b).

Beside the number of conclusions that can be derived from the structure of subterranean middle ears, species-specific characteristics of their ossicles prove also to be a useful diagnostic and taxonomic tool (Burda 1979; Burda et al. 1989; Lange et al. 2005).

9.4.2 Inner Ear

There are two approaches to dissect and examine the organ of Corti of the inner ear: the technique of histological serial sectioning and subsequent reconstruction (cf. Bruns et al. 1988) and the method of surface specimens (cf. Burda et al. 1989). Both methods are technically demanding and hence, it is no surprise that the cochlear duct and the Corti organ have been examined up to now only in a few subterranean rodent species (and specimens) (Pye 1977; Bruns et al. 1988; Burda et al. 1989; Müller et al. 1989, 1992; Raphael et al. 1991; Begall and Burda 2006; Lange 2006; Schleich et al. 2006).

The cochleae of subterranean mammals show a tower shape with more than three turns (cf. Table 9.2, Fig. 9.3). From this aspect the cochlea of subterranean rodents (including *Spalax* and *Geomys*) studied to date resembles that of caviomorphs and differs from much less coiled cochleae found in murids (Fleischer 1973; Pye 1977; Burda et al. 1988). Cochleae in subterranean rodents (and caviomorphs) exhibit, compared to skull size, longer cochlear ducts (basilar membranes) than in murids. Since the hair cell densities are roughly comparable in the studied mammalian species (see papers cited above, references therein and own unpublished data), an absolutely longer basilar membrane accommodates more hair cells. This may result in better frequency and/or intensity discrimination. The number of hair cells per octave is on average higher in subterranean rodents and the guinea pig than in the rat (Table 9.2), suggesting that mean discrimination capacity in these species is also better. It should be noted, that in cochlear regions, where the best audible frequencies are received, the hair cell density is generally higher in all the considered species. Furthermore, in *Fukomys anselli*, a clear over-representation of the frequencies between 0.6 and 1 kHz (denoted as "acoustic fovea") was revealed; in this frequency range the slope of the place-frequency map amounted to 5.3 mm/octave (Müller et al. 1992), which means that in this species actually 3000 hair cells are dedicated to process one octave (Lange 2006)! It would be interesting to study innervation pattern and see whether the higher number of hair cells is reflected also in a higher number of nerve fibres carrying the same input to the brainstem thereby increasing neuronal substrate for auditory processing or whether the outer and inner hair cells along the organ of Corti show different degrees of auditory nerve fibre convergence.

As in other mammals, the basilar membrane width in subterranean rodents increased and its thickness decreased from the cochlear base towards the cochlear apex. However, remarkably, there was no or little change in the basilar membrane width and thickness between 40 and 85% of the length of the cochlear spiral in *F. anselli* (Müller et al. 1992), further evidence of cochlear specialization. Noteworthy also that many aspects of the cochlear morphology, like low scalae (especially scala tympani) thin spiral ligament, less regular geometric pattern of the reticular lamina of the Corti organ, higher Corti organ, and "immature" outer hair cells (i.e. those that resemble in their appearance hair cells found in immature Corti organs, cf. Burda 1985), characterizing in other mammals only the most apical, i.e. low frequency, regions, are displayed in the studied subterranean rodents throughout most of the cochlear spiral (Figs. 9.3 and 9.4).

Table 9.2. Comparison of inner ear parameters in subterranean rodents (CBL Condylo-basal length; CT, cochlear turns; BML, basilar membrane length. IHC, inner hair cells; OHC, outer hair cells) [a]Values for *Cryptomys* sp. are averages over *C. hottentotus*, *C. pretoriae*, and *C. natalensis* [b]*Fukomys* sp. comprises averages over *F. anselli*, *F. damarensis*, and *F. darlingi* [c]Values for *Spalax ehrenbergi* are averages over the four chromosomal species ($2n = 52$, $2n = 54$, $2n = 58$, $2n = 60$)

Species/genus	CBL (mm)	No. CT	BML (mm)	IHC density (no/mm)	OHC dens. (no/mm)	OHC/ 100 IHC	Total no. IHC+OHC	Source
Subterranean rodents								
Cryptomys sp.[a]	36.6	3.33	9.7	119	457	383	5597	Lange 2006
Ctenomys talarum	37.4	3.25	10.58	101	400	396	5311	Schleich et al. 2006
Fukomys sp.[b]	36.3	3.42	10.53	117	457	392	6450	Lange 2006
Spalacopus cyanus	30.8	3.5	11.68	109	397	364	5920	Begall and Burda 2006
Spalax ehrenbergi[c]	41.3	4.0	12.6	93	386	415	6035	Burda et al. 1989, Lange 2006
Mean		3.5	11.0	107.8	419.4	389.8	5862.6	
Murids								
Apodemus sylvaticus	23	1.75	6.7	111	388	351	3343	Burda et al. 1988
Mus musculus	20.5	1.95	6	121	411	340	3192	Burda et al. 1988
Rattus norvegicus	48	2.2	10.7	98	364	372	4940	Burda et al. 1988
Rattus rattus	40	2.125	12.1	107	372	348	5802	Burda et al. 1988
Mean		2.0	8.9	109.3	383.8	352.8	4319.3	
Cavia porcellus		4.25	20.5	100.4	360.6	359.2	9450	Burda 1984
Mammalian mean				103	378	367.0		Burda et al. 1988

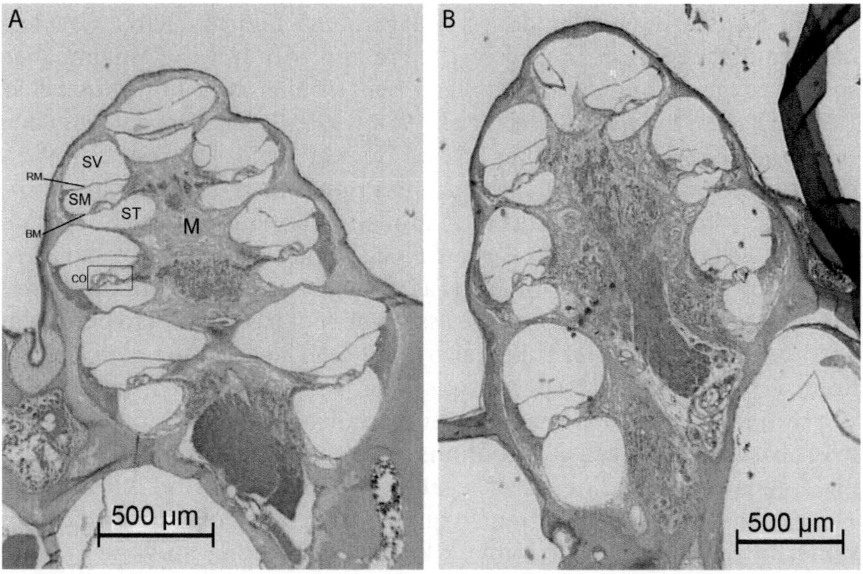

Fig. 9.3. Midmodiolar section of the cochlea of **A** *Georychus capensis* (right ear) and **B** *Fukomys damarensis* (left ear). BM, basilar membrane; CO, organ of Corti; M, modiolus; RM, Reissner's membrane; SM, scala media; ST, scala tympani, SV, scala vestibuli. Micrograph modified from Lange (2006)

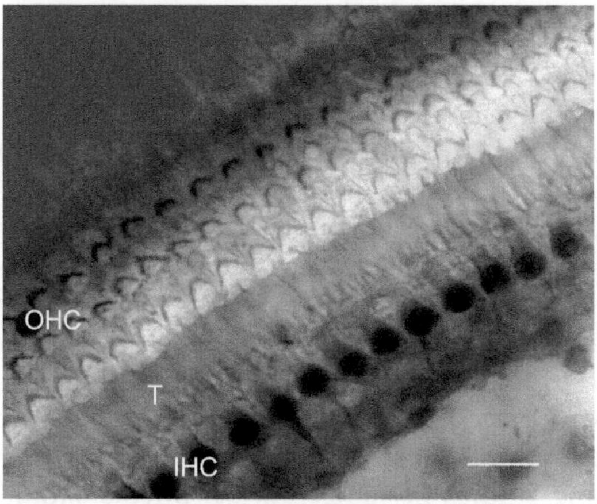

Fig. 9.4. Reticular lamina of the organ of Corti of *Spalacopus cyanus*. IHC, inner hair cells; T, cuticular plates of outer pillars forming the roof of the tunnel of Corti; OHC, outer hair cells. Focus is on stereociliar formations on cuticular plates of hair cells. Scale bar= 20 μm. Micrograph from Begall and Burda (2006)

Inner ear and higher neural structures also play a major role in determining the hearing bandwidth (cf. Ruggero and Temchin 2002). Cochlear tuning and the auditory brain centres were studied only in a few mammalian species so far, among them *Fukomys anselli* (Müller et al. 1992; Kössl et al. 1996), *Spalax ehrenbergi* (Heil et al. 1988, 1991; Bronchti et al. 1989) and *Heterocephalus glaber* (Heffner RS and Heffner 1993). The results of these studies support the idea of specialization rather than degeneration of the auditory system in subterranean mammals. Due to the cochlear specialization (well documented in *F. anselli* – cf. Müller et al. 1992), established models of place-frequency maps (Fay 1992) are generally not applicable to subterranean mammals and the assignment of frequencies to particular positions is not possible. This is also due to the fact that the high hearing limit, usually estimated by power functions, is overrated which in turn leads to a shift in the tonotopic map. However, available data on place-frequency maps based on hair cell distributions in coruros reflect perfectly the animals' behavioural audiograms (Begall et al. 2004; Begall and Burda 2006). The establishment of a position-frequency map based on a new model for subterranean rodents is desirable. Likewise, audiograms and inner ears of other (and particularly of sciurognath) subterranean rodents should be investigated in future studies to evaluate such models and to be able to assess better the role of phylogenetic constraints vs sensory-ecological moulding and functional significance of certain features (like cochlear coiling).

9.5 Synthesis

Hearing in subterranean rodents generally matches the acoustics in burrows, as it is not responsive to high frequencies (an aspect which can be considered regressive due to lack of stimulation). However, the hearing range is (in terms of octaves) not reduced but shifted towards lower frequencies. Best hearing occurs at lower frequencies which are transmitted best in burrows. Hearing sensitivity in the lower frequency range is (in comparison to rats) conserved or even improved. Reduction of best hearing sensitivity (compared to that in other rodents, though occurring at other frequencies) may compensate the amplification of sounds at the given frequencies by the "stethoscope effect" in the tunnel. To prove or falsify this assumption, hearing studies in subterranean rodents should also be conducted in tunnels. We suggest that their best hearing sensitivity in their natural environment, in burrows, is comparable to that of other rodents above the ground.

The similar ossicular morphologies of the studied subterranean rodents (and of the guinea-pig vs laboratory rat) suggest hearing tuned to low frequencies. From the quantitative and qualitative points of view, none of the components of the middle ear transmitting chain is reduced or degenerated. On the contrary, some parts, particularly the incus and the stapedial footplate, are markedly enlarged, leading to reduced sensitivity. This feature of hearing can thus be also considered an adaptation with a clearly defined morphological substrate.

The cochlea in all its aspects is specialized for sensitive perception and high resolution of low frequencies.

We can conclude that both low-frequency and low-sensitivity hearing in subterranean mammals can be considered adaptive specializations. The known similarities in ear morphology and hearing of diverse species belonging to diverse genera, families and orders of subterranean mammals provide thus a good example for convergence due to similar acoustic environments.

Acknowledgements. SB was funded by a Lise-Meitner-Habilitations-stipendium. CES was supported by a DAAD travel grant. We thank Manfred Kössl for critical comments on an earlier draft of the manuscript.

References

Aitkin LM, Horseman BG, Bush BMH (1982) Some aspects of the auditory pathway and audition in the European mole, *Talpa europaea*. Brain Behav Evol 21:49–59

Begall S, Burda H (2006) Acoustic communication and burrow acoustics are reflected in the ear morphology of the coruro (*Spalacopus cyanus*, Octodontidae), a social fossorial rodent. J Morphol 267:382–390

Begall S, Burda H, Schneider B (2004) Hearing in coruros (*Spalacopus cyanus*): special audiogram features of a subterranean rodent. J Comp Physiol A 190:963–969

Bronchti G, Heil P, Scheich H, Wollberg Z (1989) Auditory pathway and auditory activation of primary visual targets in the blind mole-rat (*Spalax ehrenbergi*): I. 2-Deoxyglucose study of subcortical centers. J Comp Neurol 284:253–274

Brückmann G, Burda H (1997) Hearing in blind subterranean Zambian common mole-rat (*Cryptomys* sp., Bathyergidae, Rodentia). J Comp Physiol A 181:83–88

Bruns V, Müller M, Hofer V, Heth G, Nevo N (1988) Inner ear structure and electrophysiological audiograms of the subterranean mole-rat, *Spalax ehrenbergi*. Hear Res 33:1–10

Burda H (1979) Morphology of the middle and inner ear in some species of shrews (Insectivora, Soricidae). Acta Sc Nat Brno 13:1–48

Burda H (1984) Guinea pig cochlear hair cell density; its relation to frequency discrimination. Hear Res 14:315–317

Burda H (1985) Qualitative assessment of postnatal maturation of the organ of Corti in two rat strains. Hear Res 17:201–208

Burda H, Ballast L, Bruns V (1988) Cochlea in old world mice and rats (Muridae). J Morphol 198:269–285

Burda H, Bruns V, Nevo E (1989) Middle ear and cochlear receptors in the subterranean mole-rat, *Spalax ehrenbergi*. Hear Res 39:225–230

Burda H, Bruns V, Müller M (1990) Sensory adaptations in subterranean mammals. In: Nevo E, Reig OA (eds) Evolution of subterranean mammals at the organismal and molecular levels. Wiley-Liss, New York, pp 269–293

Burda H, Bruns V, Hickman GC (1992) The ear in subterranean Insectivora and Rodentia in comparison with ground-dwelling representatives. I. Sound conducting system of the middle ear. J Morphol 214:49–61

Fay RR (1992) Structure and function in sound discrimination among vertebrates. In: Webster DB, Fay RR, Popper AN (eds) The evolutionary biology of hearing. Springer Berlin Heidelberg New York, p 229–263

Fleischer G (1973) Studien am Skelett des Gehörorgans der Säugetiere, einschließlich des Menschen. Säugetierk Mitt 21:131–239

Fleischer G (1978) Evolutionary principles of the mammalian middle ear. Adv Anat Embryol Cell Biol 55:1–70

Heffner H, Masterton, RB (1980) Hearing in glires: domestic rabbit, cotton rat, feral house mouse, and kangaroo rat. J Acoust Soc Am 68:1584–1599

Heffner HE, Heffner RS, Contos C, Ott T (1994) Audiogram of the hooded Norway rat. Hear Res 73:244–247

Heffner RS, Heffner HE (1990) Vestigial hearing in a fossorial mammal, the pocket gopher (*Geomys bursarius*). Hear Res 46:239–252

Heffner RS, Heffner HE (1992) Hearing and sound localization in blind mole rats (*Spalax ehrenbergi*). Hear Res 62:206–216

Heffner RS, Heffner HE (1993) Degenerate hearing and sound localization in naked mole-rats (*Heterocephalus glaber*), with an overview of central auditory structures. J Comp Neurol 331:418–433

Heffner RS, Heffner HE, Contos C, Kearns D (1994) Hearing in prairie dogs: transition between surface and subterranean rodents. Hear Res 73:185–189

Heil P, Bronchti G, Wollberg Z, Scheich H (1988) Deoxyglucose evidence for intermodal compensation. In: Elsner N, Barth FG (eds) Sense organs. Interfaces between enviroment and behaviour, G Thieme, Stuttgart, p 186

Heil P, Bronchti G, Wollberg Z, Scheich H (1991) Invasion of visual cortex by the auditory system in the naturally blind mole rat. NeuroReport 2:735–738

Heth G, Frankenberg E, Nevo E (1986) Adaptive optimal sound for vocal communication in tunnels of a subterranean mammal (*Spalax ehrenbergi*). Experientia 42:1287–1289

Knotková E, Šumbera R, Sedláček F (2005) Vocalization and behavioral hearing range in silvery mole-rat (*Heliophobius argenteocinereus*). Ninth International Mammalogical Congress (IMC9), Sapporo (Japan), p 222

Kock D, Ingram CM, Frabotta LJ, Honeycutt RL, Burda H (2006) On the nomenclature of Bathyergidae and *Fukomys* n. gen. (Mammalia: Rodentia). Zootaxa 1142:51–55

Kössl M, Frank G, Burda H, Müller M (1996) Acoustic distortion products from the cochlea of the blind African mole-rat, *Cryptomys* spec. J Comp Physiol A 178:427–434

Lange S (2006) Sinnesökologie afrikanischer Sandgräber (Bathyergidae) am Beispiel von Hör- und Geruchssinn. PhD Thesis, Univ of Duisburg-Essen

Lange S, Burda H (2005) Comparative and functional morphology of the middle ear in Zambian mole-rats (*Coetomys – Cryptomys*, Bathyergidae). Belg J Zool 135(suppl):5–10

Lange S, Stalleicken S, Burda H (2004) Functional morphology of the ear in fossorial rodents, *Microtus arvalis* and *Arvicola terrestris*. J Morphol 262:770–779

Lange S, Burda H, Bennett NC, Němec P (2005) Middle ear ossicles as a diagnostic trait in African mole-rats (Rodentia: Bathyergidae). In: Huber BA, Sinclair BJ, Lampe KH (eds) African biodiversity: molecules, organisms, ecosystems. Proceedings of the 5th International Symposium on Tropical Biology, Museum König, Bonn. Springer, Berlin Heidelberg New York, pp 329–337

Lange S, Burda H, Wegner RE, Dammann P, Begall S, Kawalika M (2007) Living in a stethoscope: burrow-acoustics promotes auditory specializations in subterranean rodents. Naturwiss 94:134–138

Lay DM (1972) The anatomy, physiology, functional significance and evolution of specialized hearing organs of gerbilline rodents. J Morphol 138:41–120

Mason MJ (2001) Middle ear structures in fossorial mammals: a comparison with non-fossorial species. J Zool Lond 255:467–486

Mason MJ (2003a) Morphology of the middle ear of golden moles (Chrysochloridae). J Zool Lond 260:391–403

Mason MJ (2003b) Bone conduction and seismic sensitivity in golden moles (Chrysochloridae). J Zool Lond 260:405–413

Mason MJ (2004) The middle ear apparatus of the tuco-tuco *Ctenomys sociabilis* (Rodentia, Ctenomyidae). J Mammal 85:797–805

Mayer A, O'Brian G, Sarmiento EE (1995) Functional and systematic implications of the ear in golden moles (Chrysochloridae). J Zool Lond 236:417–430

Müller M, Burda H (1989) Restricted hearing range in a subterranean rodent, *Cryptomys hottentotus* (Bathyergidae). Naturwiss 76:134–135

Müller M, Burda H, Bruns V (1989) Structure and function of the inner ear in subterranean mammals. In: Erber J, Menzel R, Pflüger HJ, Todt D (eds.) Neural mechanisms of behavior. Thieme, Stuttgart, p 144

Müller M, Laube B, Burda H, Bruns V (1992) Structure and function of the cochlea in the African mole rat (*Cryptomys hottentotus*): evidence for a low frequency acoustic fovea. J Comp Physiol A 171:469–476

Pye A (1977) The structure of the cochlea in some myomorph and caviomorph rodents. J Zool Lond 182:309–321

Quilliam TA (1966) The mole's sensory apparatus. J Zool Lond 149:76–88

Rado R, Himelfarb M, Arensburg B, Terkel J, Wollberg Z (1989) Are seismic communication signals transmitted by bone conduction in the blind mole-rat? Hear Res 41:23–30

Raphael Y, Lenoir M, Wroblewski R, Pujol R (1991) The sensory epithelium and its innervation in the mole rat cochlea. J Comp Neurol 314:367–382

Ruggero MA, Temchin AN (2002) The roles of the external, middle, and inner ears, in determining the bandwidth of hearing. Proc Natl Acad Sci USA 99:13206–13210

Ryan A (1976) Hearing sensitivity of the mongolian gerbil *Meriones unguiculatis*. J Acoust Soc Am 59:1222–1226

Schleich CE, Busch C (2004) Functional morphology of the middle ear of *Ctenomys talarum* (Rodentia: Octodontidae). J Mammal 85:290–295

Schleich CE, Vassallo AI (2003) Bullar volume in subterranean and surface-dwelling caviomorph rodents. J Mammal 84:185–189

Schleich CE, Begall S, Burda H (2006) Morpho-functional parameters of the inner ear in *Ctenomys talarum* (Rodentia, Ctenomyidae). Folia Zool 55:264–272

Webster DB, Webster M (1975) Auditory systems of Heteromyidae: functional morphology and evolution of the middle ear. J Morphol 146:343–376

10 Acoustic Communication in Subterranean Rodents

Cristian E. Schleich, Silke Veitl, Ema Knotková, Sabine Begall

10.1 Introduction

Why study acoustic communication? And why in subterranean rodents? There are several reasons for studying acoustic communication in this particular group of mammals. First, acoustic communication is of great importance to the organization of animal societies. Animals use vocal signals to elicit specific behaviors, initiate contacts with conspecifics, identify individuals or species, signalize status (reproductive, dominance, territorial), solicit food, warn against predators, and coordinate reproductive efforts (courtship, copulation solicitation, physiological synchrony). Beyond the intra-specific level, acoustic signals are also an important factor in reproductive isolation and speciation processes.

Furthermore, vocal communication can be a useful instrument for explaining general evolutionary principles. Signal structure and function are shaped by several factors, such as characteristics of the environmental channel in which the signal is transmitted and the physiological and morphological properties of both, the sender and the receiver. Thus, the characteristics of the vocal signals usually demonstrate the kind, intensity and process of natural selection.

Why in subterranean rodents? The answer to this question is intimately linked to the particular environment in which they live. The subterranean ecotope has influenced the evolution of the sensory biology of subterranean rodents. As seen in the previous chapter, the dark and monotonous subterranean environment limits the transmission of most signals and cues,

Cristian E. Schleich: Departamento de Biología, Facultad de Ciencias Exactas y Naturales, CC 1245, Universidad Nacional de Mar del Plata, Funes 3250, Mar del Plata (7600), Argentina, E-mail: cschleic@mdp.edu.ar

Silke Veitl: Dept. General Zoology, Institute for Biology, Faculty of Biology and Geography, University of Duisburg-Essen, D-45117 Essen, Germany

Ema Knotková: Faculty of Biological Sciences, University of South Bohemia, CZ-370 05 České Budějovice, Czech Republic

Sabine Begall: Dept. General Zoology, Institute for Biology, Faculty of Biology and Geography, University of Duisburg-Essen, D-45117 Essen, Germany

Subterranean Rodents: News from Underground
S. Begall, H. Burda, C.E. Schleich (Eds.)
© Springer-Verlag Berlin Heidelberg 2007

leading to morphological and physiological modifications in the sensory systems of this group. Therefore, studying vocal and seismic communication in subterranean rodents is a useful tool for understanding the adaptive responses of these species to the constraints imposed by life in underground burrows.

In this chapter we review the ecological and evolutionary conditions that shaped the vocal repertoire of subterranean rodents. First, we describe two of the main factors that shape the vocalization characteristics of this group: the underground environment and the social system. Second, we examine seismic signals for long-distance communication in some species of solitary subterranean rodents, detailing the main hypotheses about the evolution of these signals. Finally, we try to elucidate the role of care-elicitation calls in pups of *Ctenomys talarum*, in a first attempt to understand the significance of the begging behaviors in this and other species of subterranean rodents.

10.2 Adult Vocalizations

10.2.1 Acoustic Environment and Vocalization Structure

Subterranean rodents evolved in different geographical regions but under similar environmental conditions. Therefore, convergent morphological, physiological and ecological adaptations of these small mammals are expected to evolve. Among these adaptations, those involving vocalization characteristics are remarkable in their consistency. As demonstrated in the previous chapter, low-frequency sounds (around 200–800 Hz) are transmitted more efficiently in burrows than sounds of higher or lower frequencies (Heth et al. 1986; Lange et al. 2006). Concomitant with the acoustical characteristic of subterranean environments, vocalizations in all subterranean rodents studied to date are shifted towards the middle to low-frequency range (Heth et al. 1988; Pepper et al. 1991; Credner et al. 1997; Francescoli 1999; Veitl et al. 2000; Schleich and Busch 2002a; Knotková and Veitl, unpublished; Table 10.1). Despite some small differences in the frequency range of the vocalizations among the different species of subterranean rodents examined and among the different types of calls that they produce, the main energy components of all vocalizations occur within the low frequency range. The fact that this physical attribute of vocalizations is shared by all subterranean species, which is also congruent with hearing being tuned to low frequencies (Chap. 9), suggests that vocalization parameters in subterranean rodents are the result of convergent evolution.

Table 10.1. Vocal repertoire of solitary and social species of hystricognath subterranean rodents. The range of the main frequency (kHz) of each vocalization is indicated in *brackets*. Data are taken from Pepper et al. (1991)[a], Credner et al. (1997)[b], Francescoli (1999)[c], Veitl et al. (2000)[d], Schleich and Busch (2002)[e], Knotková and Veitl[f], unpubl. and Veitl[g], unpubl.

	Solitary			Social			
	Ctenomys talarum[e]	*Ctenomys pearsoni*[c]	*Heliophobius argenteocinereus*[f]	*Heterocephalus glaber*[a]	*Fukomys anselli*[b]	*Fukomys mechowii*[g]	*Spalacopus cyanus*[d]
Mating vocalizations	Male call (0.2–0.4) Female call (0.4–1)	C-signal (1–1.5)	High-cluck (0.5–4.1) Chirp (3.35) Female clucks (1.37) Gabbling (1.1–2.2)	V-trill (2–4)	Cluck (1.6–2.5) Shriek (1.6–2.5) Cry (0.6–0.9; 1.6–2.5; 4–8; 12)	Low trill (0.4) High trill (0.4–0.7) Cry (0.7–3.6)	Creaking (0.17–0.47) Scream (1.55)
Agonistic vocalizations	Tuc-tuc (0.2–0.4) Grunts (0.2–0.4)	S-signal (0.2–0.3) G-signal (0.5–0.6)	Low-cluck (0.3–1.5) Hiss (0.34–0.9) Grunt (0.17–3.45)	Hiss (0.4–1.3) Grunt Upsweep trill (1–9) Loud chirp (8–14)	Whistle (0.5–0.6) Trill (1.6–2.5) Trill II (2–3) Hiss (8–14) Grunt (1–2; 3) Grunt II (0.5–0.8; 2; 6)	Hiss (0.9) Grunt II (0.4–4.3)	Cluck (0.56–1.03) Cluck II (0.34–0.52)

Table 10.1. continued

	Solitary			Social			
	Ctenomys talarum[e]	*Ctenomys pearsoni*[c]	*Heliophobius argenteocinereus*[f]	*Heterocephalus glaber*[a]	*Fukomys anselli*[b]	*Fukomys mechowii*[g]	*Spalacopus cyanus*[d]
Distress vocalizations			Squeaks (1.6–5) Scream (2.8–4.6)		Loud calls (1.25–16)	Shriek (3.3–8.5) High cluck I (0.4–1.2)	Cluck III (0.34–0.69)
Contact vocalizations					Grunt (5) Twitter (0.5–3)	Low whistle (0.6–1) High whistle (0.6–2.9) Chirp (0.6–1.6) High chirp (1.5–2.8) Twitter I (1.4–3.3) Twitter II (0.6–2.6) Squeal (0.5–0.9) Purr (0.2–7.4) Grunt I (0.6–3.3) High cluck II (1.5–2.3) Cluck (0.3–1.3)	Cooing (0.34–0.52) Twitter (1.03–4.31) Twitter II (1.55) Squeak (0.34)

Table 10.1. continued

	Solitary			Social			
	Ctenomys talarum[e]	*Ctenomys pearsoni*[c]	*Heliophobius argenteocinereus*[f]	*Heterocephalus glaber*[a]	*Fukomys anselli*[b]	*Fukomys mechowii*[g]	*Spalacopus cyanus*[d]
Alarm vocalizations				Tap (0.2–1.2) Sneeze (0.2–1.2) Low-pitched chirp (1–3) Scream (1–1.5)			Trill (0.8–1.2)
Routine vocalizations				Soft chirp (2–4) Toilet call (1.5–4)			

Certainly, the effect of the acoustic environment is different for long or short-distance calls. While long-distance calls are severely constrained by the acoustical properties of the subterranean burrows, short-distance vocalizations (like contact or mating calls) suffer less attenuation and degradation, presenting therefore wider fundamental frequency ranges, although also in the low-frequency range (Table 10.1).

Another hint supporting the hypothesis of convergent evolution is seen in the fact that surface-dwelling hystricognath rodents of comparable body size, like *Cavia* sp., *Octodon degus* and *Octodontomys gliroides* emit not only low and middle frequency calls, but also several high frequency or ultrasonic vocalizations (Eisenberg 1974), which is in contrast with the predominant low-frequency range of vocalizations of subterranean rodents and the absence of high frequency or ultrasonic calls in these rodents. Therefore, the shared physical characteristics of vocalizations of these species can be considered as one of the clearest examples of convergent adaptation in the sensory biology of subterranean rodents.

10.2.2 Social System and Vocal Repertoire

One of the generally accepted facts regarding animal communication is the correlation between the degree of social organization and the diversity of the vocal repertoire. The integration of individuals into groups, maintaining group cohesion, or the arrangement of communal activities, requires the development of signals necessary to organize those activities (Bradbury and Vehrencamp 1998).

This relationship between the social system and the vocal repertoire is also observed in subterranean rodents, where solitary species emit fewer vocalizations than social ones (Table 10.1). While the solitary species *Ctenomys pearsoni*, *C. talarum*, *Spalax ehrenbergi* and *Heliophobius argenteocinereus* possess an adult repertoire size of three, four, six and eight main types of calls, respectively (Francescoli 1999; Schleich and Busch 2002a; Capranica et al. 1973; Knotková and Veitl, unpublished), the adult vocal repertoire of the social species *Heterocephalus glaber*, *Fukomys anselli*, *F. mechowii* (formerly known as *Cryptomys anselli* and *C. mechowi*, see Kock et al. 2006) and *Spalacopus cyanus* include eleven, twelve, eighteen and ten calls. (Pepper et al. 1991; Credner et al. 1997; Veitl et al. 2000). As can be seen in Table 10.1, solitary species of subterranean rodents usually present vocalizations associated with agonistic and sexual contexts that reflect their (main) social activities which are related with territorial defense and mating coordination. The vocal repertoire of *Heliophobius*, e.g., reveals variability in its distress and agonistic calls to avoid close contact with conspecifics (Knotková and Veitl, unpublished).

The situation is quite different in social species. The eusocial *F. mechowii* has a large vocal repertoire with partly polysyllabic contact calls which probably serve for individual identification (Veitl, unpublished). While the vocal repertoire of *S. cyanus* is characterized by well differentiated contact calls (Veitl et al. 2000), the vocal repertoire of *H. glaber* presents several vocalizations involved in cooperative predator avoidance and colony defense (Pepper et al. 1991). Surprisingly, naked mole-rats emit only one mating call which may reflect the lack of evident competitive behaviors around estrus and copulation (Pepper et al. 1991). In contrast to this, the vocal repertoire of *Fukomys anselli* shows a greater variety of vocalizations associated with sexual behaviors, although aggressive calls are also well differentiated (Credner et al. 1997). The differences in vocal repertoire composition of *Fukomys* and *H. glaber* have been ascribed to different predation pressures on these species and to differences in the maintenance of the eusocial structure between them (behavioral suppression in *H. glaber* vs incest avoidance in *Fukomys*) (Credner et al. 1997; Veitl et al. 2000; Burda 2003).

Although the relationship between the social system and the vocal repertoire in subterranean rodents seems to be real, the contribution of other factors, besides social complexity, to the evolution of repertoire sizes cannot be discarded. When analyzing the contribution of social complexity in explaining variation in alarm repertoire size in ground-dwelling sciurid rodents, Blumstein and Armitage (1997) found that a significant variation in repertoire size remained unexplained by social complexity. Therefore, it would be interesting to investigate (and identify) if other factors, like morpho-physiological or behavioral differences, could have influenced the evolution of communicative complexity in the different species of subterranean rodents.

Regarding the relationship between the physical structure of the calls and the motivation underlying their emission, it appears that the vocalizations of both solitary and social subterranean rodents follow the motivational-structural rules proposed by Morton (1977) for the vocal repertoires of some species of birds and mammals, which proposed that harsh, low-frequency sounds are used in "hostile" situations and pure tonelike, relatively high-frequency sounds in "friendly" situations. Figure 10.1 shows calls of solitary and social subterranean rodents, and it can be seen that sounds emitted during hostile situations (i.e. agonistic calls) usually consisted of low and atonal bands of noise, contrary to the physical characteristics of the vocalizations emitted during appeasing contexts (i.e. mating calls), which are characterized by a main band with high frequency modulation and generally with multiple bands or harmonics.

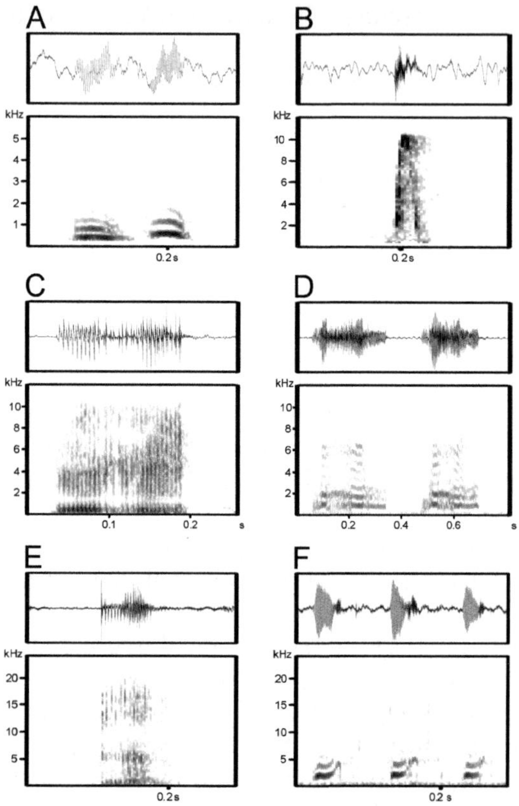

Fig. 10.1. Sonagrams and envelope curves of agonistic and mating vocalizations of: **A,B** *Spalacopus cyanus*; **C,D** *Fukomys mechowii*; **E,F** *Heliophobius argenteocinereus*, (all showing the relationship between the physical structure of the calls and the motivation underlying their emission. All sonagrams by Silke Veitl, (*Spalacopus cyanus*: Veitl et al. 2000, *Fukomys mechowii* and *Heliophobius argenteocinereus*: unpublished)

10.2.3 Territorial Signaling: To Knock or to Shout

One of the first steps when colonizing a territory involves conflict resolution with its coupled short range agonistic signals. Subsequently, the maintenance and defense of that territory raise different problems and are frequently associated with the emission of long-range signals (Bradbury and Vehrencamp 1998) which usually convey information about the identity and location of the sender. Most mammals, including subterranean rodents, defending territories use one or more signals to inform potential intruders about the owner's presence in that territory. Two types of channels are used by these animals for long-distance communication: seismic and acoustic.

Creating mechanical vibrations by drumming on the substrate with the feet or thumping by the head to transfer information between individuals has been described and deeply studied for two unrelated species of subterranean rodents, the Cape mole-rat, *Georychus capensis* (Narins et al. 1992) and the blind mole-rat, *S. ehrenbergi* (Heth et al. 1987, 1991; Rado et al. 1987; Nevo et al. 1991). Cape mole-rats create sexually dimorphic patterns of drumming using their hind feet and it is likely that these signals are used for communication between individuals and maintenance of their territories. Heth et al. (1987) and Narins et al. (1992) suggested that these signals could serve for mate attraction. *Spalax ehrenbergi*, one of the thoroughly studied cases of seismic communication, creates vibration signals by knocking its head against the tunnel ceiling, a behavior supposedly derived from digging activities. The produced low frequency signals travel further than airborne sounds providing an efficient long-distance communication channel between the solitary blind mole rats. Apart from being an important means of communication between individuals within populations (including attracting and recognizing the appropriate partners), these seismic signals are also considered an important pre-mating isolating mechanism between populations (Heth et al. 1987). Moreover, Heth et al. (1991) also found that the number of thumps and the duration of these seismic signals increase with increasing territory size and decreasing density across the species of the *S. ehrenbergi* superspecies, suggesting that this type of communication has an active role in the process of speciation in blind mole-rats.

Despite their efficiency, seismic signals are not the only long-distance communication channel used by subterranean rodents. Some species seem to rely on vocalizations for long-distance inter-burrow communication, e.g. members of South American rodents of the genus *Ctenomys*. At least three solitary species of this genus, *C. pearsoni*, *C. mendocinus* and *C. talarum*, give vocalizations associated with aggressive spacing behaviors which might be used as territorial signals (Francescoli 1999, personal communication; Schleich and Busch 2002a). The communicative function of these calls is supported by laboratory and field observations, in which males of *C. talarum* were found to emit alternating vocalizations in a "duet" (Schleich and Busch 2002a).

However, as we have seen in the previous chapter (Begall et al., this volume), acoustic signals attenuate starkly inside the burrows limiting thus transmission through the soil, although transmission efficiency also depends on the physical characteristics of the signals, since sounds with a frequency of about 440 Hz are better transmitted than sounds of lower and higher frequencies (Heth et al. 1987). The above-mentioned species of ctenomyids seem to overcome this difficulty by means of two features related to the physical characteristics of the territorial vocalizations and to the behavioral contexts. First, loudness, long duration, and low frequency characteristics of their territorial vocalizations (Francescoli 1999; Schleich and Busch 2002a) make these calls suitable for long-distance, inter-burrow communication. These physical characteristics of the vocalizations seem to assure a sufficiently large spatial range, necessary to be accurately localized by receivers. The second aspect that improves the transmission of these signals between the burrows of these rodents is the behavior of the individuals while vocalizing. In the laboratory, when chasing another individual away from its own (artificial) burrow, males of *C. talarum* usually started to vocalize at the entrance of their burrows (Schleich and Busch 2002a). This behavior can also be observed in the field, where sometimes vocalizing individuals can be observed at the entrance of their burrows. The conjunctions of these two features allow these species to use acoustic signals for long-distance communication.

As can be observed, some species of subterranean rodents seem to have solved the problem of long-distance communication by means of two different strategies: seismic and acoustic. However, why have some species developed one or the other? Burda et al. (1990) and Narins et al. (1992) linked the two communication modalities with the social system of the given subterranean species, with social species using vocal communication while solitary species have extended on seismic communication. Francescoli (2000) suggested that solitary species that rarely leave their burrows use vocal signals for short-distance and seismic signals for long-distance communication and solitary species that are active aboveground and maintain good hearing and vision abilities use vocalizations for both short and long-distance communication. Recently, Burda (2003) when discussing seismic communication in solitary African mole-rats, suggested that seismic signaling could have evolved for mate seeking in those species that disperse and look for potential mates underground. The observation that solitary silvery mole-rats do not show signs of seismic communication and probably seek mates above ground (Šumbera et al., this volume) is consistent with this hypothesis. In spite of several evolutionary motivated hypotheses and the information available at present about seismic communication in this group of mammals, further behavioral studies on other subterranean species are necessary to enlighten the general picture of the evolution of seismic communication in subterranean rodents.

10.3 Juvenile Vocalizations: Are Care-elicitation Calls Honest Advertisements of Offspring Need?

Several hypotheses have been proposed with the purpose of explaining the possible meaning of the begging behaviors observed in several species of birds and mammals. Recent theoretical signaling models suggest that begging can be an honest signal of offspring condition, provided that the signal is costly to produce (in terms of fitness) and the benefits of receiving a food item decrease with increasing physical condition (Godfray 1991, 1995; Kilner and Johnstone 1997).

Pups of several species of solitary and social subterranean rodents emit vocalizations during their nestling period. In at least four species (*H. glaber*, *S. cyanus*, *C. pearsoni* and *C. talarum*) the presence of care-elicitation calls prior to and during the weaning process has been described (Pepper et al. 1991; Veitl et al. 2000; Francescoli 2001; Schleich and Busch 2002b). Although the physical structure of these calls differs, the behavioral context associated with the emission of these vocalizations is similar among these subterranean rodents. In all these studies, the observed solicitations have been mediated by means of vocal behavior, usually accompanied by distress to which the parents respond by providing resources (food or heat) to their offspring. Although this behavior seems to be widespread among subterranean rodents, pups of some species of this group, like *Fukomys anselli* and *F. mechowii* (Burda, personal communication), only seldom emit spontaneously vocalizations during the nestling period, even when leaving the nest and staying in the cold for a while. However, while in all the previously mentioned studies the behavioral contexts associated with the emission of those juvenile vocalizations have been analyzed, only recent works have also investigated the possible significance of the juvenile vocal signals towards their parents (Schleich and Busch 2002b, 2004).

In a first attempt to learn more about the significance of the care-elicitation calls emitted by pups of *C. talarum*, Schleich and Busch (2002b) studied the ontogeny of the pups' vocal behavior. Pups emit care-elicitation cries during the first weeks after birth, when they are not capable of maintaining their body temperature and eating solid food (Schleich and Busch 2002b; Zenuto et al. 2002). As pups develop, they become able to regulate their body temperature alone and to move around the (artificial) burrow in search of solid food. The number of care-elicitation calls of *C. talarum* pups tends to decrease from the day of birth until the beginning of the sixth week concomitant with the decrease in the time spent nursing, and with the increase in the time spent eating solid foods by the pups. Weaning also coincides with an increased level of aggression of the mother towards the pups. The developmental pattern of care-eliciting calls suggests that begging calls in *C. talarum* pups reflect indeed the offspring's need.

However, signaling models also predict that begging signals must be costly to a certain extent to maintain honest advertisement of need. Two factors, predation and energetic expenditure, are usually related to signal cost. Because of low risk of predation and absence of *C. talarum* pups at the surface, Schleich and Busch (2002b, 2004) have suggested that energetic expenditure of begging is the crucial factor related to signal cost. In a series of experiments, Schleich and Busch (2004) measured the oxygen consumption associated with the emission of care-elicitation calls and found no differences in individual oxygen consumption between resting and calling *C. talarum* pups (Schleich and Busch 2004).

These negative results lead to the conclusion that begging calls of *C. talarum* cannot be considered as honest advertisements of offspring's need. However, the apparent contradiction between the experimental results and the theoretical predictions raised several questions about the validity of honest signaling models and the possible significance of the care-elicitation calls of *C. talarum* pups.

First, is increased energy expenditure one of the main costs needed to maintain the honesty of begging signals? The results obtained from *C. talarum* pups (Schleich and Busch 2004) and several species of birds (McCarty 1996; Leech and Leonard 1996; Soler et al. 1999) showed that the energetic expenditure while begging is low or negligible, suggesting that begging behaviors are not as expensive as would be expected from signaling models (but see Bergstrom and Lachmann 1998).

Second, is there a growth cost of begging? Recently, some studies searched not for the metabolic cost of begging but for the link between begging and offspring growth. However, results on different bird species have not been consistent (Kilner 2001; Rodriguez-Gironés et al. 2001; Leonard et al. 2003). Future studies on *C. talarum* should elucidate whether, despite not showing any detectable energetic cost, the emission of vocalizations in pups of this subterranean rodent could have an effect on the growth rate or other physiological parameters that could affect the fitness of the vocalizing animal.

Finally, can an increased aggressive behavior of the parents towards their offspring be the cost of begging calls? As is the case in the black-capped chickadee (Leonard et al. 1988, 1991) and in *C. talarum* (Schleich and Busch 2002b) there is an aggressive behavior of the parents towards the juveniles that vocalize persistently soliciting food, suggesting that this parental behavior could cause the development of a cost associated with the emission of these calls. However, until now, few cases of this punishment cost have been reported (Roulin 2001). Thus, the role of this kind of behavior in the maintenance of begging as an honest signal is still under debate.

In conclusion, based on the current signaling models, the vocalizations of *C. talarum* pups cannot be considered as honest signals of need. However, although the work on *C. talarum* constitutes an important contribution to the knowledge of begging behaviors, more studies, both at the theoretical and experimental level, are necessary to elucidate the role of care-elicitation calls in this and other species of subterranean rodents.

10.4 Conclusion

In this chapter, we describe different factors influencing the evolution of acoustic signals used by subterranean rodents to communicate. It is clear that vocalizations have been an important component of subterranean rodents' adaptation to the underground environment, and selection has optimized their design in the context of improving signal efficacy. However, despite the great quantity of studies on the topic, there are still many unanswered questions about the evolution and functional meaning of the different signals employed by subterranean rodents to communicate. One of the factors that impede obtaining a general picture of the evolution of the vocal behavior of subterranean rodents is our biased knowledge. As clearly pointed out by Burda (2003), only a small fraction of all the genera of subterranean rodents have been studied. Moreover, within the most speciose genera, only a few species belonging to a narrow geographical range have been investigated. Only the extension of behavioral studies to other species and genera of subterranean rodents will help us to understand and obtain a general picture of the evolution of the acoustic behavior in these species of mammals.

References

Bergstrom CL, Lachmann M (1998) Signaling among relatives. III. Talk is cheap. Proc Natl Acad Sci USA 95:5100–5101

Blumstein DT, Armitage BK (1997) Does sociality drive the evolution of communicative complexity? A comparative test with ground-dwelling sciurid alarm calls. Am Nat 150:179–200

Bradbury JW, Vehrencamp SL (1998) Principles of animal communication. Sinauer, Sunderland, Mass

Burda H (2003) Adaptations for subterranean life. In: Kleiman DG, Geist V, Hutchins M, McDade MC (eds) Mammals I. Vol. 12 of Grzimek's animal life encyclopedia. Farmington Hills, Mich, Gale Group, pp 69–78

Burda H, Bruns V, Müller M (1990) Sensory adaptations in subterranean mammals. In: Nevo E, Reig OA (eds) Evolution of subterranean mammals at the organismal and molecular levels. Wiley-Liss, New York, pp 269–293

Capranica RR, Moffat J, Nevo E (1973) Vocal repertoire of a subterranean rodent (*Spalax*). In: Abstr Acoust Soc Am Ann Meeting, Los Angeles

Credner S, Burda H, Ludescher F (1997) Acoustic communication underground: vocalization characteristics in subterranean social mole-rats (*Cryptomys* sp., Bathyergidae). J Comp Physiol A 180:245–255

Eisenberg JF (1974) The function and motivational basis of hystricomorph vocalizations. In: Rowlands IW, Weir BJ (eds.) The biology of hystricomorph rodents. Zool Soc Lond, pp 211–247

Francescoli G (1999) A preliminary report on the acoustic communication in Uruguayan *Ctenomys* (Rodentia: Octodontidae): Basic sound types. Bioacoustics 10:203–218

Francescoli G (2000) Sensory capabilities and communication in subterranean rodents. In: Lacey EA, Patton JL, Cameron G (eds) Life underground: the biology of subterranean rodents. Univ Chicago Press, Chicago, pp 111–144

Francescoli G (2001) Vocal signals from *Ctenomys pearsoni* pups. Acta Theriol 46:327–330

Godfray HCJ (1991) Signalling of need by offspring to their parents. Nature 352:328–330

Godfray HCJ (1995) Evolutionary theory of parent–offspring conflict. Nature 376:133–138

Heth G, Frankenberg E, Nevo E (1986) Adaptive optimal sound for vocal communication in tunnels of a subterranean mammal (*Spalax ehrenbergi*). Experientia 42:1287–1289

Heth G, Frankenberg E, Raz A, Nevo E (1987) Vibrational communication in subterranean mole rats (*Spalax ehrenberghi*). Behav Ecol Sociobiol 21:31–33

Heth G, Frankenberg E, Nevo E (1988) "Courtship" call of subterranean mole rats (*Spalax ehrenbergi*): physical analysis. J Mammal 69:121–125

Heth G, Frankenberg E, Pratt H, Nevo E (1991) Seismic communication in the blind subterranean mole-rat: patterns of head thumping and their detection in the *Spalax ehrenbergi* superspecies in Israel. J Zool Lond 224:633–638

Kilner R (2001) A growth cost of begging in captive canary chicks. Proc Natl Acad Sci USA 98:11394–11398

Kilner R, Johnstone RA (1997) Begging the question: are offspring solicitation behaviours signals of need? Trends Ecol Evol 12:11–15

Kock, D, Ingram CM, Frabotta LJ, Honeycutt RL, Burda H (2006) On the nomenclature of Bathyergidae and *Fukomys* n. gen. (Mammalia: Rodentia). Zootaxa 1142:51–55

Lange S, Burda H, Wegner R, Dammann P, Begall S, Kawalika M (2007) Living in a "stethoscope": burrow-acoustics promotes auditory specializations in subterranean rodents. Naturwiss 94:134–138

Leech S, Leonard ML (1996) Is there an energetic cost to begging in nestling tree swallows (*Tachycineta bicolor*)? Proc R Soc Lond B 263:983–987

Leonard ML, Horn AG, Eden SF (1988) Parent-offspring aggression in moorhens. Behav Ecol Sociobiol 23:265–270

Leonard ML, Horn AG, Ratcliff LM (1991) Parental aggression in black-capped chickadees. Behav Ecol 2:228–233

Leonard ML, Horn AG, Porter J (2003) Does begging affect growth in nestling tree swallows, *Tachycineta bicolor*? Behav Ecol Sociobiol 54:573–577

McCarty JP (1996) The energetic cost of begging in nestling passerines. Auk 113:178–188

Morton ES (1977) On the occurrence and significance of motivation-structural rules in some birds and mammals. Am Nat 111:855–869

Narins PM, Reichman OJ, Jarvis JUM, Lewis ER (1992) Seismic signal transmission between burrows of the cape mole-rat, *Georychus capensis*. J Comp Physiol 170:13–21

Nevo E, Heth G, Pratt H (1991) Seismic communication in a blind subterranean mammal: a major somatosensory mechanism in adaptive evolution underground. Proc Natl Acad Sci USA 88:1256-1260

Pepper JW, Stanton HB, Lacey EA, Sherman PW (1991) Vocalizations of the naked mole-rat. In: Sherman PW, Jarvis JUM, Alexander RD (eds) The biology of the naked mole-rat. Princeton Univ Press, Princeton, NJ, pp 243–274

Rado R, Levi N, Hauser H, Witcher J, Adler N, Intrator N, Wollberg Z, Terkel, J (1987) Seismic signalling as a means of communication in a subterranean mammal. Anim Behav 35:1249–1251

Rodriguez-Girones MA, Zuñiga JM, Redondo T (2001) Effects of begging on growth rates of nestling chicks. Behav Ecol 3:269–274

Roulin A (2001) On the cost of begging vocalization: implications of vigilance. Behav Ecol 12:506–511

Schleich CE, Busch C (2002a) Acoustic signals of a solitary subterranean rodent *Ctenomys talarum* (Rodentia: Ctenomyidae): physical characteristics and behavioural correlates. J Ethol 20:123–131

Schleich CE, Busch C (2002b) Juvenile vocalizations of *Ctenomys talarum* (Rodentia: Octodontidae). Acta Theriol 47:25–33

Schleich CE, Busch C (2004) Energetic expenditure during vocalization in pups of the subterranean rodent *Ctenomys talarum*. Naturwiss 91:548–551

Soler M, Soler JJ, Martinez JG, Moreno J (1999) Begging behaviour and its energetic cost in great spotted cuckoo and magpie host chicks. Can J Zool 77:1794–1800

Veitl S, Begall S, Burda H (2000) Ecological determinants of vocalisation parameters: the case of the coruro *Spalacopus cyanus* (Octodontidae), a fossorial social rodent. Bioacoustics 11:129–148

Zenuto R, Antinuchi D, Busch C (2002) Bioenergetics of reproduction and pup development in a subterranean rodent (*Ctenomys talarum*). Physiol Biochem Zool 75:469–478

11 Visual Systems and the Role of Vision in Subterranean Rodents: Diversity of Retinal Properties and Visual System Designs

Pavel Němec, Pavla Cveková, Hynek Burda, Oldřich Benada, Leo Peichl

11.1 Introduction

The visual system of subterranean mammals is assumed to be regressed and rudimentary in response to their lightless ecotope. Among rodents this has been exemplified by detailed studies of the blind mole-rat *Spalax ehrenbergi* (see Nevo 1999, this volume). However, recent studies involving a larger spectrum of subterranean mammalian species have 'unearthed' an unexpected diversity of ocular and retinal features. Similarly, it was shown that not all visual brain nuclei across different species are equally degenerate. Taken together, these findings suggest different visual capabilities and adaptations in different subterranean rodents. They challenge the widely held view that vision is an expendable sense underground. Here we summarize the current knowledge and discuss the potential role of vision for some subterranean mammals.

11.2 Eye Morphology

The eye is the input stage to the visual system and its properties determine what information becomes available to the visual brain centres for further processing. The size of the eye determines the image size on the retina,

Pavel Němec: Department of Zoology, Charles University, Viničná 7, CZ-128 44 Praha 2, Czech Republic; Department of Anatomy, Johann Wolfgang Goethe University, Theodor Stern-Kai 7, D-60590 Frankfurt a. M., Germany, E-mail: pgnemec@natur.cuni.cz

Pavla Cveková: Department of Zoology, Charles University, Viničná 7, CZ-128 44 Praha 2, Czech Republic

Hynek Burda: Department of General Zoology, University of Duisburg-Essen, Universitätsstr. 5, D-45117 Essen, Germany

Oldřich Benada: Institute of Microbiology, Academy of Sciences of the Czech Republic, CZ-14220 Praha 4, Czech Republic

Leo Peichl: Max Planck Institute for Brain Research, Deutschordenstrasse 46, D-60528 Frankfurt a. M., Germany

Subterranean Rodents: News from Underground
S. Begall, H. Burda, C.E. Schleich (Eds.)
© Springer-Verlag Berlin Heidelberg 2007

and the properties of the optical apparatus, comprising cornea and lens, determine the image quality. Eye sizes vary substantially across subterranean rodents (Fig. 11.1). At one end of the spectrum are minute eyes like those of *Spalax*. The regressed, subcutaneous *Spalax* eye has lost the ability of image formation, but can still detect ambient light levels for photoperiodic perception (see Nevo, this volume). It should be noted here that the eye of the European mole *Talpa europaea* is equally small but superficially positioned and normally developed (Fig. 11.1A; Quilliam 1966). At the other end of the spectrum are eyes of normal size, like those of the coruro *Spalacopus cyanus* with a diameter of nearly 6 mm (Peichl et al. 2005). This is well within the range of eye dimensions in surface-dwelling rodents of similar body size (Howland et al. 2004). Most subterranean rodents have normal ocular properties: eyelids, clear cornea, lens, and vitreous, iris with a pupillary aperture and a well-developed retina lining the back of the eye (Fig. 11.1). This indicates the capability of image-forming vision.

Small eye size limits the image size on the retina, resulting in poor image quality and visual acuity. This limitation is shared by all small-eyed (microphthalmic) mammals, e.g. shrews and microchiropteran bats. An additional corollary of small eye size is a small pupil limiting the photon flux to the retina. Furthermore, the lens in many subterranean species (but not in the naked mole-rat *Heterocephalus glaber*) is small in relation to the eye size (Fig. 11.1; Nikitina et al. 2004). This is surprising because nocturnal surface-dwelling mammals, like rat and mouse, have relatively large lenses to collect light effectively. Relatively small lenses are a characteristic of diurnal mammals. Assuming that the eyes of subterranean rodents are adapted to make use of the little light available in their environment (just as eyes of nocturnal species), their optical properties appear paradoxical. The small lens in non-visual species may be dismissed as a regressive, metabolic cost-saving trait, but only if one ignores a potentially relevant sensory adaptation. A related paradox is found in the photoreceptor arrangements (see below).

11.3 Retina

The general morphology and layering of the retina is preserved in subterranean mammals, but the thickness of the layers varies greatly across species, indicating different processing capacities (Fig. 11.2). The types and packing densities of neurons determine the retinal processing charac-

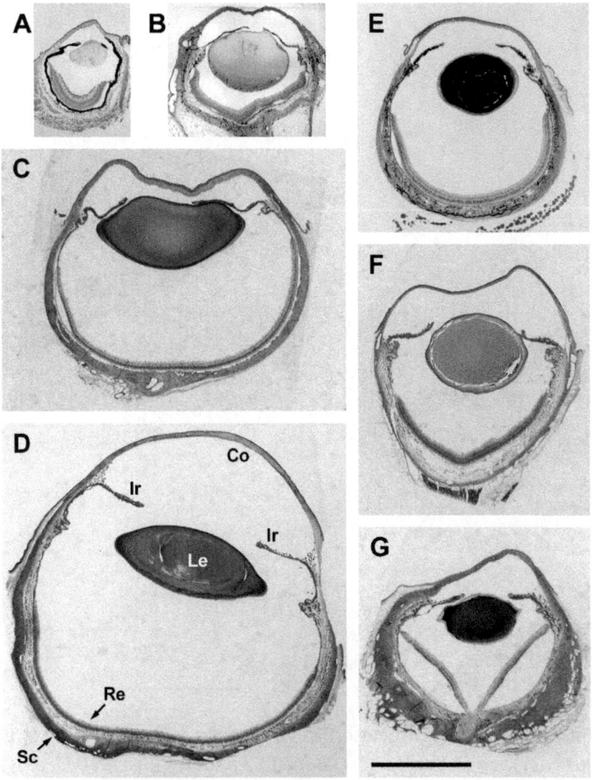

Fig. 11.1. Axial sections of the eyes of various subterranean mammals: **A** *Talpa europaea*; **B** *Heterocephalus glaber*; **C** *Georychus capensis*; **D** *Bathyergus suillus*; **E** *Fukomys anselli*; **F** *Cryptomys hottentotus*; **G** *Heliophobius argenteocinereus*. Eye sizes differ greatly between species, but all have a well-structured optical apparatus consisting of cornea, lens and pupil. Note that *T. europaea* features a cellular lens composed of nucleated cells, a peculiarity with unknown consequences for image-forming properties. Corneal deformations and retinal detachment in some eyes are histological artefacts of fixation and dehydration. Toluidine blue stained semithin sections of Epon embedded eyes. Co, cornea; Ir, iris with pupillary opening; Le, lens; Re, retina; Sc, Sclera. All eyes shown at same magnification, scale 1 mm. A, kindly provided by M. Glösmann; B,E, from Peichl et al. (2004)

teristics. Together with the optical resolution (set by eye size), the neural resolution (set by retinal neuron density) determines the visual acuity. The photoreceptor properties determine the sensitivity to the light intensity and spectral composition.

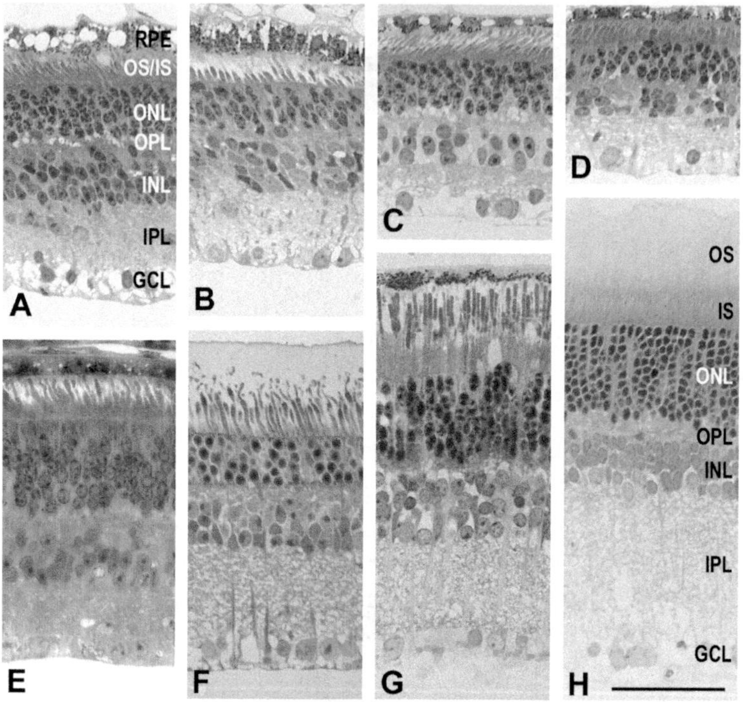

Fig. 11.2. Retinal cross-sections at higher magnification: **A** *Cryptomys hottentotus*; **B** *Heterocephalus glaber*; **C** *Georychus capensis*; **D** *Bathyergus suillus*; **E** *Fukomys anselli*; **F** *Geomys sp.*; **G** *Spalacopus cyanus*; **H** laboratory rat *Rattus norvegicus*. Semithin sections stained with toluidine blue. All sections taken from central parts of the eye and shown at same magnification, scale 50 µm. RPE, retinal pigment epithelium; OS/IS, photoreceptor outer and inner segments; ONL, outer nuclear layer; OPL, outer plexiform layer; INL, inner nuclear layer; IPL, inner plexiform layer; GCL, ganglion cell layer. B,E, modified from Peichl et al. (2004); G, modified from Peichl et al. (2005)

11.3.1 Photoreceptors

The photoreceptors are the first stage of visual processing, translating the light signals into neural signals. Mammalian photoreceptors comprise two categories, rods and cones. The rods are more light-sensitive and used for scotopic (night) vision at low light levels. The cones operate at higher light levels and are used for photopic (daylight and colour) vision. At intermediate light levels, both rods and cones are involved. Rod/cone ratios vary across mammals, roughly correlating with the light intensities at which the animal is active. Generally, nocturnal species have less cones (<1–3%) among their photoreceptors, diurnal species have 8–95% cones, and cathemeral species have intermediate (2–10%) proportions of cones

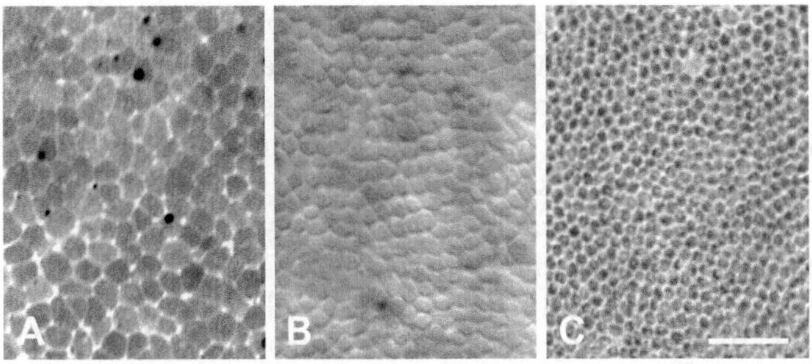

Fig. 11.3. Flat views of the retinal photoreceptor mosaic at the level of the inner segments: **A** *Fukomys anselli*; **B** *Spalacopus cyanus*; **C** laboratory rat, all shown at same magnification. In both subterranean species, the photoreceptors have larger diameters and hence a lower packing density than in rat. Most of the photoreceptors are rods. A,C, toluidine blue stained sections; B, phase image taken in an unstained, unsectioned retina. Scale 10 μm. A, modified from Peichl et al. (2004); B, modified from Peichl et al. (2005)

(recent overviews: Ahnelt and Kolb 2000; Peichl 2005). It is notable that even in most diurnal mammals the rods greatly outnumber the cones.

Retinal thickness is determined to a large extent by the thickness of the outer nuclear layer (ONL) which contains the photoreceptor somata (Fig. 11.2). The photoreceptor inner and outer segments have smaller diameters than the somata, so higher packing densities of outer segments require more rows of somata in the ONL. This is obvious in the rat retina, where the photoreceptor somata are stacked in about ten rows (Fig. 11.2H). In contrast, the retinae of the subterranean species have a thinner ONL, indicating lower packing densities of photoreceptors (Fig. 11.2). This becomes even more obvious in en-face views at the level of the photoreceptor inner segments, directly showing the photoreceptor mosaic (Fig. 11.3). Both Ansell's mole-rat *Fukomys anselli* (denoted as *Cryptomys anselli* in previous papers) and *Spalacopus cyanus* have larger inner segment diameters than the rat and hence a lower packing density. Photoreceptor densities amount to 100,000–150,000/mm^2 in *F. anselli* and *F. mechowii* (Peichl et al. 2004), to 160,000–224,000/mm^2 in *S. cyanus* (Peichl et al. 2005), and to 70,000–125,000/mm^2 in the pocket gopher *Thomomys bottae* (Williams GA et al. 2005). In comparison, photoreceptor densities average 400,000/mm^2 in the laboratory hooded rat (Hallett 1987) and 450,000/mm^2 in the house mouse (Jeon et al. 1998).

It is not clear why, compared to nocturnal murids, subterranean rodents have larger and less densely packed rods, since both groups are thought to be adapted to low-light vision. It is unknown whether higher retinal light sensitivity is better achieved by lower densities of larger rods with a potentially higher content of visual pigment per cell, or by higher densities of smaller rods allowing a higher convergence onto postsynaptic neurons and hence good signal-to-noise ratio. Assuming that the length and diameter of the rod outer segment is an indicator of visual pigment content, we may conclude that the rods in different subterranean species contain different amounts of pigment. Across species, the outer segments are similarly thick but of very different length, being rather short in bathyergids (Fig. 11.2A–E), and longer in *Geomys* and *S. cyanus* (Fig. 11.2F,G). Overall, it is obvious that subterranean rodents and nocturnal surface-dwellers have responded to the sensitivity challenge in different ways or that they had to respond to different challenges.

While all rods contain the same visual pigment rod opsin, the cones comprise different spectral types characterized by different cone opsins. The basic mammalian pattern is dichromatic colour vision based on two cone types, one containing a short-wave-sensitive (S) opsin (blue- to ultraviolet-sensitive, depending on species), and the other containing a middle-to-long-wave-sensitive (L) opsin (commonly green- or yellow-sensitive).

In three bathyergid species and *S. cyanus*, we have identified the spectral cone types and analyzed their distribution (Fig. 11.4; Peichl et al. 2004, 2005). In *F. anselli* nearly all cones express the S opsin in their outer segments (Fig. 11.4A). Many of these S cones co-express small amounts of L opsin, but there are only few pure L cones expressing exclusively L opsin. Cone densities in *F. anselli* are 8000–15,000/mm^2, and similar cone densities are found in *F. mechowii* and *Heterocephalus glaber*, i.e. in these species about 10% of the photoreceptors are cones. The coruro has also about 10% of cones among the photoreceptors, but their density is higher (10,000–30,000/mm^2) exhibiting a centro-peripheral gradient across the retina. Here, however, a majority of cones are pure L cones and a minority pure S cones, opsin co-expression does not occur (Fig. 11.4B). Highest cone densities of 15,000–34,000/mm^2 and highest cone proportions of about 26% are found in *T. bottae* (Williams GA et al. 2005). Here, most cones of dorsal retina co-express L and S opsin, whereas practically all cones in ventral retina are pure S cones. In the retina of *Spalax ehrenbergi* L cone opsin but no S cone opsin is expressed (David-Gray et al. 2002). Interestingly, the European mole also has about 10% cones. In its dorsal retina, there are about twice as many L cones as S cones; in the ventral retina, this ratio is reversed (Glösmann et al. 1999).

The S opsins of the coruro and the pocket gopher have their spectral sensitivity maximum near 365 nm ('ultraviolet'), and the L opsins near 505 nm ('green') (Peichl et al. 2005; Williams GA et al. 2005). UV-sensitive S cones are found in a number of rodent species, whereas most other mammals have blue-sensitive S cones (reviewed in Peichl 2005). In *S. ehrenbergi*, the L opsin is unusually red-shifted to 534 nm, presumably associated with the subcutaneous and hence hemoglobin-dominated light environment of the

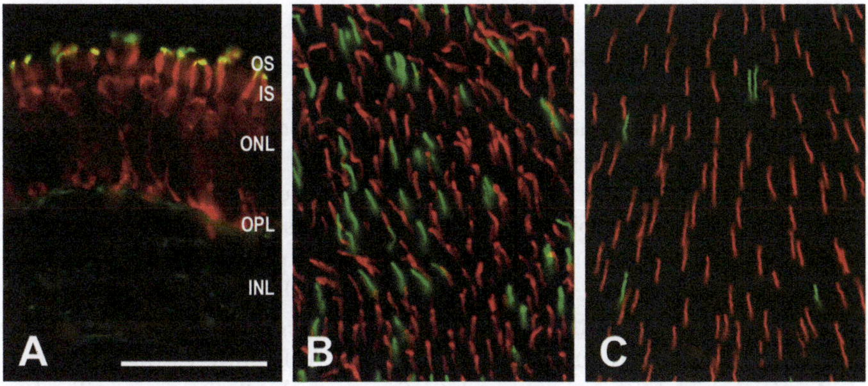

Fig. 11.4. Cone photoreceptor mosaics, shown by immunocytochemical labelling of the cones: **A** *Fukomys anselli*: retina, cross-section labelled for cone transducin, a general cone marker (*red*) and for S cone opsin (*green*). Transducin labelling reveals the complete cone morphology; nearly all cones have the S opsin in their outer segments (yellow merge of the red and green label). Retinal layers indicated as in Fig. 11.2; **B,C** *Spalacopus cyanus* and laboratory rat retina, flat views of the cone mosaic at the level of the opsin-containing outer segments, labelled for L cone opsin (*red*) and S cone opsin (*green*). *Spalacopus cyanus* has a rather high cone density with a majority of L cones and a substantial minority of S cones. Rat has a lower cone density and a lower S cone proportion. Rat cone outer segments are slimmer than those of *F. anselli* and *S. cyanus*. All micrographs shown at same magnification, scale 50 µm. A, modified from Peichl et al. (2004); B, modified from Peichl et al (2005); C, kindly provided by M. Glösmann

eye (David-Gray et al. 1998). This may also explain the loss of the *Spalax* S opsin. For bathyergids, the exact spectral tuning of the cone opsins is not known.

The laboratory rat has a markedly sparser cone population (about 1% of the photoreceptors) than the studied subterranean mammals, with a majority of L cones and a minority of S cones (Fig. 11.4C). Thus, the subterranean species (having cone/rod ratios of 1:10 or higher) appear more similar to diurnal than to nocturnal surface-dwellers. This and the low absolute rod densities suggest that the photoreceptor arrangements are more adapted to higher light levels than to the lightless underground ecotope. While all the studied subterranean species share high cone proportions, they markedly differ in the pattern of cone opsin expression. The range includes a conventional dominance of L cones over S cones (e.g. *S. cyanus*), a regional or retina-wide dominance of S opsin (e.g. *F. anselli*), and a complete absence of S opsin (*S. ehrenbergi*). This suggests evolutionary adaptations to different, species-specific visual needs. The species with two cone types have the potential for dichromatic colour vision, provided that the appropriate retinal and cortical circuits for colour processing are also preserved.

11.3.2 Retinal Interneurons and Circuitry

The thickness of the inner nuclear layer (INL) correlates with the number of interneurons (bipolar cells, horizontal cells and amacrine cells) available per unit area of retina. The thickness of the inner plexiform layer (IPL), where the processes of bipolar and amacrine cells contact the dendrites of ganglion cells, indicates the amount of synapses per retinal unit area and hence the potential processing complexity. The thickness and 'regularity' of the INL and IPL differ markedly across species (Fig. 11.2), even at intrageneric and intrafamilial levels (Fig. 11.2A–E). Interestingly, eye size and retinal thickness do not correlate. For example, *Bathyergus suillus* has a larger eye but a thinner retina (Figs. 11.1D and 11.2D) than *Fukomys anselli* (Figs. 11.1E and 11.2E). Obviously, these two determinants of visual performance do not go hand in hand. The rat IPL (Fig. 11.2H) is much thicker than that of any of the studied subterranean rodents.

The retinae in *Spalax ehrenbergi* and *Fukomys anselli* exhibit relatively normal cellular and synaptic ultrastructural characteristics (Cernuda-Cernuda et al. 2002, 2003). In *Heterocephalus glaber*, horizontal cells, rod bipolar cells, several types of cone bipolar cells, and several types of common amacrine cells were identified (Fig. 11.5; Mills and Catania 2004). Apparently most of the basic mammalian retinal neuron types are conserved in this species, but their structural organization is considerably less regular than in more visual mammals (Fig. 11.5). It is expected that disorganization of the retinal architecture has severe consequences for visual processing. Congruently, the corneal electroretinogram (ERG) exhibits typical features reflecting activity in photoreceptors, ON-bipolar cells, and inner retinal feedback circuitry. However, the ERG response of the retina is greatly attenuated (Hetling et al. 2005). Future studies will have to clarify whether a similar disorganization occurs in other subterranean rodents.

11.3.3 Retinal Ganglion Cells (RGCs)

The retinal ganglion cells are the output stage of the retina. Their axons form the optic nerve and optic tract, and terminate in the primary visual centres. Since the axons of the RGCs provide the only link between the eye and the brain, the total number of RGCs limits the amount of visual information that can be transmitted per time unit. Furthermore, their density, and not the density of the photoreceptors, limits visual acuity, because commonly there is a considerable convergence from photoreceptors to ganglion cells (reviewed in Wässle 2004).

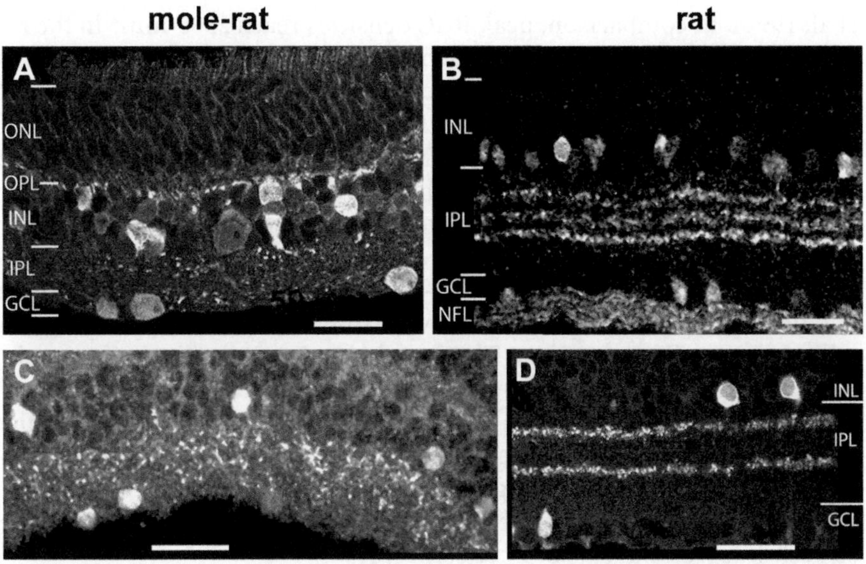

Fig. 11.5. Retinal interneurons: **A,C** in *Heterocephalus glaber*; **B,D** in the rat . A,B – immunolabelling for calretinin. In the laboratory rat, the label shows several types of amacrine cells with somata in the INL and narrowly stratified processes in the IPL, as well as somata in the GCL and axons in the NFL. In *H. glaber*, the label shows diffusely stratified amacrine cells, horizontal cells in the INL/OPL, and some somata in the GCL. C,D – immunolabelling for choline acetyl transferase, a marker of cholinergic amacrine cells. In the rat, there are two populations with somata in the INL and GCL, respectively, forming two narrow strata in the IPL. In *H. glaber*, the two populations also exist, but their IPL stratification is diffuse. Both markers reveal a less organized IPL in *H. glaber* than in the rat. All scales 25 μm. Abbreviations as in Fig. 11.2; NFL, optic nerve fibre layer. Modified from Mills and Catania (2004) with kind permission by the authors and the publisher

The total number of RGCs in mole-lemmings *Ellobius lutescens* and *E. talpinus* was estimated to be at least ∼3700 and ∼20,000, respectively (Herbin et al. 1994); estimates for mole-rats may be derived from optic nerve fibre counts (see below). The distribution of the RGCs is fairly homogenous in both mole-lemmings and mole-rats. Central regions of elevated ganglion cell density that serve high acuity vision (commonly referred to as the visual streak and the area centralis) are not discernible. However, the overall RGC density differs significantly between spalacids, bathyergids and mole-lemmings. In *Spalax ehrenbergi*, the RGCs are sparsely distributed throughout the retina with a mean density of $\sim 630/mm^2$ (Cooper HM et al. 1993a). Among bathyergids, mean RGC densities range between $\sim 1000/mm^2$ and $1600/mm^2$, and peak RGC densities between $\sim 1600/mm^2$ and $2500/mm^2$(Cveková et al., in preparation). The mole-lemmings have even higher RGC densities (mean $\sim 3400/mm^2$, peak $\sim 5000/mm^2$; Herbin et al. 1994). In comparison, peak RGC densities reach 2500/mm^2 in the rat (Perry 1981), 6000/mm^2 in the golden hamster (Tiao and Blakemore 1976) and 10,000 in the house mouse (Jeon et al. 1998). Thus, mole-lemmings and at least some bathyergid species have RGC densities comparable to those of nocturnal surface-dwelling rodents. Nevertheless, their visual acuities are low due to low optical resolution of their small eyes.

The upper limits of visual acuity, estimated from peak RGC density and eye size, are 0.3–0.5 cycles/degree in bathyergid mole-rats, 0.4–0.9 cycles/degree in mole-lemmings and 1–1.75 cycles/degree in the mentioned nocturnal rodents. These values are overestimates, because not all ganglion cell types contribute to acuity. The actual visual acuities in the above subterranean species most likely are lower than the behaviourally determined visual acuities of the mouse (~0.5 cycles/degree; Jellali et al. 2005) and the rat (~1 cycle/degree; Prusky et al. 2002). It is clear that in all these subterranean and surface-dwelling rodents, the brain only receives a coarsely 'pixelated' image of the outside world. In comparison, humans can resolve 60 cycles/degree and cats 6–9 cycles/degree.

In well-studied mammals, including the mouse and the rat, about 15 types of RGCs have been identified. They serve different tasks in image analysis, e.g. form discrimination, movement detection, and colour vision and thus represent different visual processing channels (reviewed in Wässle 2004). Very little is known about the RGC types in subterranean mammals. Mills and Catania (2004) found a variety of RGC types in *Heterocephalus glaber*. Myelinated axons with a wide range of diameters and an additional population of unmyelinated fibres (see below) indicate the presence of several types of ganglion cells also in other bathyergid species (Omlin 1997; Němec et al. 2004; Cveková et al., in preparation). It remains to be shown whether there are as many RGC types and hence processing channels as in more visual mammals, and what the situation is in other subterranean mammals. We expect that the diversity of RGC types may vary as much across subterranean species as does the number of RGCs and optic nerve fibres (see below).

Interestingly, a substantial fraction (~20%) of *Spalax* RGCs contains the photopigment melanopsin (Hannibal et al. 2002). In contrast to 'regular' ganglion cells, these recently discovered melanopsin-containing RGCs are intrinsically photosensitive (hence termed ipRGCs) and provide a major input to the central circadian pacemaker, the suprachiasmatic nucleus, and to the pretectum (reviewed in Berson 2003). In the surface-dwelling mammals, the ipRGCs represent only a few percent of the RGCs. Their abundance in *Spalax* is undoubtedly associated with the marked predominance of the retinohypothalamic pathway (Bronchti et al. 1991; Cooper HM et al. 1993a,b).

11.4 Optic Nerve (ON)

The diameter and the number of fibres of the optic nerve vary considerably across subterranean species. Not surprisingly, the ON diameter of the large-eyed species like *Spalacopus cyanus*, ctenomyids and geomyids is comparable to that of surface-dwelling rodents. In contrast, the ON is thin and macroscopically barely visible in microphthalmic species. The structure of the ON has been studied in *Talpa europaea* (Quilliam 1966; Cveková et al., unpublished data), in *Spalax ehrenbergi* (Herbin et al. 1995) and in bathyergid mole-rats (Omlin, 1997; Cernuda-Cernuda et al. 2003; Němec et al. 2004; Hetling et al. 2005; Cveková et al., in preparation). However, only two electron-microscopic studies provide reliable total fibre counts (Fig. 11.6A; Herbin et al. 1995; Cveková et al., in preparation). A significant reduction in the total number of optic fibres is evident when compared with rodents such as mouse (45,000–65,000; Williams RW et al. 1996; Jeon et al. 1998), hamster (65,000–114,000; Rhoades et al. 1979; Tay et al. 1986) and rat (100,000–120,000; Fukuda et al. 1982; Perry et al. 1983). Other characteristics shared by mole-rats and the European mole are a high proportion (if not dominance) of unmyelinated fibres, the absence of a radial gradient of fibre gauge and/or packing density across the nerve, and a high proportion of glial tissue. The latter is a reliable index of the degree of ON regression/degeneration (Fig. 11.6B).

In *Spalax ehrenbergi*, the ON measures 50–80 µm in diameter and is composed exclusively of unmyelinated axons; less than 1000 fibres comprise only 4.8% of the cross sectional area of the nerve, with astrocytes, connective tissue and a prominent nerve sheath accounting for the remainder. In *Talpa europaea*, the ON has ~50 µm in diameter and contains ~3000 axons, of which 15% are myelinated. In bathyergids, ON diameters range between ~40 and 110 µm, total fibre counts between ~2000 and 6000. Stark interspecific differences also occur in the ON texture and degree of myelination (Figs. 11.6 and 11.7). The ON is rather well-organized in *Georychus capensis, Bathyergus suillus, Cryptomys hottentotus* and *C. natalensis*, where myelinated axons predominate and are tightly packed in fascicules, and where optic fibres comprise ~50–70% of the nerve. Other species exhibit a less accentuated fasciculation with many individual optic fibres scattered among abundant glia. A severe regression is found in *Fukomys darlingi* and *Heliophobius argenteocinereus*, in which optic fibres comprise 20% and 14% of the ON, respectively. Less than 7% of the optic fibres are myelinated in the latter species. An almost complete absence of myelinated axons (<0.5%) is reported in *Heterocephalus glaber*. In com-

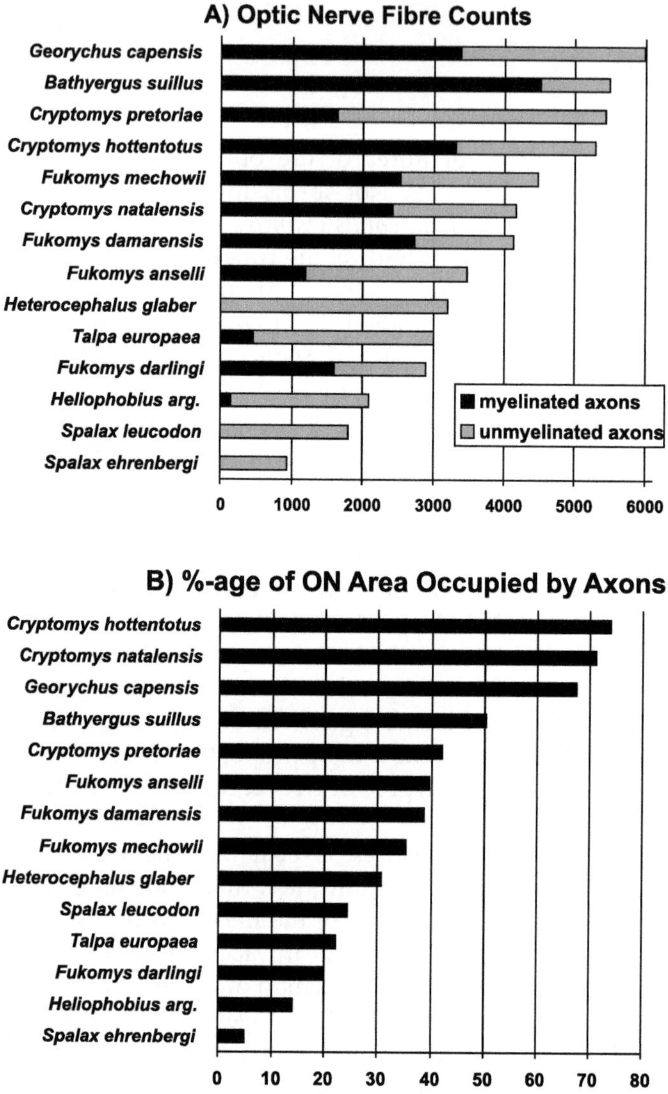

Fig. 11.6. A The number of optic nerve fibres per eye in bathyergids, spalacids and the European mole. Note the striking interspecific differences in the proportion of unmyelinated fibres. **B** The proportion of optic nerve cross-sectional area occupied by axons in these species. Data pooled from Herbin et al. (1995) and Cveková et al. (in preparation)

parison, there are only 3.5% and 5% unmyelinated fibres in the ON of the golden hamster and house mouse, respectively (Rhoades et al. 1979; Williams RW et al. 1996).

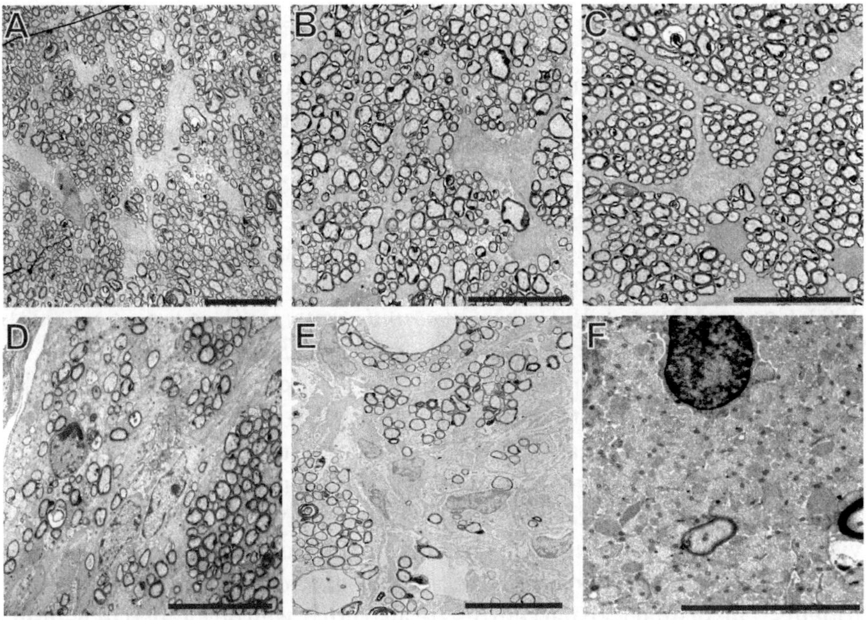

Fig. 11.7. Electron micrographs of transverse sections of optic nerves showing the appearance of typical tissue in six African mole-rat species: **A** *Bathyergus suillus*; **B** *Georychus capensis*; **C** *Cryptomys natalensis*; **D** *Cryptomys pretoriae*; **E** *Fukomys anselli*; **F** *Heterocephalus glaber*. Note the marked interspecific differences in texture and the near-absence of myelinated fibres in *H. glaber*. Scales 10 μm in A–E, 5 μm in F. Data from Cveková et al. (in preparation)

11.5 Subcortical Visual System

The projections from the retina to the brain are organized in parallel pathways, and the fibres of the optic tract terminate in five major subcortical visual centres – the suprachiasmatic nucleus (SCN), the lateral geniculate complex (LG), the pretectum, the superior colliculus (SC), and the accessory optic system (AOS). These areas receive input from different types of retinal ganglion cells and hence are provided with different types of visual information (Wässle 2004). They act as the relays that feed this information into different circuits having distinct roles in visual processing. Thus, these subcortical areas are key constituents of individual functional subsystems.

The relative amount of retinal input to the primary visual centres may be estimated from the distribution and the density of the retinal projections. Technically, the projections are visualized by injection of an anterogradely transported dye (tracer) into the eye, and computer-assisted image analysis can be used to measure the integral optical density of the terminal label within each primary visual structure (Magnin et al. 1989; Cooper HM et al. 1993b). It has been assumed that this measure reflects the relative number of ganglion cells projecting to a given structure (Cooper HM et al. 1993b). Thus it can be used as a criterion for evaluation of the relative degree of development and the relative importance of individual functional subsystems in a given species.

To date, retinal projections have been studied in five species of subterranean rodents: *Spalax ehrenbergi* (Bronchti et al. 1991; Rado et al. 1992; Cooper HM et al. 1993a,b), *Cryptomys pretoriae* (Negroni et al. 2003), *Fukomys anselli* (Němec et al. 2004), *Heterocephalus glaber* (Crish et al. 2006), and *Ellobius talpinus* (Herbin et al. 1994) (cf. Fig. 11.8). Data on the relative magnitude of retinal projections are also available for *Spalacopus cyanus* (Cooper HM et al. 1993a,b). The distribution of retinal input to the primary visual centres differs starkly between species, a clear indication that different subterranean rodents have evolved remarkably different visual systems. In *S. cyanus*, the visual system is not reduced and the distribution pattern of the retinal projections is similar to that of surface-dwelling rodents. The primary visual centres are also fairly well developed in *E. talpinus*. A slight hypoplasy of the LG is the only regressive feature observed in this species. In bathyergids, by contrast, all subcortical visual centres are cytoarchitecturally poorly developed and reduced in size, while the degree of reduction differs between nuclei. The LG and the pretectal nuclei are moderately reduced, whereas the superficial visual layers of the SC and the AOS are extremely reduced. Finally, in *S. ehrenbergi* all subcortical visual centres, except for the SCN, are severely reduced. The corollary of this extreme reduction is that as much as ~20% and ~17% of retinal projections are channelled to the SCN and the nuclei of the stria terminalis/anterior thalamic nuclei, respectively. These relative proportions are oddly high compared to the minor retinal input (<1%) which these structures receive in other mammals. Thus, the degree and the pattern of regression of functional visual subsystems differ tremendously among subterranean species. In spite of that, a conspicuous common trend can be observed: a strong selective reduction of the retinal input to the SC.

11.5.1 Suprachiasmatic Nucleus, Hypothalamus and Basal Telencephalon

The SCN is a central circadian pacemaker that governs a wide range of behavioural, physiological, and biochemical rhythms and synchronizes them with the ambient 24-h light–dark cycle. It also mediates effects of photoperiod on breeding and other seasonally recurring phenomena.

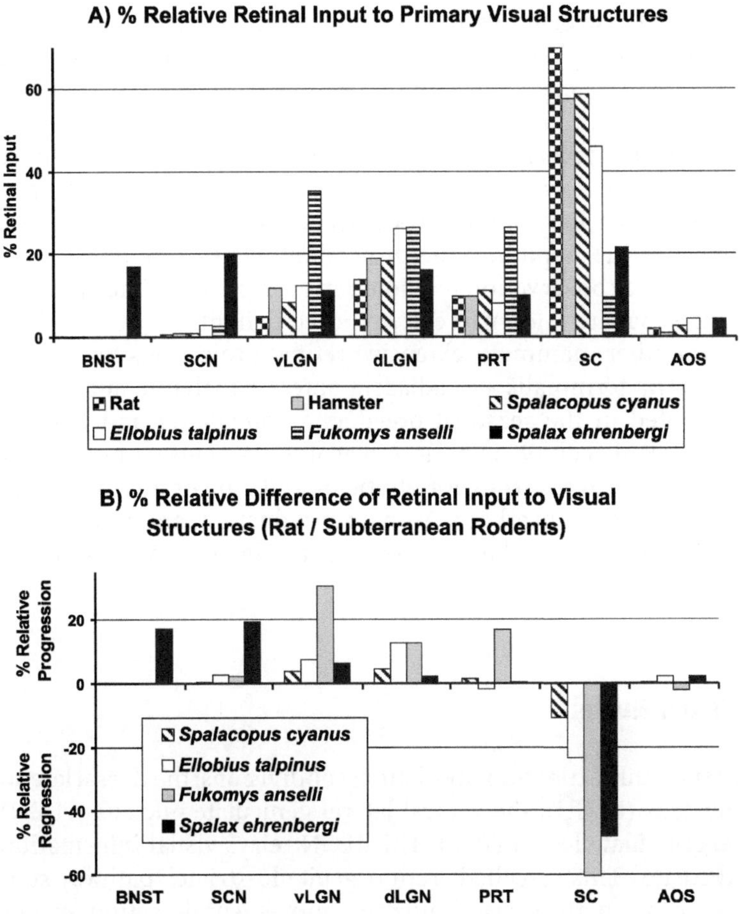

Fig. 11.8. Relative magnitude of retinal projections to primary visual centres in laboratory rat, hamster and four subterranean rodent species, *Spalacopus cyanus*, *Ellobius talpinus*, *Fukomys anselli* and *Spalax ehrenbergi*. There are marked interspecific differences in the distribution of retinal inputs. **B** Relative degree of change in the proportions of retinal input to different primary visual centres in the above subterranean species compared to the rat. The most conspicuous common trend is a strong selective reduction of the retinal input to the superior colliculus (SC). AOS, accessory optic system; BNST, bed nucleus of the stria terminalis; dLGN, dorsal lateral geniculate nucleus; PRT, pretectum; SCN, suprachiasmatic nucleus; vLGN, ventral lateral geniculate nucleus. Data pooled from Toga and Collins (1981), Cooper HM et al. (1993a,b), Herbin et al. (1994), and Němec et al. (2004)

All the studied subterranean rodents as well as insectivore moles possess a large, cytoarchitecturally well developed paired SCN that is bilaterally innervated by the retina (Bronchti et al. 1991; Kudo et al. 1991; Rado et al. 1992; Cooper HM et al. 1993a,b; Herbin et al. 1994; Negroni et al. 1997, 2003;

Němec et al. 2004; Crish et al. 2006). The size, cytoarchitecture and pattern of retinal innervation of their SCNs are similar to those in surface-dwelling rodents (for review see, e.g., Card and Moore 1991). The distribution pattern of neuropeptides in the SCN allows identification of two functionally distinct subdivisions typical of other rodents (for review see, e.g., Moore et al. 2002) in both *Spalax ehrenbergi* and *Cryptomys pretoriae*, albeit with some deviations from the common rodent pattern (Negroni et al. 1997, 2003). A robust retino-hypothalamic projection and an unreduced SCN clearly speak for the conservation of photoperiod-responsive mechanisms in these animals that are seldom, if ever, exposed to light.

As in several other mammals, extensive retinal projections to the bed nucleus of the stria terminalis and adjacent anterior thalamic nuclei have been demonstrated in the seasonal breeders *S. ehrenbergi* and *Ellobius talpinus* (Cooper HM et al. 1993a,b; Herbin et al. 1994). This pathway likely provides the anatomical substrate for photic modulation of thermoregulation and for the photic control of seasonal reproduction, and is absent in the aseasonally breeding social African mole-rats (Němec et al. 2004; Crish et al. 2006).

11.5.2 Thalamic Visual Nuclei

The LG consists of three functionally distinct subdivisions: the dorsal lateral geniculate nucleus (dLGN), the ventral lateral geniculate nucleus (vLGN) and the intergeniculate leaflet (IGL). The dLGN relays visual information to the visual cortex. This so-called “retino-geniculo-striate” pathway subserves image vision (pattern discrimination, movement detection, colour vision, etc.). The precise function of the vLGN is not known, although it has been implicated in brightness discrimination, the pupillary light reflex, a variety of visuomotor functions such as the optokinetic nystagmus, and saccadic eye movements, perhaps also in the detection of self-motion vs object-motion and the monitoring of eye movements (for review see Harrington 1997). It has also been suggested that the vLGN has a role in the regulation of circadian rhythms, a function that is more commonly associated with the adjacent intergeniculate leaflet (IGL) (Harrington 1997). A major role of the IGL seems to be mediating photic and non-photic phase shifts of circadian rhythms. In addition, its connectivity suggests that it may also contribute to the regulation of sleep and arousal, and to visuomotor functions (for review see Morin and Allen 2006).

In *Spalacopus cyanus*, the LG is large and heavily innervated by the retina. In terms of the size and relative amount of retinal input it is comparable to that of other rodents such as hamster and rat (Cooper HM et al. 1993a,b). In

Ellobius talpinus, the LG is normally organized and bilaterally innervated, but somewhat reduced in size (Herbin et al. 1994). Its volume averages about 60% of the mouse LG (*E. talpinus* 0.28 mm^3, mouse 0.46 mm^3, but note that *E. talpinus* is about twice as large as mouse). The LG of *E. talpinus* receives a dense retinal projection, comprising ~40% of all retinal projections. The dLGN has clearly defined binocular and monocular segments with partially segregated regions receiving ipsilateral or contralateral retinal input, an innervation pattern that suggests conserved binocular vision. The vLGN is also relatively well developed and can be divided into medial and lateral divisions as in other rodents (for reviews see Sefton and Dreher 1995; Harrington 1997). The IGL is cytoarchitecturally distinct and receives an equivalent bilateral retinal innervation (Herbin et al. 1994).

Bathyergid mole-rats possess a much smaller LG (Negroni et al. 2003; Němec et al. 2004; Crish et al. 2006; Xiao et al. 2006). For example, the volume of the *Heterocephalus glaber* LG (0.05 mm^3, Crish et al. 2006; 0.075 mm^3, Xiao et al. 2006) averages 10–15% of the mouse LG and 20–25% of the *E. talpinus* LG. Despite a significant size reduction, the LG is the main target of retinal projections, dwarfing the tectal structures by comparison. In *Fukomys anselli* for instance, about 62% of all retinal projections terminate within the LG (Němec et al. 2004). The dLGN is poorly differentiated, lacks a distinct lamination or well-defined cytoarchitectural subdivisions, and receives almost exclusively contralateral retinal input (Fig. 11.9). The vLGN is less reduced than the dLGN and, at least in *Fukomys* mole-rats, can be divided into medial and lateral divisions. Only the lateral division receives retinal input, and features high cytochrome oxidase activity (Němec et al. 2004) and expresses c-Fos in response to light exposure (Oelschläger et al. 2000). Thus, in *Fukomys*, the typical rodent organization of the vLGN is conserved, in contrast to the situation in other strictly subterranean mammals (Lund and Lund 1965; Kudo et al. 1988; Bronchti et al. 1991, Cooper HM et al. 1993a,b). The IGL is very small and poorly differentiated, but identifiable on the basis of neuropeptidergic features and a differential concentration of the ipsilateral retinal projection (Negroni et al. 2003; Němec et al. 2004). In *Cryptomys pretoriae*, the IGL contains few neuropeptide Y immunoreactive (NPY) neurons, and no enkephalin immunoreactive (ENK) neurons (Negroni et al. 2003). In the laboratory rat, the NPY neurons project to the SCN, the ENK neurons to the contralateral IGL (Moore and Card 1994). Hence, the geniculo-hypothalamic tract is likely conserved, whereas commissural connections between the two IGLs may be lacking in bathyergids.

The degree of LG regression is maximal in *Spalax ehrenbergi*. The volume of the *S. ehrenbergi* LG (0.066 mm^3; Cooper HM et al. 1993a,b) is about the same as in *H. glaber* (note that *S. ehrenbergi* is about three times larger than the naked mole-rat and its brain is more than four times larger). About 27% of all retinal projections terminate within the LG in *Spalax*. Retinal

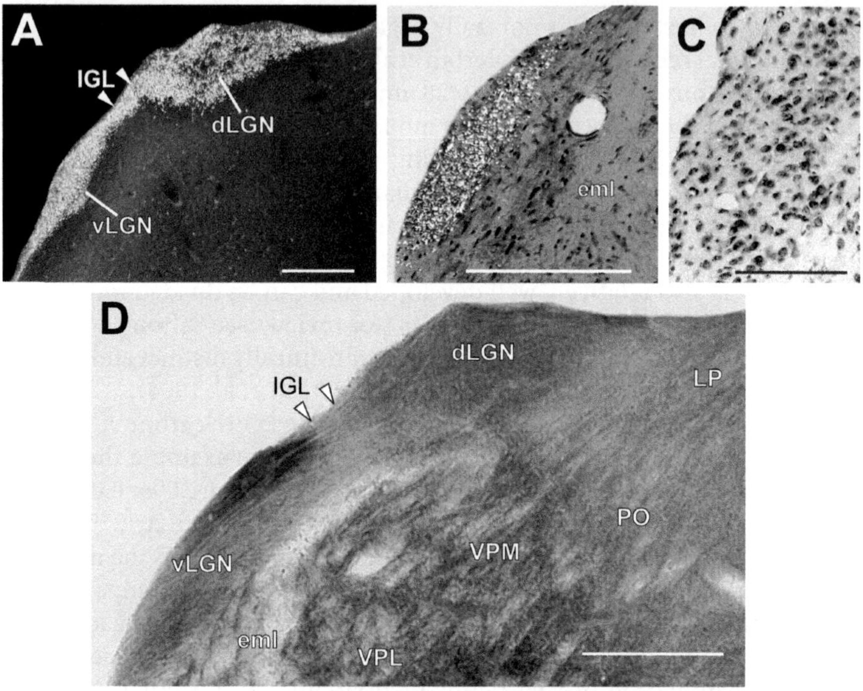

Fig. 11.9. The lateral geniculate complex in *Fukomys anselli*: **A** frontal section through the dorsal thalamus showing the distribution of retinal projection in the contralateral LG (darkfield illumination). *Arrowheads* indicate approximate borders of the IGL; **B** the distribution of retinal projection within the contralateral vLGN, Nissl counterstaining (polarized light). Note that the retinal input is confined to the lateral part of the nucleus; **C** cytoarchitecture of the vLGN, note the differences in cell size and packing density between the lateral and medial parts of the vLGN; **D** in this section stained for cytochrome oxidase activity, the dLGN and vLGN stand out as densely stained nuclei, while the IGL is only faintly labelled. The lateral part of the vLGN exhibits striking cytochrome oxidase activity. Abbreviations: eml, external medullary lamina; LP, lateral posterior thalamic nucleus; PO, posterior thalamic nuclear group; VPL, ventral posterolateral thalamic nucleus; VPM, ventral posteromedial thalamic nucleus. All scales 200 μm. All micrographs modified from Němec et al. (2004)

input is bilateral, although the ipsilateral projection is scanty. The dLGN is reduced to a narrow sheet 3–5 neurons thick, located on the surface of the dorsal thalamus. The vLGN is also cytoarchitecturally undifferentiated and reduced in size. Neither intrinsic subdivisions of the vLGN nor an IGL can be distinguished. A highly regressed LG has also been reported in insectivore moles (Lund and Lund 1965; Kudo et al. 1988).

11.5.3 Midbrain Nuclei: Pretectum, Superior Colliculus and Accessory Optic System

The pretectum is involved in luminance detection, the pupillary light reflex, the generation of optokinetic responses, and smooth pursuit eye movements (for review see Büttner-Ennever and Horn 1997). Robust, contralaterally predominant retinal projections to the pretectum have been demonstrated in all subterranean mammals studied so far (Lund and Lund 1965; Cooper HM et al. 1993a,b; Herbin et al. 1994; Němec et al. 2004; Crish et al. 2006). Retinal input mainly reaches the olivary pretectal nucleus (OPN) and the nucleus of the optic tract (NOT). The OPN contains neurons that act as luminance detectors and mediates the pupillary light reflex; the NOT generates the horizontal optokinetic nystagmus. The size and the degree of cytoarchitectural differentiation of these nuclei vary significantly among subterranean species. Both are well developed in *Ellobius talpinus* (Herbin et al. 1994). In bathyergids, the NOT is reduced in size and cytoarchitecturally indistinct, whereas the OPN is rather well developed (Němec et al. 2004; Crish et al. 2006). Although reduced in absolute size, it retains its characteristic olive-shaped morphology, receives a prominent retinal input (Fig. 11.10) and expresses c-Fos in response to light exposure (Oelschläger et al. 2000). In *Fukomys anselli*, the OPN and dLGN receive similar proportions of retinal input (each ∼26%). This is in line with a normal iris and pupil (cf. Fig. 11.1). In *Spalax ehrenbergi* both the NOT and the OPN are vestigial.

The SC is a prominent subcortical multimodal sensorimotor integrator that plays an important role in the control of directing eye and head movements. In the context of vision, it is involved in the detection of and orientation toward objects in the peripheral visual field, allowing approach (prey) versus avoidance (predator) decisions. Moreover, the SC constitutes a key component of the gaze control network that plays a crucial role in saccade generation and the maintenance of active visual fixation, and participates in the execution of microsaccades, smooth pursuit and vergence eye movements.

As in other rodents (Linden and Perry 1983), the superficial SC layers of *Spalacopus cyanus* and *E. talpinus* are well developed and receive a robust retinal input (Fig. 11.11) (Cooper HM et al. 1993a,b; Herbin et al. 1994). As much as 59% and 46% of all retinal projections terminate within the SC in *S. cyanus* and *E. talpinus*, respectively. In *E. talpinus*, the strata zonale, griseum superficiale et opticum are cytoarchitecturally distinct and attain a maximum width of about 380 μ m. The pattern of retinal innervation conforms to the typical non-primate pattern (cf. Kaas and Preuss 1993): ganglion cells that project to the SC are distributed throughout the

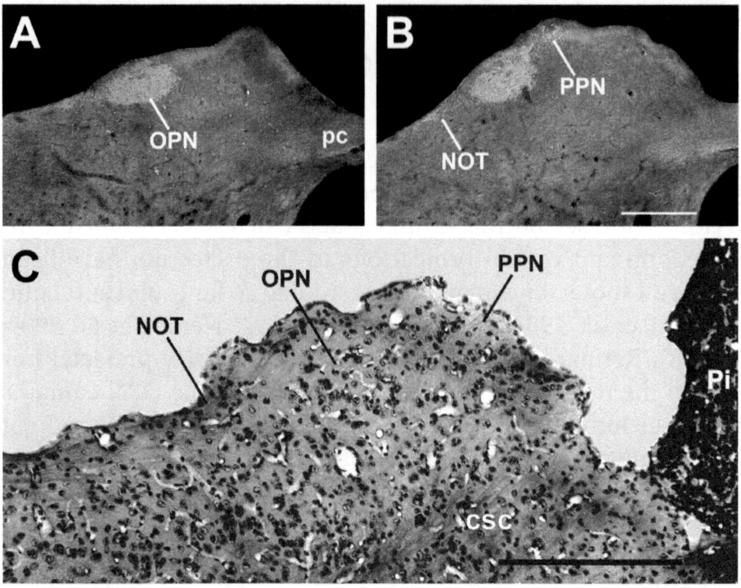

Fig. 11.10. The olivary pretectal nucleus in *Fukomys anselli*: **A,B** frontal sections through the mid and caudal pretectal complex, respectively, showing the distribution of retinal projection in the contralateral pretectum. Note the densely labelled OPN; **C** cytoarchitecture of the mid-pretectal complex, Klüver-Barrera staining. Note the discrete appearance of the OPN. csc, commissure of the superior colliculus; pc, posterior commissure; Pi, pineal gland; PPN, posterior pretectal nucleus. Scales 300 μm. All micrographs modified from Němec et al. (2004)

contralateral retina, while the ipsilateral projection arises from ganglion cells restricted to the temporal retina. This projection pattern suggests the presence of a topographically organized representation of the entire contralateral retinal field. In mole-rats, by contrast, the superficial, retinorecipient SC layers are collapsed to a single ill-differentiated layer attaining a maximum width of about 50 μm in *Fukomys anselli* and *S. ehrenbergi*, and of about 100 μm in *Heterocephalus glaber* (Fig. 11.11) (Cooper HM et al. 1993a,b; Němec et al. 2004; Crish et al. 2006). Moreover, no c-Fos expression can be detected in these layers after light stimulation (Vuillez et al. 1994; Oelschläger et al. 2000). The SC of mole-rats is thus composed almost entirely of well-developed multimodal intermediate and deep layers. In bathyergid mole-rats, the retinal projection is almost exclusively contralateral and in its medio-lateral extent it is largely confined to the central portion of the SC (Němec et al. 2004; Crish et al. 2006). In *S. ehrenbergi*, by contrast, the retinal projection is bilateral and spans the entire medio-lateral extent of the SC. The diffuse distribution of ipsilateral retinal afferents suggests a lack of precise retinotopy (Cooper et al. HM 1993a,b).

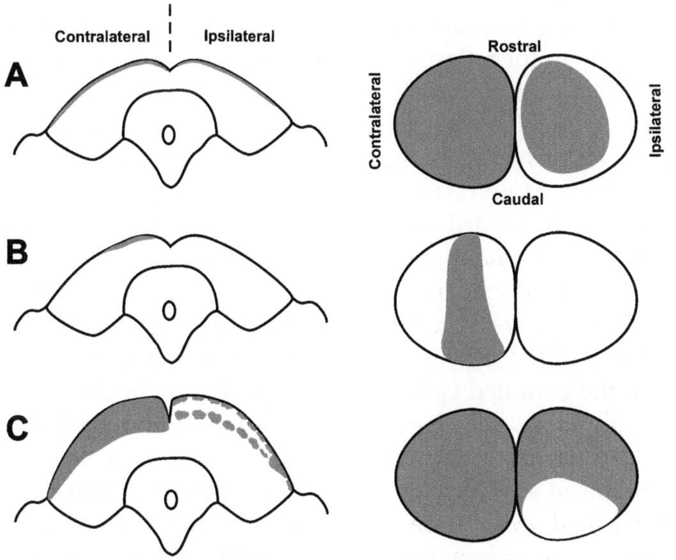

Fig. 11.11. Schematic drawings of the retinal projection to the superior colliculus in three subterranean rodent species: **A** *Spalax ehrenbergi*; **B** *Fukomys anselli*; **C** *Ellobius talpinus*. The *left column* shows the extent of the retinal projection as seen in frontal sections through the rostral SC. The *right column* shows the overall distribution of the retinal input as seen in dorsal view. Superficial, retinorecipient layers are severely regressed in *S. ehrenbergi* and *F. anselli* but normally developed in *E. talpinus*. These species also differ starkly in the distribution and the pattern of termination of the retino-tectal projection. The typical mammalian discontinuous ('patchy') distribution of the ipsilateral input is only conserved in *E. talpinus*. The figure is based on data reported by Cooper HM et al. (1993a,b), Herbin et al. (1994) and Němec et al. (2004)

Moles have been reported to lack any retinal innervation of the SC (Lund and Lund 1965; Kudo et al. 1988). Correspondingly, no visual responses were recorded from the mole SC (Crish et al. 2003).

The AOS is a set of three mesencephalic nuclei: the dorsal, the lateral and the medial terminal nuclei. It mediates optokinetic responses in reaction to a slow motion of large visual stimuli across the retina. Via detection of whole visual field motion, it likely serves to signal self-motion and to coordinate eye and head movements in relation to the animal's own movement within the visual field (Simpson 1984; Cooper HM and Magnin 1986). The AOS is complete and well developed in *E. talpinus* (Herbin et al. 1994) and, as far as we can judge from available quantitative data (Cooper HM et al. 1993a,b), the same is true for *S. cyanus*. In contrast, the AOS is vestigial and incomplete in mole-rats (Cooper HM et al. 1993a,b; Němec et al. 2004; Crish et al. 2006) and absent in insectivore moles (Lund and Lund 1965; Kudo et al. 1988).

Taken together, the above evidence shows that the brain centres involved in coordination of visuomotor reflexes (NOT, SC, AOS) are well developed in *Spalacopus* and *Ellobius*, but severely reduced in mole-rats and moles. This indicates that only the former taxa are well equipped for orientation in the three-dimensional visual environment that is encountered by a rodent active on the surface. Mole-rats and moles, by contrast, would find it difficult to detect and orient towards objects in the visual field and to track moving objects. This would render above-ground visually guided navigation, foraging, seeking for conspecifics, and predator avoidance ineffective if not impossible. We therefore speculate that the visual system of these animals is rather designed for 'underground' vision. Many visuomotor reflexes are useless in the confined space of a burrow. As the body largely obstructs the animal's backward view, food, conspecifics and predators could only be seen in a narrow space in front of the animal, provided that enough light is available at all. Detection and visual tracking of objects in the peripheral visual field seem to be irrelevant in these circumstances. Moreover, in contrast to an animal moving in an extended environment with many visual landmarks, an animal moving in a narrow tunnel does not experience the 'visual flow field' required to modulate AOS neuronal activity. Finally, other visuomotor functions such as saccadic and vergence eye movements and active visual fixation seem superfluous in animals that lack a central retinal region specialized for high acuity vision and that have tiny laterally positioned eyes. In view of these arguments, the extensive convergent regression of visuomotor midbrain centres in strictly subterranean mammals comes as no surprise.

In contrast, the OPN is conserved in all subterranean mammals that possess superficially positioned eyes with a normal iris and pupil. It has been suggested that occasional exposures to drastic changes in light level (the darkness of the burrow versus full intensity daylight) impose a selection pressure for the retention of the protective pupillary light reflex (Crish et al. 2006). It seems odd that no single ophthalmologic examination has been conducted to check for a pupillary reflex or other visuomotor reflexes in subterranean mammals.

11.6 Visual Cortex

Very little is known about the visual cortex of subterranean mammals. A search for the primary visual cortex has been conducted in mole-rats and moles, but the available evidence remains controversial. Retrograde tracing unambiguously demonstrated the presence of a rudimentary geniculostriate pathway in *Spalax ehrenbergi* (Cooper HM et al. 1993b). However, the topographic organization of connections between the dLGN and the visual cortex is imprecise, as distinct regions of area V1 receive projections from neurons located in widespread and overlapping regions of the dLGN. Contrary to this anatomical evidence, no visually evoked potentials could be recorded in the occipital cortex (Haim et al. 1983; Necker et al. 1992). Instead, electrophysiological and 2-deoxiglucose mapping experiments revealed that occipital regions, which are occupied by visual cortex in sighted mammals, can be activated by somatosensory (Necker et al. 1992) and auditory stimuli (Heil et al. 1991; Bronchti et al. 2002). Likewise, electrophysiological mapping provided no evidence for the existence of visually responsive cortical areas in *Heterocephalus glaber* (Catania and Remple 2002; Henry et al. 2006). In contrast, a recently completed anatomical study utilizing retrograde tracing and the technique of flattening and sectioning the cortex to visualize area boundaries has evidenced the presence of a primary visual cortex in two other African mole-rat species, *Fukomys anselli* and *F. mechowii* (Wielkopolska et al., in preparation). Moreover, as in sighted mammals (for review see Kaczmarek and Caudhuri 1997), light stimulation after a period of dark adaptation elicits c-Fos expression in the occipital cortex of *F. anselli* (Oelschläger et al. 2000), another indication of the presence of a definable visual cortex. Finally, in the star-nosed mole *Condylura cristata*, but not in the Eastern mole *Scalopus aquaticus*, a small region of cortex exhibiting elevated cytochrome oxidase activity has been observed in the expected location of the primary visual cortex (Catania and Kaas 1995, 1997). Thus, area V1 seems to be present only in the former species. Clearly, further studies are needed to resolve this issue.

Interestingly, light exposure elicits c-Fos expression also in the retrosplenial cortex of *F. anselli* (Oelschläger et al. 2000). In addition to the processing of emotional experience and behavioural reactions such as arousal and attentiveness (del Cano et al. 2000 and citations therein), this structure is involved in spatial memory (e.g. Vann and Aggleton 2002) and path integration (Cooper BG and Mizumori 1999). More specifically, it provides mnemonic associations of the visual and nonvisual environment that can be used to correct for cumulative errors that occur during path integration. Noteworthy, retrosplenial cortex inactivation selectively impairs navigation in darkness (Cooper BG and Mizumori 1999; Cooper BG et al. 2001). Given that path integration is a dominant modus operandi for orientation in subterranean burrows (Burda et al. 1990; Kimchi and Terkel 2002), it is tempting to speculate that visual information is integrated with multimodal sensory information at the level of the retrosplenial cortex and thus contributes to spatial behaviour and/or predator avoidance in certain circumstances, for instance when a tunnel is opened. In any case, light entrained c-Fos expression in the retrosplenial cortex suggests that *F. anselli* is attentive to visual stimuli.

11.7 Oculomotor Nuclei

To our knowledge, neither the oculomotor nuclei nor extraocular muscles have been studied systematically in subterranean rodents. A surprisingly well-developed oculomotor nucleus was described in *Fukomys anselli* (Němec et al. 2004), and cholinergic oculomotor and trochlear nuclei were described in *Cryptomys pretoriae* (Da Silva et al. 2006). The trochlear nucleus is reported to be composed of very few motoneurons and partly merged with the posterior portion of the oculomotor nucleus. It is likely that most of the oculomotor nucleus motoneurons innervate the well-developed musculus retractor bulbi, which may subserve the protective retraction of the eye during digging (Němec et al. 2004).

11.8 Role of Vision

Data on visual capacities of subterranean rodents remain scarce (for review see Burda et al. 1990; Francescoli 2000). At present, most of what we know about vision in subterranean rodents has been inferred from the above-reviewed anatomical evidence, from anecdotal observations, and from a limited number of electrophysiological and behavioural studies. Generally, the eye size is a very good predictor of visual capabilities. The large-eyed species like the coruro, ctenomyids and pocket gophers have reasonably good visual acuity and are able to detect moving objects. The markedly dorsal position of the eyes in these taxa suggests that vision is used when the animals are active at burrow entrances (cf. Burda et al. 1990; Francescoli 2000). Correspondingly, the visual system of the coruro remains unreduced and in all respects appears well suited for normal 'surface' vision.

Likewise, in *Ellobius talpinus* the eye is only moderately reduced (axial length 2.9 mm), and the central visual system is fairly well developed and comparable to that of surface-dwelling rodents (Herbin et al. 1994). Actually, a moderately reduced number of retinal ganglion cells, their homogenous distribution, and a slight hypoplasy of the dLGN are the only regressive features observed in this species. Thus, a low visual acuity appears to be the only factor constraining visual capacity in the mole-lemmings.

The situation is less conclusive in the African bathyergid mole-rats. Anecdotal observations suggest that they are oblivious to light stimuli such as moving objects or full-field light flashes (Eloff 1958; Oelschläger et al. 2000, but see Poduschka 1978; Hetling et al. 2005). This is in line with a severe regression of the midbrain structures subserving coordination of visuomotor reflexes (see above). Actually, the apparent behavioural blindness of the African mole-rats might be caused by the fact that these animals are unable to generate spatially appropriate orientation responses. While the lack of an overt behavioural reaction to light suggests a reduced reliance on visual cues, it does not per se imply that these animals are blind. Indeed, recent behavioural studies in three bathyergid species utilizing a preferential nesting assay have clearly demonstrated the capability to distinguish between light and darkness (*Fukomys anselli* and *F. kafuensis*, Wegner et al. 2006; *Heliophobius argenteocinereus*, Kott, unpublished BSc thesis). In addition, physiological responses of the retina to bright light flashes and light avoidance behaviour have been reported in *Heterocephalus glaber* (Hetling et al. 2005). This result is again in good accord with the anatomical evidence, as the structures implicated in brightness discrimination (vLGN and OPN) are relatively well-developed, receive robust retinal input, and express c-Fos in response to light exposure. The latter is also true for the occipital cortex. The moderately reduced dLGN and the conserved retino-geniculo-striate pathway further suggest that the African mole-rats may have retained the capacity for crude form and/or movement discrimination.

In contrast, abundant evidence indicates a complete lack of visual capabilities in *Spalax ehrenbergi*. The minute, subcutaneous eye lacks any image forming capability. Accordingly, the retinal input to structures involved in form and movement discrimination and visuomotor coordination is dramatically reduced (Bronchti et al. 1991; Cooper HM et al. 1993a,b). These thalamic, tectal and tegmental nuclei are highly hypoplasic, lack a distinct topographical organization and do not express c-Fos in reaction to light stimulation (Leder 1974; Cooper HM et al. 1993a,b; Rehkämper et al. 1994; Vuillez et al. 1994). The lack of evidence for a visually responsive cortex (see above) together with the fact that *Spalax* fail to learn avoidance of shocks associated with a light stimulus (Rado et al. 1992), also support the tentative conclusion that they are entirely blind. Nevertheless, a preferential nesting assay has shown that they are, just as the African mole-rats, able to distinguish between light and darkness (Rado et al. 1992). Rudimentary luminance discrimination abilities are also conserved in *Talpa europaea* (Lund and Lund 1965; Johannesson-Gross 1988).

The adaptive significance of vision in the underground ecotope remains enigmatic. Nevertheless, the above-reviewed evidence sheds some light on this puzzle. The visual system of mole-rats and moles is not suited for above-ground orientation. Thus, rare surface activities (e.g. dispersal, mate-seeking excursions, exceptional foraging above ground) can hardly be visually guided. Likewise, vision seems useless in a dark subterranean environment since light does not penetrate into sealed underground burrows effectively enough to provide a perceivable cue. Paradoxically, the photoreceptor arrangements suggest that subterranean species are adapted to day-light conditions rather than to a dark subterranean environment. Why should the visual system be designed for 'underground' vision, but the photoreceptor mosaic adapted to perceive daylight intensities? As has been recently suggested (Hetling et al. 2005; Wegner et al. 2006), residual vision may enable subterranean mammals to localize breeches in the burrows that let in light. Incidence of light may well indicate a tunnel being opened by a predator and may thus warn the animal not to approach the opening too closely but instead to plug it. Plugging of illuminated tunnels has been reported in *Fukomys* and *Heliophobius* mole-rats (own field observations) and *Thomomys* pocket gophers (laboratory experiment, Werner et al. 2005), and light-evoked escape behaviour in *H. glaber* (Hetling et al. 2005). Alternatively, light may signal an accidental collapse of the burrow and induce its maintenance.

11.9 Chronobiology

Circadian rhythms are a fundamental adaptation of living cells and organisms to the daily and seasonal fluctuation in light and temperature. To secure the proper phasing to the sequence of external changes, circadian rhythms are phase-adjusted (entrained) by day-night cycles. Abundant evidence from anatomical, physiological, behavioural and ecological studies shows that this mechanism is conserved in subterranean mammals. An extensive review of the literature is beyond the scope of this chapter. The following section is therefore restricted to detailed studies on mole-rat circadian rhythms.

In *Spalax ehrenbergi*, prominent retinal input to the SCN and the bed nucleus of the stria terminalis provides the anatomical substrate for photic entrainment of locomotor and thermoregulatory activity rhythms and the photic control of seasonal reproduction (Cooper HM et al. 1993a,b; Negroni et al. 1997). A light pulse given during the subjective night induces c-Fos expression in the SCN (Vuillez et al. 1994; Tobler et al. 1998), the thermoregulatory capacity is photoperiod dependant (Haim et al. 1983), and circadian activity can be entrained by the ambient light-dark cycle (Rado et al. 1991; Ben-Shlomo et al. 1995; Goldman et al. 1997; Tobler et al. 1998). Taken together with the arguments reviewed above, the available evidence strongly suggests that the blind mole-rat's visual system has retained one major function, the photic entrainment of circadian and circannual physiology.

In the African bathyergid mole-rats, the SCN is also large, cytoarchitecturally well developed and bilaterally innervated by the retina. c-Fos expression in the SCN is gated according to the phase of the circadian clock in seasonally breeding, solitary *Georychus capensis*, but not in social species (Oelschläger et al. 2000; Oosthuizen et al. 2005). Nevertheless, both solitary and social species exhibit light entrained circadian rhythms of melatonin secretion (Richter et al. 2003; Hart et al. 2004; Gutjahr et al. 2004; Vasicek et al. 2005a), body temperature (Lovegrove and Muir 1996; Riccio and Goldman 2000a) and locomotor activity (Lovegrove and Papenfus 1995; Riccio and Goldman 2000b; Oosthuizen et al. 2003; Hart et al. 2004; Vasicek et al. 2005b; Schöttner et al. 2006). In contrast to these results obtained when mole-rats were individually housed, circadian rhythms of locomotor activity were typically not observed in social species housed in colony settings (Bennett 1992; Davis-Walton and Sherman 1994; Riccio and Goldman 2000b).

Why should herbivore mammals inhabiting lightless, thermally stable ecotopes entrain their rhythms to the outside photoperiod? One possibility is that light-entrained rhythmicity is a useless relict from ancestral above-ground existence. This notion is supported by a large intra- and interindividual variability in many aspects of circadian function observed in all mole-rat species. However, the very fact that the SCN is the only visual structure escaping the evolutionary regression proves that circadian capacities were retained because they serve some specific functions. For instance, circadian rhythms may be used to optimize timing of rare above-ground forays or removing loose earth from the burrows - activities that may expose animals to predation. Furthermore, photoperiod detection may be useful for synchronizing circannual rhythms in higher latitudes. Seasonally breeding species need to be synchronized to the annual environmental cycle to anticipate the breeding season, thus providing another adaptive rationale for retention of the circadian system in subterranean species (for further discussion see, e.g., Riccio and Goldman 2000b).

Acknowledgements. We thank Dr. Martin Glösmann for providing unpublished material and Walter Hofer and Jitka Koláčková-Sedláčková for excellent histological assistance. Our work was supported by grants from the DFG (BU 717/10-3 to HB; OE 217/4-2 to PN), the Czech Science Foundation (206/03/0638 to PN), the Ministry of Education, Youth and Sport of the Czech Republic (0021620828 to PN) and Institutional Scientific Project (AV0Z50200510 to OB).

References

Ahnelt PK, Kolb H (2000) The mammalian photoreceptor mosaic – adaptative design. Progr Ret Eye Res 19:711–777

Bennett NC (1992) The locomotory activity patterns of a functionally complete colony of *Cryptomys hottentotus* (Rodentia: Bathyergidae). J Zool Lond 228:435–443

Ben-Shlomo R, Ritte U, Nevo E (1995) Activity pattern and rhythm in the subterranean mole rat superspecies *Spalax ehrenbergi*. Behav Gen 25:239–245

Berson DM (2003) Strange vision: ganglion cells as circadian photoreceptors. Trends Neurosci 26:314–320

Bronchti G, Rado R, Terkel J, Wollberg Z (1991) Retinal projections in the blind mole rat: a WGA-HPR tracing study of a natural degeneration. Dev Brain Res 58:159–170

Bronchti G, Heil P, Sadka R, Hess A, Scheich H, Wollberg Z (2002) Auditory activation of 'visual' cortical areas in the blind mole rat (*Spalax ehrenbergi*). Eur J Neurosci 16:311–329

Büttner-Ennever JA, Horn AKE (1997) Anatomical substrates of oculomotor control. Curr Opin Neurobiol 7:872–879

Burda H, Bruns V, Müller M (1990) Sensory adaptations in subterranen mammals. In: Nevo E, Reig OA (eds) Evolution of subterranean mammals at the organismal and molecular levels. Wiley-Liss, New York, pp 269–294

Card JP, Moore RY (1991) The organization of visual circuits influencing the circadian activity of the suprachiasmatic nucleus. In: Klein DC, Moore RY, Reppert SM (eds) Suprachiasmatic nucleus: the mind's clock. Oxford Univ Press, New York, pp 51–76

Catania KC, Kaas JH (1995) Organization of the somatosensory cortex of the star-nosed mole. J Comp Neurol 351:549–567

Catania KC, Kaas JH (1997) Organization of somatosensory cortex and distribution of corticospinal neurons in the eastern mole (*Scalopus aquaticus*). J Comp Neurol 378:337–353

Catania KC, Remple MS (2002) Somatosensory cortex dominated by the representation of teeth in the naked mole-rat brain. Proc Natl Acad Sci USA 99:5692–5697

Cernuda-Cernuda R, DeGrip WJ, Cooper HM, Nevo E, García-Fernández JM (2002) The retina of *Spalax ehrenbergi*: novel histologic features supportive of a modified photosensory role. Invest Ophthalmol Vis Sci 43:2374–2383

Cernuda-Cernuda R, García-Fernández JM, Gordijn MCM, Bovee-Geurts PHM, DeGrip WJ (2003) The eye of the African mole-rat *Cryptomys anselli*: to see or not to see? Eur J Neurosci 17:709–720

Cooper BG, Mizumori SJ (1999) Retrosplenial cortex inactivation selectively impairs navigation in darkness. Neuroreport 10:625–630

Cooper BG, Manka TF, Mizumori SJ (2001) Finding your way in the dark: the retrosplenial cortex contributes to spatial memory and navigation without visual cues. Behav Neurosci 115:1012–1028

Cooper HM, Magnin M (1986) A common mammalian plan of accessory optic system organization revealed in all primates. Nature 324:457–459

Cooper HM, Herbin M, Nevo E (1993a) Ocular regression conceals adaptive progression of the visual system in a blind subterranean mammal. Nature 361:156–159

Cooper HM, Herbin M, Nevo E (1993b) Visual system of a naturally microphthalmic mammal: the blind mole rat, *Spalax ehrenbergi*. J Comp Neurol 328:313–350

Crish SD, Comer CM, Marasco PD, Catania KC (2003) Somatosensation in the superior colliculus of the star-nosed mole. J Comp Neurol 464:415–425

Crish SD, Dengler-Crish CM, Catania KC (2006) Central visual system of the naked mole-rat (*Heterocephalus glaber*). Anat Rec A 288:205–212

Da Silva JN, Fuxe K, Manger PR (2006) Nuclear parcellation of certain immunohistochemically identifiable neuronal systems in the midbrain and pons of the highveld molerat (*Cryptomys hottentotus*). J Chem Neuroanat 31:37–50

David-Gray ZK, Janssen JWH, DeGrip WJ, Nevo E, Foster RG (1998) Light detection in a 'blind' mammal. Nat Neurosci 1:655–656

David-Gray ZK, Bellingham J, Munoz M, Avivi A, Nevo E, Foster RG (2002) Adaptive loss of ultraviolet-sensitive/violet-sensitive (UVS/VS) cone opsin in the blind mole rat (*Spalax ehrenbergi*). Eur J Neurosci 16:1186–1194

Davis-Walton J, Sherman PW (1994) Sleep arrhythmia in the eusocial naked mole-rat. Naturwiss 81:272–275

del Cano G, Gerrikagoitia I, Martínez-Millián L (2000) Morphology and topographical organization of the retrospleniocollicular connection: a pathway to relay contextual information from the environment to the superior colliculus. J Comp Neurol 425:393–408

Eloff G (1958) Functional and structural degeneration of the eye of the south African rodent mole, *Cryptomys bigalkei* and *Bathyergus maritimus*. S Afr J Sci 54:292–302

Francescoli G (2000) Sensory capabilities and communication in subterranean mammals. In: Lacey EA, Patton JL, Cameron GN (eds) Life underground: the biology of subterranean rodents. Univ Chicago Press, Chicago, pp 111–144

Fukuda Y, Sugimoto T, Shirokawa T (1982) Strain differences in quantitative analysis of the rat optic nerve. Exp Neurol 75:525–532

Glösmann M, Fliesser M, Ahnelt PK (1999) Cone opsin topography in the retina of the European mole. Invest Ophthalmol Vis Sci 40:S158

Goldman BD, Goldman SL, Riccio AP, Terkel J (1997) Circadian patterns of locomotor activity and body temperature in blind mole-rats, *Spalax ehrenbergi*. J Biol Rhythms 14:348–361

Gutjahr GH, van Rensburg LJ, Malpaux B, Richter TA, Bennett NC (2004) The endogenous rhythm of plasma melatonin secretion and its regulation by light in the Highveld mole-rat (*Cryptomys hottentotus pretoriae*): a microphthalmic, seasonally breeding rodent. J Pineal Res 37:185–192

Haim A, Heth G, Pratt H, Nevo E (1983) Photoperiodic effects on thermoregulation in a "blind" subterranean mammal. J Exp Biol 107:59–64

Hallett PE (1987) The scale of the visual pathways of mouse and rat. Biol Cybern 57:275–286

Hannibal J, Hindersson P, Nevo E, Fahrenkrug J (2002) The circadian photopigment melanopsin is expressed in the blind subterranean mole rat *Spalax*. Neuroreport 13:1411–1414

Harrington ME (1997) The ventral lateral geniculate nucleus and the intergeniculate leaflet: interrelated structures in the visual and circadian systems. Neurosci Biobehav Rev 21:705–727

Hart L, Bennett NC, Malpaux B, Chimimba CT, Oosthuizen MK (2004) The chronobiology of the Natal mole-rat, *Cryptomys hottentotus natalensis*. Physiol Behav 82:563–569

Heil P, Bronchti G, Wollberg Z, Scheich H (1991) Invasion of visual cortex by the auditory system in the naturally blind mole rat. Neuroreport 2:735–738

Henry EC, Remple MS, O'Riain MJ, Catania KC (2006) Organization of somatosensory cortical areas in the naked mole-rat (*Heterocephalus glaber*). J Comp Neurol 495:434–452

Herbin M, Repérant J, Cooper HM (1994) Visual system of the fossorial mole-lemmings, *Ellobius talpinus* and *Ellobius lutescens*. J Comp Neurol 346:253–275

Herbin M, Rio J-P, Repérant J, Cooper HM, Nevo E, Lemire M (1995) Ultrastructural study of the optic nerve in blind mole-rats (Spalacidae, *Spalax*). Vis Neurosci 12:253–261

Hetling JR, Baig-Silva MS, Comer CM, Pardue MT, Samaan DY, Qtaishat NM, Pepperberg DR, Park TJ (2005) Features of visual function in the naked mole-rat *Heterocephalus glaber*. J Comp Physiol A 191:317–730

Howland HC, Merola S, Basarab JR (2004) The allometry and scaling of the size of vertebrate eyes. Vision Res 44:2043–2065

Jellali A, Meziane H, Ouagazzal A, Rousseau S, Romand R, Auwerx J, Sahel J, Chambon P, Picaud S (2005) The optomotor response: a robust first-line visual screening method for mice. Vision Res 45:1439–1446

Jeon C-J, Strettoi E, Masland RH (1998) The major cell populations of the mouse retina. J Neurosci 18:8936–8946

Johannesson-Gross K (1988) Brightness discrimination of the mole (*Talpa europaea* L) in learning experiments applying a modified tube-maze method. Z Säugetierk 53:193–201

Kaas JH, Preuss TM (1993) Archontan affinities as reflected in the visual system. In: Szalay FS, Novacek MJ, McKenna MC (eds) Mammal phylogeny: placentals. Springer, Berlin Heidelberg New York, pp 115–128

Kaczmarek L, Caudhuri A (1997) Sensory regulation of immediate-early gene expression in mammalian visual cortex: implications for functional mapping and neural plasticity. Brain Res Rev 23:237–256

Kimchi T, Terkel J (2002) Seeing or not seeing. Curr Opin Neurobiol 12:728–734

Kudo M, Nakamura Y, Moriizumi T, Tokuno H, Kitao Y (1988) Direct retinal projections to the lateroposterior thalamic nucleus (LP) in the mole. Neurosci Lett 93:176–180

Kudo M, Yamamoto M, Nakamura Y (1991) Suprachiasmatic nucleus and retinohypothalamic projections in moles. Brain Behav Evol 38:332–338

Leder ML (1974) Zur mikroskopischen Anatomie des Gehirns eines blinden Nagers (*Spalax leucodon*, Nordmann 1840) mit besonderer Berücksichtigung des visuellen Systems. Teil 1: Atlas. Zool Jb Anat 93:353–410

Linden R, Perry VH (1983) Massive retinotectal projection in rats. Brain Res 272:145–149

Lovegrove BG, Muir A (1996) Circadian body temperature rhythms of the solitary Cape mole rat *Georychus capensis* (Bathyergidae). Physiol Behav 60:991–998

Lovegrove BG, Papenfus ME (1995) Circadian activity rhythms in the solitary Cape molerat (*Georychus capensis*: Bathyergidae) with some evidence of splitting. Physiol Behav 58:679–685

Lund RD, Lund JS (1965) The visual system of the mole, *Talpa europaea*. Exp Neurol 13:302–316

Magnin M, Cooper HM, Mick G (1989) Retinohypothalamic pathway: a break in the law of Newton-Müller-Gudden? Brain Res 488:390–397

Mills SL, Catania KC (2004) Identification of retinal neurons in a regressive rodent eye (the naked mole rat). Vis Neurosci 21:107–117

Moore RY, Card JP (1994) Intergeniculate leaflet: an anatomically and functionally distinct subdivision of the lateral geniculate complex. J Comp Neurol 344:403–430

Moore RY, Speh JC, Leak RK (2002) Suprachiasmatic nucleus organization. Cell Tissue Res 309:89–98

Morin LP, Allen CN (2006) The circadian visual system, 2005. Brain Res Rev 51:1–60

Necker R, Rehkämper G, Nevo E (1992) Electrophysiological mapping of body representation in the cortex of the blind mole rat. Neuroreport 3:505–508

Negroni J, Nevo E, Cooper HM (1997) Neuropeptidergic organization of the suprachiasmatic nucleus in the blind mole rat (*Spalax ehrenbergi*). Brain Res Bull 44:633–639

Negroni J, Bennett NC, Cooper HM (2003) Organization of the circadian system in the subterranean mole rat, *Cryptomys hottentotus* (Bathyergidae). Brain Res 967:48–62

Němec P, Burda H, Peichl L (2004) Subcortical visual system of the African mole-rat *Cryptomys anselli*: to see or not to see? Eur J Neurosci 20:757–768

Nevo E (1999) Mosaic evolution of subterranean mammals: regression, progression and global convergence. Oxford Univ Press, Oxford

Nikitina NV, Maughan-Brown B, O'Riain MJ, Kidson SH (2004) Postnatal development of the eye in the naked mole rat (*Heterocephalus glaber*). Anat Rec 277:317–337

Oelschläger HHA, Nakamura M, Herzog M, Burda H (2000) Visual system labeled by c-Fos immunohistochemistry after light exposure in the 'blind' subterranean Zambian mole-rat (*Cryptomys anselli*). Brain Behav Evol 55:209–220

Omlin FX (1997) Optic disc and optic nerve of the blind Cape mole-rat (*Georychus capensis*): a proposed model for naturally occurring reactive gliosis. Brain Res Bull 44:627–632

Oosthuizen MK, Cooper HM, Bennett NC (2003) Circadian rhythms of locomotor activity in solitary and social species of African mole-rats (Family: Bathyergidae). J Biol Rhythm 18:481–490

Oosthuizen MK, Bennett NC, Cooper HM (2005) Fos expression in the suprachiasmatic nucleus in response to light stimulation in a solitary and social species of African mole-rat (family Bathyergidae). Neuroscience 133:555–560

Peichl L (2005) Diversity of mammalian photoreceptor properties: adaptations to habitat and lifestyle? Anat Rec A 287:1001–1012

Peichl L, Němec P, Burda H (2004) Unusual cone and rod properties in subterranean African mole-rats (Rodentia, Bathyergidae). Eur J Neurosci 19:1545–1558

Peichl L, Chávez AE, Ocampo A, Mena W, Bozinovic F, Palacios AG (2005) The eye and vision in the subterranean rodent cururo (*Spalacopus cyanus*, Octodontidae). J Comp Neurol 486:197–208

Perry VH (1981) Evidence for an amacrine cell system in the ganglion cell layer of the rat retina. Neuroscience 6:931–944

Perry VH, Henderson Z, Linden R (1983) Postnatal changes in retinal ganglion cell and optic axon populations in the pigmented rat. J Comp Neurol 219:356–368

Poduschka W (1978) Zur Frage der Wahrnehmung von Lichtreizen durch die Mullratte, *Cryptomys hottentotus* (Lesson, 1826). Säugetierk Mitt 1978:269–274

Prusky GT, Harker KT, Douglas RM, Whishaw IQ (2002) Variation in visual acuity within pigmented, and between pigmented and albino rat strains. Behav Brain Res 136:339–348

Quilliam TA (1966) The problem of vision in the ecology of *Talpa europaea*. Exp Eye Res 5:63–78

Rado R, Gev H, Goldman BD, Terkel J (1991) Light and circadian activity in the blind mole-rat. In: Riklis E (ed) Photobiology. Plenum Press, New York, pp 581–589

Rado R, Bronchti G, Wollberg Z, Terkel J (1992) Sensitivity to light of the blind mole rat: behavioural and neuroanatomical study. Isr J Zool 38:323–331

Rehkämper G, Necker R, Nevo E (1994) Functional anatomy of the thalamus in the blind mole rat *Spalax ehrenbergi*: an architectonic and electrophysiologically controled tracing study. J Comp Neurol 347:570–584

Rhoades RW, Hsu L, Parrfett G (1979) An electron microscopic analysis of the optic nerve in the golden hamster. J Comp Neurol 186:491–504

Riccio AO, Goldman BD (2000a) Circadian rhythms of body temperature and metabolic rate in naked mole-rats. Physiol Behav 71:15–22

Riccio AO, Goldman BD (2000b) Circadian rhythms of locomotor activity in naked mole-rats (*Heterocephalus glaber*). Physiol Behav 71:1–13

Richter TA, Malpaux B, Fleming PA, Molteno AJ, Bennett NC (2003) Melatonin secretion in a strictly subterranean mammal, the Damaraland mole-rat (*Cryptomys damarensis*). J Zool Lond 261:313–319

Schöttner K, Oosthuizen MK, Broekman M, Bennett NC (2006) Circadian rhythms of locomotor activity in the Lesotho mole-rat, Cryptomys hottentotus subspecies from Sani Pass, South Africa. Physiol Behav 89:205–212

Sefton AJ, Dreher B (1995) Visual system. In: Paxinos G (ed) The rat nervous system. Academic Press, San Diego, pp 833–880

Simpson JI (1984) The accessory optic system. Ann Rev Neurosci 7:13–41

Tay D, So K-F, Jen LS, Lau KC (1986) The postnatal development of the optic nerve in hamsters: an electron microscopic study. Dev Brain Res 30:268–273

Tiao TC, Blakemore C (1976) Regional specializations in the golden hamster's retina. J Comp Neurol 168:439–458

Tobler I, Herrmann M, Cooper HM, Negroni J, Nevo E, Achermann P (1998) Rest-activity rhythm of the blind mole rat *Spalax ehrenbergi* under different lighting conditions. Behav Brain Res 96:173–183

Toga AW, Collins RC (1981) Metabolic response of optic centers to visual stimuli in the albino rat: anatomical and physiological considerations. J Comp Neurol 199:443–464

Vann SD, Aggleton JP (2002) Extensive cytotoxic lesions of the rat retrosplenial cortex reveal consistent deficits on tasks that tax allocentric spatial memory. Behav Neurosci 116:85–94

Vasicek CA, Malpaux B, Fleming PA, Bennett NC (2005a) Melatonin secretion in the Mashona mole-rat, *Cryptomys darlingi* – influence of light on rhythmicity. Physiol Behav 83:689–697

Vasicek CA, Oosthuizen MK, Cooper HM, Bennett NC (2005b) Circadian rhythms of locomotor activity in the subterranean Mashona mole rat, *Cryptomys darlingi*. Physiol Behav 84:181–191

Vuillez P, Herbin M, Cooper HM, Nevo E, Pevet P (1994) Photic induction of Fos immunoreactivity in the suprachiasmatic nuclei of the blind mole rat (*Spalax ehrenbergi*). Brain Res 654:81–84

Wässle H (2004) Parallel processing in the mammalian retina. Nature Rev Neurosci 5:747–757

Wegner RE, Begall S, Burda H (2006) Light perception in "blind" subterranean Zambian mole-rats. Anim Behav 72:1021–1024

Werner SJ, Nolte DL, Provenza FD (2005) Proximal cues of pocket gopher burrow plugging behavior: influence of light, burrow openings, and temperature. Physiol Behav 85:340–345

Williams GA, Calderone JB, Jacobs GH (2005) Photoreceptors and photopigments in a subterranean rodent, the pocket gopher (*Thomomys bottae*). J Comp Physiol A 191:125–134

Williams RW, Strom RC, Rice DS, Goldowitz D (1996) Genetic and environmental control of variation in retinal ganglion cell number in mice. J Neurosci 16:7193–7205

Xiao J, Levitt JB, Buffenstein R (2006) The use of a novel and simple method of revealing neural fibers to show the regression of the lateral geniculate nucleus in the naked mole-rat (*Heterocephalus glaber*). Brain Res 1077:81–89

12 Magnetic Compass: A Useful Tool Underground

Regina E. Moritz, Hynek Burda, Sabine Begall, Pavel Němec

12.1 Introduction

Living aboveground offers animals diverse cues for spatial orientation, but the underground picture is less colourful: besides darkness, subterranean rodents face restriction in useful orientation cues such as odors or sounds (Burda et al. 1990a). Target and landmarks are not directly perceivable and, as a consequence, distant orientation is heavily impeded. Subterranean mammals need to solve several major tasks relevant to spatial orientation. 1) They have to orientate quickly and efficiently in their dynamically changing three-dimensional burrow system, in order to frequent nest, food chambers, latrines, and harvesting grounds. 2) They must keep the course when digging longer foraging and dispersing tunnels. Straight tunnelling conserves energy because the animals do not search in the same area twice. 3) They need to restore and interconnect damaged burrows, effectively bypass obstacles, etc. 4) Animals temporarily leaving their burrows while foraging or searching for mates aboveground need to find their way back home.

Landmark-independent navigation, such as path integration (also called dead reckoning), is likely used for orientation within a familiar burrow system or for short-distance tasks. In this form of navigation, an animal uses idiothetic cues, i.e. internal movement cues based on proprioceptive and vestibular information from sensory flow, or efferent copies of movement commands. However, depending exclusively on self-generated signals, path

Regina E. Moritz: Department of General Zoology, Institute for Biology, University of Duisburg-Essen, Campus Essen, Universitaetsstrasse 5, 45117 Essen, Germany, E-mail: regina.moritz@uni-due.de

Hynek Burda: Department of General Zoology, Institute for Biology, University of Duisburg-Essen, Campus Essen, Universitaetsstrasse 5, 45117 Essen, Germany

Sabine Begall: Department of General Zoology, Institute for Biology, University of Duisburg-Essen, Campus Essen, Universitaetsstrasse 5, 45117 Essen, Germany

Pavel Němec: Department of Zoology, Charles University, Viničná 7, Praha 2, 12844 Czech Republic; Dr. Senckenbergische Anatomie (Institute for Anatomy III), J. W. Goethe University, Theodor Stern-Kai 7, D-60590 Frankfurt a. M., Germany

Subterranean Rodents: News from Underground
S. Begall, H. Burda, C.E. Schleich (Eds.)
© Springer-Verlag Berlin Heidelberg 2007

integration is severely constrained by the rapid accumulation of errors (Etienne et al. 1988; Benhamou et al. 1990). Therefore, an external directional reference is inevitable for navigation over longer tracks. For successful navigation within the complex burrow maze, a subterranean rodent thus requires, besides idiothetic cues, a compass sense and also a mental representation of its environment (a cognitive map) that can be used for spatial navigation and spatial memory (reviewed in Etienne and Jeffery 2004). The questions a subterranean mammal must address when navigating are "Where am I?", "Where do I want to go?", and "How do I get there from here?" In a monotonous, dark world, the Earth's magnetic field provides a relatively constant and reliable source of directional, and perhaps also positional, information. Hence, it may not be surprising that the first evidence for a mammalian magnetic compass comes from a subterranean rodent, the Ansell's mole-rat (*Fukomys anselli*) (Burda et al. 1990b; in previous studies denoted as *Cryptomys hottentotus* or *C. anselli*, cf. Kock et al. 2006). Here, we show that this species is still used as a model for refining and continuously expanding the knowledge on the mechanisms of magnetic orientation and magnetoreception in mammals.

12.2 Available Information: From Earth to Animal

Magnetoreception is defined by the ability to sense/perceive magnetic cues (intensity and/or the orientation of the local geomagnetic field) and transfer them to the nervous system, which extracts, processes, and interprets the relevant information. An individual can obtain differential information either from the spatially varying gradients of *intensity* of the magnetic field (strength expressed in μT) or *inclination* (i.e. the angle between the magnetic field vector and the horizontal plane), or from directional cues provided by horizontal *polarity* (i.e. the direction of the magnetic field lines pointing towards magnetic North, as visualised by a compass needle).

By means of an *inclination compass*, an animal can perceive the course of the magnetic axis by detecting the field lines and by using the inclination angle to determine the direction the animal is heading to, i.e. polewards or equatorwards. With a *polarity compass*, an animal can distinguish between magnetic North and South by directly detecting the field lines' direction. With both compass types, an animal can determine the *azimuth*, i.e. the deviation between magnetic North and the horizontal direction of its own movement. Keeping the azimuth constant thus secures keeping the movement direction. Combining the information on the locally varying gradients in inclination and magnetic field intensity, an animal can determine *magnetic coordinates* of any particular location within a given corridor. While total intensity and/or inclination may provide positional (map) information, it is the magnetic vector that provides the necessary directional (compass) information (reviewed in, e.g. Wiltschko and Wiltschko 1995, 2005, 2006; Bingman and Cheng 2005; Lohmann and Lohmann, 2006; Phillips et al. 2006).

In terrestrial vertebrates, magnetoreception has been reported in, e.g. newts, turtles, migratory and homing birds (reviewed in, e.g. Wiltschko and Wiltschko 1995). Among mammals, magnetic compass orientation has been unambiguously demonstrated in two species of subterranean rodents, the African Ansell's mole-rat (*F. anselli*) (Burda et al. 1990b; Marhold et al. 1997a,b) and the blind mole-rat (*Spalax ehrenbergi* superspecies) from Israel (Burda et al. 1991; Kimchi and Terkel 1999, 2001; Kimchi et al. 2004), and in two species of epigeic rodents: the Siberian hamster (*Phodopus sungorus*) (Deutschlander et al. 2003) and the laboratory mouse (Muheim et al. 2006). Earlier studies of magnetic orientation in rodents provided ambiguous and in some cases questionable results (Mather and Baker 1981; Madden and Phillips 1987; Sauvé 1988; August et al. 1989).

12.3 Compass Mode: From Behavioural Experiment to Proof

12.3.1 Experimental Design

Studying a putative magnetic compass, the researcher receives confirmation from animals' responses to artificial shifts of magnetic North with respective predictable directional changes. Classical experiments make use of *homing orientation* (reviewed in, e.g. Wiltschko and Wiltschko 1995). Whereas plenty of evidence exists for homing abilities in animals of diverse taxa, this experimental approach poses severe limits when examining magnetic compass orientation: homing is realized over longer distances; negative results need to be considered in a motivational context; and being a complex multifactor task, homing can hardly provide evidence for the exclusive use of a particular orientation mechanism. Experimental refinement of this method, practicable in smaller organisms also under laboratory conditions, is based on "simulated magnetic displacements", i.e. when the animal is exposed to a defined artificial magnetic field characterizing the opposite side of its "home", corresponding changes in the direction of its homing orientation are expected (cf. Fischer et al. 2001; Phillips et al. 2002; Boles and Lohmann 2003). At least males of seasonally breeding solitary subterranean mammals, which seek for mates over longer distances (e.g. the silvery mole-rat, *Heliophobius argenteocinereus*; Šumbera et al., this volume), show good homing abilities presumably based on a magnetic compass. Homing abilities have been demonstrated also in diverse other species of subterranean mammals (cf. Burda et al. 1990a). Nevertheless, examining the proximate mechanisms of homing would be technically very difficult both in the field and in the laboratory, for instance because of the limited availability and probably seasonally limited motivation of these animals.

The real treat for scientists is *spontaneous behaviour*, specifically innate preference for a certain direction displayed by a migratory direction or positioning of a nest. The experimental set-up in the magnetic orientation assay of rodents, first applied in Ansell's mole-rats (Burda et al. 1990b), is simple: mole-rats are placed in a *circular arena* with scattered nesting material and food items. In an undisturbed geomagnetic field (in the given case characterized by 66° inclination and 46 μT), the mole-rats preferably place their nest in the south-eastern sector of the arena. Magnetic intensity and polarity in the arena can be manipulated by a pair of *Helmholtz coils*, by e.g. shifting magnetic north by a specified angle. The mole-rats place their nests according to the altered magnetic field and relative to their southern magnetic preference direction. This experimental design, which in some studies has been modified to a radial 8-armed-maze with terminal nest boxes, has proven to be useful to demonstrate magnetic compass orientation, also in the blind mole-rat (Burda et al. 1991; Kimchi and Terkel 2001) and in the Siberian hamster (Deutschlander et al. 2003).

A modified arena-assay to test whether animals use magnetic cues for orientation when digging in order to hide or escape failed in South American tuco-tucos (*Ctenomys talarum*) (Schleich and Antinuchi 2004) and coruros (*Spalacopus cyanus*) (Begall, unpublished). These results, however, do not conclusively exclude the presence of a magnetic compass in these animals, since it is possible that they do not rely on the magnetic compass in such a stressful situation. Standard nest-building experiments proved unfeasible in coruros, because no nest at all was built in most of the trials (82%).

A constructive and innovative modification of this experimental arena design involves *conditioning (learning)*: the nest position in an arena is imposed upon the animals using a fixed nest-box, and it is then tested whether the animals choose their nest place or nest box according to the previously inflicted (learned) relationship to the magnetic field, also after having been released into a new (cleaned) arena and/or after the magnetic field has been manipulated. In this way, magnetic compass orientation has been shown in the Siberian hamster (Deutschlander et al. 2003) and in the laboratory mouse (Muheim et al., submitted). This experimental paradigm certainly harbours a lot of potential, also in the study of spatial orientation in subterranean mammals.

Maze experiments have so far seldom been applied in the study of magnetic orientation of subterranean rodents. At least Ansell's mole-rats behave in a maze differently than e.g. laboratory rats (Burda, unpublished observations). They learn a maze after the first run, but start to make "false errors", driven by exploration, in later runs. Furthermore, explorative behaviour or xenophobia may be species and gender specific, making the evaluation of maze experiments a more complex task (Heth et al. 1987; Burda, unpublished observations). Nevertheless, in the blind mole-rat, magnetic cues also play a role in maze navigation and path integration (Kimchi et al. 2004).

Indications for magnetic compass orientation may also be derived from *field observations*. Subterranean mammals in natural habitats with poor food supply tend to build linearly arranged burrow systems with a long, straight main tunnel, the so-called runway, and a nest positioned frequently rather eccentrically in respect to the longest axis of the burrow system (cf. Eloff 1951; Heth et al. 2002 and literature cited therein). It can be speculated that in a uniform habitat, without geomorphologic and other features, such as water streams, slopes, rocks, roads, trees, fields or neighbours, that might potentially canalise burrowing direction, animals may project their spontaneous directional preference into a predictable orientation of the burrow system. Indeed, the first hints for possible magnetic compass orientation in the Ansell's mole-rat (Burda 1987) were derived from a few then available maps of burrow systems, which exhibited - apparently incidentally - a north-south orientation of their longest axis. Lovegrove et al. (1992) and Schleich and Antinuchi (2004) could not confirm any prevailing directionality of burrows in *Fukomys damarensis* and *Ctenomys talarum*, respectively. Apart from the fact that the studies did not include circular statistics as a method to assess the orientation, the random or directional patterns observed may be due to the problem that apparently established old systems were examined, where the original primary tunnel might not have been recognizable anymore (or perhaps not even existed anymore). Burrow systems are not static, but are steadily reworked and rearranged (cf. Šumbera et al. 2003). We suggest the following field experiment: subterranean rodents are released in a uniform habitat (see above), and the direction of single primary tunnels is checked some days later. Note that there are only few published reports on mapped burrow systems, that compass orientation is usually not recorded, and that studies usually report old, established burrow systems. No one has been interested so far to uncover, map - and publish - the architecture of short burrow systems in statu nascendi.

12.3.2 Character of the Magnetic Compass

Consistent with the subterranean lifestyle, the magnetic compass of Ansell's mole-rats has been specified as a *light-independent polarity compass*, as neither light nor artificial shifts of inclination affected the preferred south-eastern nesting direction (Marhold et al. 1997a). Similarly, the magnetic compass of the blind mole-rat is light-independent (Kimchi and Terkel 2001) and, although this point has not been addressed explicitly, it is highly probable that its compass is also polarity-based.

The directional positioning of nests in mole-rats may express an alignment behaviour towards the magnetic field's polarity and per se does not prove that the animals use the magnetic compass also in navigation. However, the findings that the blind mole-rat uses magnetic cues as a stable, external reference (comparable to visual cues in sighted mammals) for orientation within complex maze systems (Kimchi et al. 2004) as well as recent neuroanatomical findings (described below) indicate that the magnetic compass of mole-rats is involved in both path integration and navigation.

12.4 Transduction Mechanisms: From Signal to Receptor

12.4.1 Physical Models and Behavioural Evidence

Extensive body of evidence demonstrates that two starkly different magnetoreceptive mechanisms have evolved in terrestrial vertebrates: (1) a light-dependent mechanism involving a photo-induced radical pair reaction occurring in a specialized photoreceptor, and (2) a light-independent mechanism involving particles of biogenic magnetite (reviewed in, e.g. Kirschvink et al. 2001; Ritz et al. 2002; Johnsen and Lohmann 2005; Wiltschko and Wiltschko 2005, 2006). It needs to be remarked, however, that despite the behavioural evidence for these two systems, the primary receptors have not yet been identified in any animal with certainty.

Light-dependent magnetoreception, described by the so-called *Radical-Pair-Model* (RPM; Ritz et al. 2000), is based on a set of biochemical reactions in specialized photoreceptor proteins involving excitation of electrons by photon absorption and consecutive formation of radical pair intermediates. Magnetic field alignments alter the electron energy states that drive phototransduction, and thus affect the response to light (Ritz et al. 2000). The inclination compass of birds and amphibians presumably bases upon this mechanism (cf. Deutschlander et al. 1999b; Wiltschko and Wiltschko 2002).

Consistent with the model, both compass orientation of birds and newts depend on wavelength and/or light intensity (Phillips and Borland 1992a,b; Wiltschko and Wiltschko 2002). In birds, recent studies strongly support the RPM-model, showing a disruption of their magnetic orientation by weak, oscillating radio frequency fields in the MHz-range; these affect energy states in radical pair systems (Ritz et al. 2004; Thalau et al. 2005). Primary receptors of this system are supposedly located in the retina of birds (cf. Wiltschko and Wiltschko 2002; Möller et al. 2004; Mouritsen et al. 2004) or in the pineal organ in newts (Deutschlander et al. 1999a,b; Phillips et al. 2001).

Light-independent magnetoreception supposedly implicates magnetite particles (cf. Kirschvink and Gould 1981; Shcherbakov and Winklhofer 1999; Kirschvink et al. 2001; Davila et al. 2005). *Single-domain magnetite* crystals ($\sim$0.50 μm) act as permanently magnetised bar magnets twisting into alignment with the magnetic field if allowed to rotate freely. A chain of single-domain crystals (as found in the rainbow trout, see below) may exert torque or pressure on secondary receptors such as stretch receptors, hair cells or mechanoreceptors. Alternatively, rotation of intracellular crystals might open ion channels directly, if cytoskeletal filaments connect the crystals to the channels (Kirschvink et al. 2001). *Superparamagnetic magnetite* crystals ($<$0.05 μm) do not have a permanent magnetic moment and so cannot physically rotate into alignment with the Earth's field; in an external field, they nevertheless develop a magnetic moment. In an Earth-strength magnetic field, clusters of superparamagnetic particles (as found in the pigeon, see below) can attract or repel each other, depending on the orientation of the external field (Davila et al. 2003). These interactions have the potential to deform the matrix (e.g. the cell membrane) in which they are embedded. Moreover, recent simulations and experiments have demonstrated that a group of superparamagnetic clusters self-assembles into a chain-like structure that behaves like a compass needle in an external field (Davila et al. 2005).

It should be noted, that evidently these two distinct magnetoreceptive systems occur complementary in newts and birds: apparently, these animals use the light-dependent radical-pair mechanism to derive compass information, and a light-independent magnetite-based mechanism to derive map information (cf. Phillips 1986; Phillips et al. 2002; Munro et al. 1997a,b; Brassart et al. 1999).

As expected in subterranean rodents, however, living and orientating in darkness, the hitherto determined characteristics of the magnetic compass in Ansell's mole-rats are consistent with the magnetite-based magnetoreception model. Evidence for this kind of sensation is supported by the following findings. Compass orientation of Ansell's mole-rats is: 1) light-independent (Marhold et al. 1997a); 2) sensitive to the magnetic field's polarity (Marhold et al. 1997a); 3) disrupted by a brief strong magnetic pulse designed to alter (re-magnetize) single-domain magnetite or to affect superparamagnetic particles (Marhold et al. 1997b); 4) not disrupted by very weak oscillating high-frequency fields disturbing avian compass orientation (Thalau et al. 2006). While the latter experiment per se does not prove that the magnetic compass of mole-rats is magnetite-based, it indicates that it is not based on RPM. In any case, conclusions on the light-independent magnetic compass in mole-rats as a representative of the mammalian magnetic compass must be drawn cautiously. The magnetite-based light-independent compass in mole-rats may alternatively display, along with the reduced visual system, an adaptation to the dark subterranean environment. An interesting parallel may be drawn with blind salamanders inhabiting aphotic caves: they are not expected to have a light-dependent magnetic compass although this type of mechanisms occurs in terrestrial salamanders like the Eastern red-spotted newt (Phillips, personal communication). It will thus be of high interest to study the mechanisms of magnetoreception also in surface-dwelling rodents.

12.4.2 Histological and Neurobiological Evidence

Magnetite particles associated with afferent trigeminal terminals, specifically the ophthalmic nerve branches, have been found in the upper beak tissue of birds (cf. Hanzlik et al. 2000; Williams and Wild 2001; Fleissner et al. 2003), and also within the olfactory lamellae of the rainbow trout (Walker et al. 1997; Diebel et al. 2000). Congruently, impairment experiments involving anaesthesia and bilateral section of the ophthalmic nerve confirmed that this nerve might well be the carrier of magnetic field information to the brain (Beason and Semm 1996; Mora et al. 2004). Electrophysiological recordings (Semm and Beason 1990; Walker et al. 1997) as well as conditioned-choice experiments (Walker et al. 1997; Mora et al. 2004) suggested further that the magnetoreceptors associated with the ophthalmic nerve do not participate in the compass, but instead yield map information.

In Ansell's mole-rats, our preliminary histological (Burda, unpublished) and experimental behavioural studies (Wegner et al. 2006) suggest that the cornea, i.e. a paired, highly mechano-sensitive ocular structure innervated by the ophthalmic nerve, may be the seat of magnetite-based receptors. In contrast, Cernuda-Cernuda et al. (2003) reported findings of crystalloid bodies in the inner segments of retinal photoreceptors of the Ansell's mole-rat. The authors interpreted these structures as potential magnetite grains, suggesting the retinal photoreceptors as the respective magnetite-based structure.

12.5 Neuronal Processing: From Receptor to Brain

In contrast to the wealth of information on the role that the magnetic sense plays in animal orientation, on its distribution across animal taxa, and on its behavioural characterization, our knowledge of the neural substrate subserving magnetic orientation remains meagre (reviewed in Johnsen and Lohmann 2005; Němec et al. 2005; Wiltschko and Wiltschko 2005): only a fistful of studies have tried to shed light on the neural aspects of magnetoreception in mammals.

Early electrophysiological studies have demonstrated the presence of magneto-responsive units in the pineal organ of the guinea pig (Semm et al. 1980), the laboratory rat (Reuss et al. 1983), and the Mongolian gerbil *Meriones unguiculatus* (Stehle et al. 1988). Magnetic stimulation was reported to affect pineal melatonin synthesis in the laboratory rat (Olcese et al. 1985; Reuss and Olcese 1986; Welker et al. 1983). Interestingly, these effects appeared to be light- and vision-dependent (Olcese et al. 1985, 1988; Reuss and Olcese 1986), indicating the involvement of a photoreceptor-based magnetoreception mechanism.

Recently, functional neuroanatomical mapping based on monitoring stimulus-evoked expression of inducible transcription factors has been introduced into magnetoreception research (Němec et al. 2001). This method can be used to identify neurons specifically activated by magnetic stimuli, and it also offers cellular resolution and the possibility to screen for neuronal activation throughout the central nervous system (reviewed in Němec et al. 2005). Respective experiments performed in Ansell's mole-rats provided evidence for magnetic input to the superior colliculus (Němec et al. 2001). Interestingly, homologous brain regions of birds and mammals seem to co-opt independently into magnetic information processing: while in birds, magneto-responsive neurons could be identified in superficial layers of the optic tectum receiving a robust visual input (Semm and Demaine 1986); magneto-responsive neurons in mole-rats, on the other hand, were found within the intermediate layers of the superior colliculus dominated by trigeminal input (Němec et al. 2001) – visual input to the superior colliculus is extremely reduced in these rodents (Němec et al. 2004). Since it has been repeatedly demonstrated that magnetite-based magnetoreception is associated with the trigeminal nerve system, the data provided by Němec et al. (2001) indirectly support the hypothesis of a magnetite-based compass mechanism in Ansell's mole-rats.

In addition, recently completed experiments (Němec et al., in preparation) show that both the hippocampus (dentate gyrus and fields CA1 and CA3) and some brain centres harbouring *head direction cells* contain neurons that are responsive to magnetic stimuli. The hippocampus is generally considered to be the site of the cognitive map of the animal's environment (O'Keefe and Nadel 1978; Sharp 2002; Jeffery 2003). Pyramidal neurons of the CA1 and the CA3 serve as *place cells*, i.e. neurons that fire when the animal occupies a specific location in a particular environment (reviewed in, e.g. Muller 1996); such cells with place correlates also lie in the dentate gyrus (Jung et al. 1994). Complementary to this, *head direction cells* fire when the animal's head faces, in all environments, a particular direction (reviewed in, e.g. Taube 1998; Sharp et al. 2001). Thus, *place cells* are the keystone of the neural machinery generating an abstract representation of the animal's spatial surroundings (a map), while *head direction cells* are involved in the moment-by-moment representation of the animal's current heading (a compass). These results indicate that subterranean mole-rats may not only have a magnetic compass that derives directional information, as has been hitherto demonstrated by behavioural experiments, but that they also use magnetic cues to acquire magnetic map information.

Despite these highly interesting results, our knowledge about magnetoreception, i.e., what happens where in the rodent neuronal circuits, remains fragmentary and is far from complete, particularly regarding the high number of (subterranean) rodent species that have not yet been examined in this respect.

Acknowledgements. Our work was supported by grants from the DFG (BU 717/10-3 to HB and REW; OE 217/4-2 to PN), the Czech Science Foundation (206/03/0638 to PN), the Ministry of Education, Youth and Sport of the Czech Republic (0021620828 to PN) and the Lise-Meitner-Habilitationsprogramm (to SB). We wish to thank R. Muheim for constructive comments on the manuscript.

References

August PV, Ayvazian SG, Anderson JTG (1989) Magnetic orientation in a small mammal, *Peromyscus leucopus*. J Mammal 70:1–9

Beason RC, Semm P (1996) Does the avian ophthalmic nerve carry magnetic navigational information? J Exp Biol 199:1241–1244

Benhamou S, Sauve JP, Bovet P (1990) Spatial memory in large-scale movements – efficiency and limitations of the egocentric coding process. J Theor Biol 145:1–12

Bingman VP, Cheng K (2005) Mechanisms of animal global navigation: comparative perspectives and enduring challenges. Ethol Ecol Evol 17:295–318

Boles LC, Lohmann KJ (2003) True navigation and magnetic maps in spiny lobsters. Nature 412:60–63

Brassart J, Kirschvink JL, Phillips JB, Borland SC (1999) A magnetite-based map component of homing in the Eastern red-spotted newt. J Exp Biol 202:3155–3160

Burda H (1987) Magnetische Navigation bei den Graumullen, *Cryptomys hottentotus* (Bathyergidae)? Z Säugetierk Suppl 61:12

Burda H, Bruns V, Müller M (1990a) Sensory adaptations in subterranean mammals. In: Nevo E, Reig OA (eds) Evolution of subterranean mammals at the organismal and molecular levels. Wiley-Liss, New York, pp 269–293

Burda H, Marhold S, Westenberger T, Wiltschko R, Wiltschko W (1990b) Magnetic compass orientation in the subterranean rodent *Cryptomys hottentotus* Bathyergidae. Experientia 46:528–530

Burda H, Beiles A, Marhold S, Simson S, Nevo E, Wiltschko W (1991) Magnetic orientation in subterranean mole rats of the superspecies *Spalax ehrenbergi*: experiments, patterns and memory. Isr J Zool 37:182–183

Cernuda-Cernuda R, Garcia-Fernandez JM, Gordijn MCM, Bovee-Geurts PHM, DeGrip WJ (2003) The eye of the african mole-rat *Cryptomys anselli*. to see or not to see? Eur J Neurosci 17:709–720

Davila AF, Fleissner G, Winklhofer M, Petersen N (2003) A new model for a magnetoreceptor in homing pigeons based on interacting clusters of superparamagnetic magnetite. Phys Chem Earth 28:647–652

Davila AF, Winklhofer M, Sheherbakov V, Petersen N (2005) Magnetic pulse affects a putative magnetoreceptor mechanism. Biophysics J 89:56–63

Deutschlander ME, Borland SC, Phillips JB (1999a) Extraocular magnetic compass in newts. Nature 400:324–325

Deutschlander ME, Phillips JB, Borland SC (1999b) The case for a light-dependent magnetoreception mechanism in animals. J Exp Biol 202:891–908

Deutschlander ME, Freake MJ, Borland SC, Phillips JB, Anderson LE, Wilson BW (2003) Learned magnetic compass orientation by the Siberian hamster, *Phodopus sungorus*. Anim Behav 65:779–786

Diebel CE, Proksch R, Green CR, Neilson P, Walker MM (2000) Magnetite defines a vertebrate magnetoreceptor. Nature 406:299–302

Eloff G (1951) Orientation in the mole-rat *Cryptomys*. Brit J Psychol 1/2:134–145

Etienne AS, Jeffery KJ (2004) Path integration in mammals. Hippocampus 14:180–192

Etienne AS, Maurer R, Saucy F (1988) Limitations in the assessment of path dependant information. Behaviour 106:81–111

Fischer JH, Freake MJ, Borland SC, Phillips JB (2001) Evidence for the use of magnetic map information by amphibian. Anim Behav 62:1–10

Fleissner G, Holtkamp-Rötzler E, Hanzlik M, Winklhofer M, Fleissner G, Petersen N, Wiltschko W (2003) Ultrastructural analysis of a putative magnetoreceptor in the beak of homing pigeons. J Comp Neurol 458:350–360

Hanzlik M, Heunemann C, Holtkamp-Rötzler E, Winkelhofer M, Petersen N, Fleissner G (2000) Superparamagnetic magnetite in the upper beak tissue of homing pigeons. BioMetals 13:325–331

Heth G, Nevo E, Beiles A (1987) Adaptive exploratory behaviour: differential patterns in species and sexes of subterranean mole rats. Mammalia 51:27–37

Heth G, Todrank J, Begall S, Braude S, Koch R, Zilbiger Y, Nevo E, Burda H (2002) Odour-guided foraging: "blind" subterranean rodents do not search "blindly". Behav Ecol Sociobiol 52:53–58

Jeffery KJ (2003) The neurobiology of spatial behaviour. Oxford Univ Press, Oxford

Johnsen S, Lohmann KJ (2005) The physics and neurobiology of magnetoreception. Nat Neurosci Rev 6:703–712

Jung M, Wiener SI, McNaughton BL (1994) Comparison of spatial firing characteristics of units in dorsal and ventral hippocampus of the rat. J Neurosci 14:7347–7356

Kimchi T, Terkel J (1999) Magnetic orientation by blind mole rats in a labyrinth. Isr J Zool 45:318

Kimchi T, Terkel J (2001) Magnetic compass orientation in the blind mole rat *Spalax ehrenbergi*. J Exp Biol 204:751–758
Kimchi T, Etienne AS, Terkel J (2004) A subterranean mammal uses the magnetic compass for path integration. Proc Natl Acad Sci USA 101:1105–1109
Kirschvink JL, Gould JL (1981) Biogenic magnetite as a basis for magnetic field detection in animals. BioSystems 13:181–201
Kirschvink JL, Walker MM, Diebel CE (2001) Magnetite-based magnetoreception. Curr Op Neurobiol 11:462–467
Kock D, Ingram CM, Frabotta LJ, Honeycutt RL, Burda H (2006) On the nomenclature of Bathyergidae and *Fukomys* n. gen. Mammalia: Rodentia. Zootaxa 1142:51–55
Lohmann KJ, Lohmann CMF (2006) Sea turtles, lobsters, and oceanic magnetic maps. Mar Freshwater Behav Physiol 39:49–64
Lovegrove BG, Körtner Gü, Körtner Ge (1992) The magnetic compass orientation of the burrows of the Damara mole-rat *Cryptomys damarensis* (Bathyergidae). J Zool Lond 226:631–633
Madden RM, Phillips JB (1987) Failure to demonstrate magnetic field sensitivity in two species of small mammals. Anim Learn Behav 15:130–134
Marhold S, Wiltschko W, Burda H (1997a) A magnetic polarity compass for direction finding in a subterranean mammal. Naturwiss 84:421–423
Marhold S, Burda H, Kreilos I, Wiltschko W (1997b) Magnetic orientation in common molerats from Zambia. In: Orientation and navigation: birds, humans and other animals. Paper No 5. Royal Institute of Navigation, Oxford
Mather JG, Baker RR (1981) Magnetic sense of direction in woodmice for route-based navigation. Nature 291:152–155
Möller A, Sagasser S, Wiltschko W, Schierwater B (2004) Retinal cryptochrome in a migratory passerine bird: a possible transducer for the avian magnetic compass. Naturwiss 91:585–588
Mora CV, Davison M, Wild JM, Walker MM (2004) Magnetoreception and its trigeminal mediation in the homing pigeon. Nature 432:508-511
Mouritsen H, Janssen-Bienhold U, Liedvogel M, Feenders G, Stalleicken J, Dirks P, Weiler R (2004) Cryptochromes and activity markers co-localize in bird retina during magnetic orientation. Proc Natl Acad Sci USA 101:14294–14299
Muheim R, Edgar NM, Sloan KA, Phillips JB (2006) Magnetic compass orientation in C57BL/6 mice. Learn Behav 34:366–373
Muller R (1996) A quarter of a century of place cells. Neuron 17:813–822
Munro U, Munro JA, Phillips JB, Wiltschko R, Wiltschko W (1997a) Evidence for a magnetite-based navigational 'map' in birds. Naturwiss 84:26–28
Munro U, Munro JA, Phillips JB, Wiltschko W (1997b) Effect of wavelength of light and pulse magnetization on different magnetoreception systems in a migratory bird. Austr J Zool 45:189–198
Němec P, Altmann J, Marhold S, Burda H, Oelschläger HA (2001) Magnetotopic organization in the superior colliculus: involvement of a brain structure in magnetoreception. Science 294:366–368
Němec P, Burda H, Peichl L (2004) Subcortical visual system of the African mole-rat *Cryptomys anselli*: to see or not to see? Eur J Neurosci 203:757–768
Němec P, Burda H, Oelschläger HA (2005) Towards the neural nature of magnetoreception: a neuroanatomical approach. Naturwiss 92:151–157
O'Keefe J, Nadel L (1978) The hippocampus as a cognitive map. Oxford Univ Press, Oxford
Olcese J, Reuss J, Vollrath L (1985) Evidence for the involvement of the visual system in mediating magnetic field effects on pineal melatonin synthesis in the rat. Brain Res 333:382–384

Olcese J, Reuss S, Stehle J, Steinlechner S, Vollrath L (1988) Responses of the mammalian retina to experimental alteration of the ambient magnetic field. Brain Res 448:325–330
Phillips JB (1986) Two magnetoreception pathways in a migratory salamander. Science 233:765–767
Phillips JB, Borland SC (1992a) Behavioral evidence for the use of a light-dependent magnetoreception mechanism by a vertebrate. Nature 359:142–144.
Phillips JB, Borland SC (1992b) Magnetic compass orientation is eliminated under near-infrared light in the eastern red-spotted newt *Notophthalmus viridescens*. Anim Behav 44:796–797
Phillips JB, Deutschlander ME, Freake MJ, Borland SC (2001) The role of extraocular photoreceptors in newt magnetic compass orientation: evidence for parallels between light-dependent magnetoreception and polarized light detection in vertebrates. J Exp Biol 204:2543–2552
Phillips JB, Freake MJ, Fischer JH, Borland SC (2002) Behavioral titration of a magnetic map coordinate J Comp Physiol A 188:157–160
Phillips JB, Schmidt-Koenig K, Muheim R (2006) True navigation: sensory basis of gradient maps. In: Brown MF, Cook RG (eds) Animal spatial cognition: comparative, neural and computational approaches. Comparative Cognition Press: *http://www.pigeon.psy.tufts.edu/index.php?content=animal_spatial_cognition*
Reuss S, Olcese J (1986) Magnetic field effects on rat pineal gland: role of retinal activation by light. Neurosci Lett 64:97–101
Reuss S, Semm P, Vollrath L (1983) Different types of magnetically sensitive cells in the rat pineal gland. Neurosci Lett 40:23–26
Ritz T, Adem S, Schulten K (2000) A model for photoreceptor-based magnetoreception in birds. Biophys J 78:707–718
Ritz T, Phillips JB, Dommer DH (2002) Shedding light on vertebrate magnetoreception. Neuron 34:503–506
Ritz T, Thalau P, Phillips J, Wiltschko R, Wiltschko W (2004) Resonance effects indicate a radical-pair mechanism for avian magnetic compass. Nature 429:177–180
Sauvé JP (1988) Analyse de l'orientation initiale dans une expérience de retour au gîte chez le mulot, *Apodemus sylvaticus*. Sci Tech Anim Lab 13:9–91
Schleich C, Antinuchi CD (2004) Testing magnetic orientation in a solitary subterranean rodent *Ctenomys talarum* (Rodentia: Octodontidae). Ethology 110:485–495
Semm P, Demaine C (1986) Neurophysiological properties of magnetic cells in the pigeon's visual system. J Comp Physiol A 159:619–625
Semm P, Beason C (1990) Responses to small magnetic variations by the trigeminal system of the bobolink. Brain Res Bull 25:735–740
Semm P, Schneider T, Vollrath L (1980) Effects of an earth-strength magnetic field on electrical activity in pineal cells. Nature 288:607–608
Sharp PE (2002) The neural basis of navigation. Evidence from single cell recording. Kluver Academic Publishers, Boston
Sharp PE, Blair HT, Cho JW (2001) The anatomical and computational basis of the rat head-direction cell signal. Trends Neurosci 24:289–294
Shcherbakov VP, Winklhofer M (1999) The osmotic magnetometer: a new model for magnetite-based magnetoreceptors in animals. Eur Biophys J 28:380–392
Stehle J, Reuss S, Schroeder H, Henschel M, Vollrath L (1988) Magnetic field effects on pineal *N*-acetyltransferase activity and melatonin content in the gerbil – the role of pigmentation and sex. Physiol Behav 44:91–94
Šumbera R, Burda H, Chitaukali WN, Kubova J (2003) Silvery mole-rats (*Heliophobius argenteocinereus*, Bathyergidae) change their burrow architecture seasonally. Naturwiss 90:370–373

Taube JS (1998) Head direction cells and the neurophysiological basis for a sense of direction. Progr Neurobiol 55:225–256

Thalau P, Ritz T, Stapput K, Wiltschko R, Wiltschko W (2005) Magnetic compass orientation of migratory birds in the presence of a 1.315 MHz oscillating field. Naturwiss 92:86–90

Thalau P, Ritz T, Burda H, Wegner RE, Wiltschko R (2006) The magnetic compass mechanisms of birds and rodents are based on different physical principles J R Soc Interface 3:583–587

Walker MM, Diebel CE, Haugh CV, Pankhurst PM, Montogomery JC, Green CR (1997) Structure and function of the vertebrate magnetic sense. Nature 390:371–376

Wegner RE, Begall S, Burda H (2006) Magnetic compass in the cornea: local anaesthesia impairs orientation in a mammal. J Exp Biol 209:4747–4750

Welker H, Semm P, Willig R, Commentz J, Wiltschko W, Vollrath L (1983) Effects of an artificial magnetic field on serotonin *N*-acetyltranferase activity and melatonin content of the rat pineal gland. Exp Brain Res 50:426–432

Williams MN, Wild JM (2001) Trigeminally innervated iron-containing structures in the beak of homing pigeons and other birds. Brain Res 889:243–246

Wiltschko R, Wiltschko W (1995) Magnetic orientation in animals, Springer, Berlin Heidelberg New York

Wiltschko W, Wiltschko R (2002) Magnetic compass orientation in birds and its physiological basis. Naturwiss 89:445–452

Wiltschko W, Wiltschko R (2005) Magnetic orientation and magnetoreception in birds and other animals. J Comp Physiol A 191:675–693

Wiltschko R, Wiltschko W (2006) Magnetoreception. BioEssays 28:157–168

13 Adaptive Neural Organization of Naked Mole-Rat Somatosensation (and Those Similarly Challenged)

Thomas J. Park, Kenneth C. Catania, Dalia Samaan, Christopher M. Comer

13.1 Introduction

The somatosensory system of subterranean mammals has received relatively little attention in comparison to their visual and auditory systems. In this chapter we will briefly summarize recent findings on somatosensory organization in naked mole-rats (*Heterocephalus glaber*). We will compare this information with previous findings from other subterranean rodents and some non-rodent subterranean mammals, and we will contrast this picture of adaptation to life underground with neural and behavioral organization in appropriate non-subterranean mammals. The analysis suggests that subterranean mammals are somatosensory specialists in the same sense that bats and barn owls are auditory specialists, and that the subterraneans will provide general insights into brain function and its evolution in a manner similar to the well-known examples from the auditory domain.

13.2 Naked Mole-rat Body Hairs

The skin of naked mole-rats is unusual in several respects. On gross inspection the skin is obviously exceptionally loose and furless (Thigpen 1940; Tucker 1981; Klauer et al. 1997; Daly and Buffenstein 1999). However,

Thomas J. Park: Laboratory of Integrative Neuroscience, Department of Biological Sciences, University of Illinois at Chicago, 840 West Taylor Street, Chicago, Illinois, 60607, USA, E-mail: TPark@uic.edu

Kenneth C. Catania: Department of Biological Sciences, Vanderbilt University, VU Station B, Nashville, Tennessee 37235, USA

Dalia Samaan: Laboratory of Integrative Neuroscience, Department of Biological Sciences, University of Illinois at Chicago, 840 West Taylor Street, Chicago, Illinois, 60607, USA

Christopher M. Comer: Laboratory of Integrative Neuroscience, Department of Biological Sciences, University of Illinois at Chicago, 840 West Taylor Street, Chicago, Illinois, 60607, USA

Subterranean Rodents: News from Underground
S. Begall, H. Burda, C.E. Schleich (Eds.)
© Springer-Verlag Berlin Heidelberg 2007

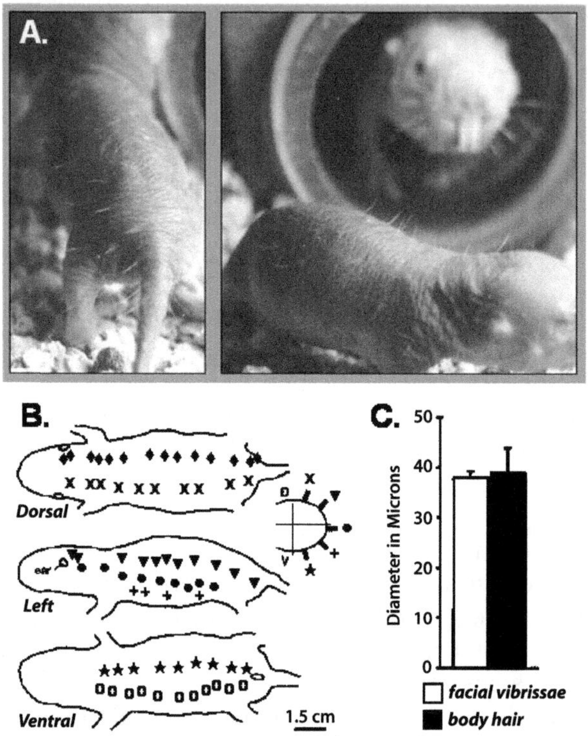

Fig. 13.1. Naked mole-rats retain a sparse array of guard hairs on the body (body hairs): **A** photographs showing body hairs on naked mole-rats. There also are hairs at some joints, on the limbs and the tail, as seen on other rodents. Note that the facial vibrissae (whiskers) can be seen on an animal in the background of the right hand panel; **B** each naked mole-rat has 10 rows of body hairs: two on the dorsal surface, three on each side, and two on the ventral surface; **C** the body hairs of naked mole rats are of a size similar to their whiskers. From Crish et al. (2003), re-published with permission

it is not entirely hairless, as it retains a grid-like pattern of sensory vibrissae (Fig. 13.1). The results of a microanatomical and immunolabeling analysis indicates that these hairs are very large guard hairs (Park et al. 2003). There are about 80 of these guard hairs on the body of a naked mole-rat, and the patterning of the hairs is consistent from individual to individual. Compared to the guard hairs found on the rat and the common mole-rat (*Cryptomys hottentotus*), the guard hairs on the naked mole-rat are far fewer in number, and the hairs and follicles are much larger in size, being almost as stout as the vibrissae found on the face (Crish et al. 2003).

13.3 Somatosensation and Behavior

Because of their lightless environment and extremely limited vision, it is no surprise that subterranean mammals rely heavily on non-visual sensory modalities to guide them through their complex tunnel systems (Burda et al. 1990; Burda 2003). We have found that naked mole-rats have poor vision (Artwohl et al. 2002; Mills and Catania 2004; Hetling et al. 2005; Crish et al. 2006a) typical of subterranean mammals (see Němec et al., this volume). Naked mole-rats also show poor sound localization ability (Heffner and Heffner 1993). So presumably their basic tasks of spatial orientation rely on non-visual and non-auditory cues which could include magnetic sensation, olfaction, vestibular input, and somatosensation. With regard to the latter, a recent study by Kimchi and Terkel (2004) using maze procedures supports the notion that blind mole-rats (*Spalax*) use touch cues in maze orientation.

We speculated that naked mole-rats might use their body hairs in maneuvering through their tunnel systems. In their natural environment, naked mole-rats must negotiate an ever changing tunnel system that can extend for kilometers, and that includes designated toilet, feeding, and nesting areas (Brett 1991). Likewise, they might use their body hairs to orient toward and maintain contact with colony mates. Based on what we know about the capacities of their sensory systems, we hypothesized that the body hairs might play a role in either of these spatial orientation tasks.

To explore this idea, we first tested naked mole-rats in an artificial tunnel system shaped like a radial maze (Fig. 13.2A). A small amount of food was placed at the end of each tunnel, and the hungry animal was allowed to travel through the maze, eating the food. In this type of procedure, non-subterranean rodents typically travel through the maze in an economical fashion, so as to visit each food area only once. Our tunnel maze was a typical eight arm radial maze design with one important exception, one of the arms was directly connected to the animals' home cage system, which was used to allow entry to the maze during test trials. The animals quickly learned to negotiate the maze by turning in a consistent direction (half of the animals turned right at each corner [e.g. Fig. 13.2A], half turned left).

We tested 14 mole-rats after removing the facial and body vibrissae from one side of the body. Each of these animals showed a significant turn preference, as did naked mole-rats with intact vibrissae. Seven of the mole-rats with hairs removed turned in the direction of the remaining, uncut hairs, and those mole-rats showed no deficit in orientation performance compared with intact mole-rats. On the other hand, seven mole-rats showed a preference to turn in the same direction as the removed hairs, and those mole-rats showed a large initial deficit compared to the mole-rats with intact hairs. Figure 2B shows the average choice

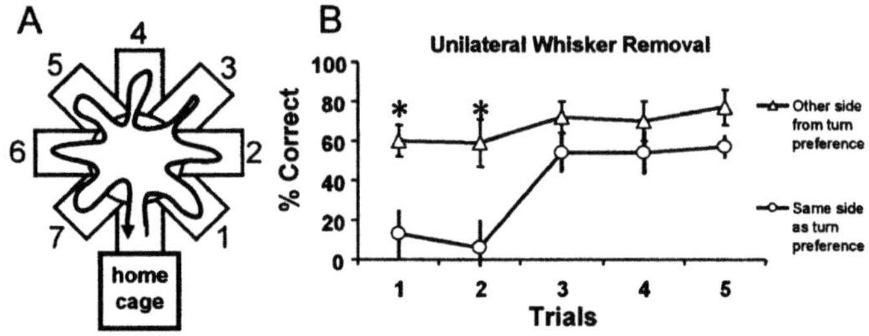

Fig. 13.2. The role of tactile hairs in orientation through a radial tunnel maze: **A** schematic of the maze with curves indicating the path taken by an experienced, intact naked mole-rat. Note that this animal consistently turned to the right. The tunnels (1–7 plus the tunnel to the home cage) were constructed of PVC pipe, 1 m long and 5 cm in diameter, and the center arena was constructed of plastic with a 26 cm diameter; **B** performance over five trials of naïve mole-rats tested after unilateral hair removal. Mole-rats that turned in the same direction as the removed hairs (*triangles*) made significantly more errors in the first two trials compared to mole-rats that turned in the same direction as the un-removed hairs (*circles*). Data from D. Samaan, PhD thesis, Univ. Illinois, Chicago

accuracy of the naked mole-rats that turned in the same direction as the removed hairs compared to average choice accuracy of those that turned in the direction of the remaining hairs (choice accuracy was calculated as the number of baited tunnels visited in the first seven choices divided by 7). All of the mole-rats eventually retrieved all seven bits of food, but the mole-rats that turned in the same direction as the cut hairs made many more errors, frequently re-visiting tunnels where food had already been eaten. Our interpretation of these data is that naked mole-rats rely to some extent on information from their vibrissae to accurately negotiate the tunnels. However, even with vibrissae removed the animals rapidly overcome the deficit after some experience in the tunnel maze. Additional experiments (not shown) suggest that the mole-rats use proprioceptive and vestibular cues (path integration) to accomplish this in a manner similar to that of other mammals (Lackner and DiZio 2005).

The basic ability to use somatosensory cues for negotiating a maze in the absence of visual cues is certainly wide-spread among mammals (Dale and Innis 1986; Toldi et al. 1994; Save et al. 1998; Kimchi and Terkel 2004). Perhaps a more clear-cut demonstration of the importance of body hairs to naked mole-rat behavior is the demonstration of their role in immediate sensory guided orientation. We have found that naked mole-rats are remarkably adept at orienting toward the point of contact when even a single body hair is deflected (Crish et al. 2003; Fig. 13.3).

Orientation was tested by manually deflecting a single hair on unrestrained animals, or by vibrating a hair via electromagnetic field, and

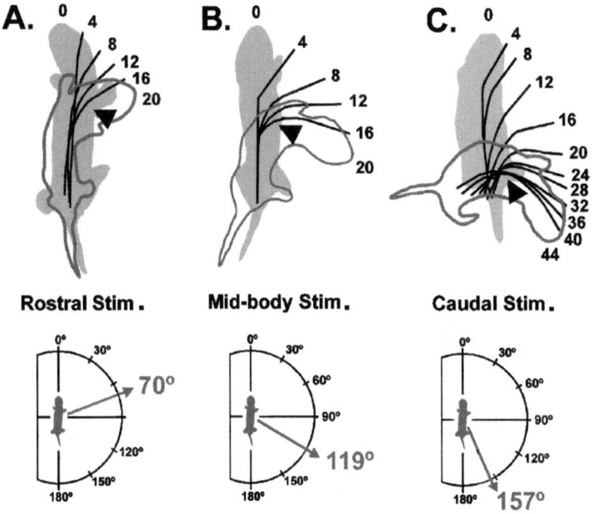

Fig. 13.3. Deflection of individual body hairs gave rise to well-directed orientation responses: **A–C** reconstructions from video tapes of three separate trials all from the same animal, which varied only in the rostro-caudal locus of the stimulated hair. In each, the *gray* form shows an animal's position just before stimulation. The *black outline* shows the animals' maximal turning position, and *black lines* show position of longitudinal body axis on every fourth video frame during the course of the response. *Arrowheads* show the rostro-caudal location at which a hair was stimulated. *Below each drawing*, the measured final angle of turn (*arrow with measured angle near tip*) is shown on a circular display so that the turn amplitudes can be compared

recording behavioral responses on video tape. Deflection of a single body hair triggered a pronounced orientation of the snout toward the point of stimulation, and the ability to evoke such an orientation response was very reliable. Clear responses were triggered on 95% of all trials (n=173 total trials; N=6 animals).

The drawings presented in Fig. 13.3 are reconstructions from video-taped trials and they illustrate the type of responses observed. Note that the angular orientation of the head axis at the completion of responses, relative to the original body axis, increased systematically as stimulus position varied along the body. Pooled data from numerous trials and animals (touched on facial whiskers or body hairs on one of five different body regions – Fig. 13.4A) show that naked mole-rats demonstrate topographically directed orienting to touch of body hairs (Fig. 13.4B). Since stimulation of skin in between body hairs did not lead to systematically directed orienting (Crish et al. 2003), the body hairs represent specialized points of touch sensitivity, and with a function that – at least for some stimuli – parallels that of facial whiskers.

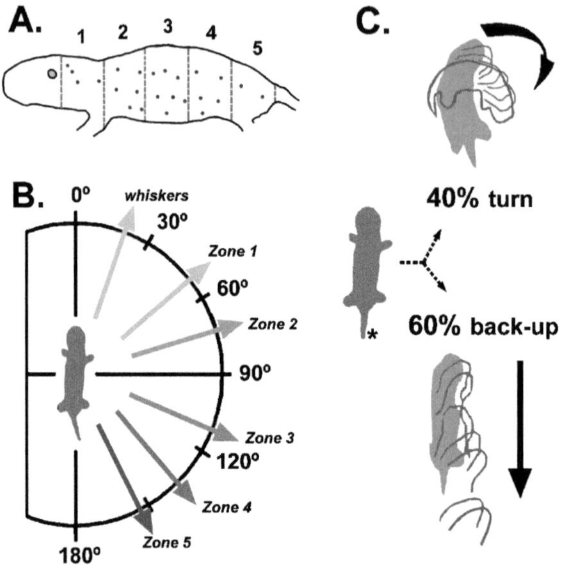

Fig. 13.4. Topography of orienting responses: **A** illustration of the five body regions used to quantify turning angles for stimulation of body hairs along the rostro-caudal axis; **B** average turning angles of orientation responses to stimulation of hairs at different rostro-caudal locations. Position of each *gray arrow* is placed so that it crosses the circular frame at a point corresponding to the mean turning angle for all responses evoked from that region. Based on pooled data from six animals (n=173 total trials). Mean turns for each locus of stimulation: facial vibrissae = 17.5°, zone 1 = 49.6°, zone 2 = 74.5°, zone 3 = 108.1°, zone 4 = 135.4°, zone 5 = 154.5°; **C** orienting responses to stimulation of hairs on the tail displayed "motor equivalence." Responses to stimulation of tail hairs fell into two categories and an example of each is shown. Both of these trials are from the same animal, and in both cases the same tail hair was stimulated. From Crish et al. (2003), re-published with permission

An interesting observation that we made during these experiments was that stimulation of hair sites on the tail also triggered orientation responses that brought the snout near the point of stimulation. However, orientation to tail sites involved a mixed repertoire of motor patterns compared to more stereotyped turning responses triggered by stimulation of hairs located on more rostral regions of the body. The reconstructions displayed in Fig. 13.4C illustrate two different orientation responses to the same hair located on the tail. In one case (top) the animal rotated nearly 180° so that the snout was pointing toward the original position of the tail. However on another trial (bottom) instead of rotating the head and torso toward the point of stimulation, the animal stepped straight backwards, stopping at a position that placed the snout in close proximity to the original position of the tail. Stimulation of tail hairs led to rotations on about 40% of all trials, but reverse locomotion on about 60% of all trials (30 trials total,

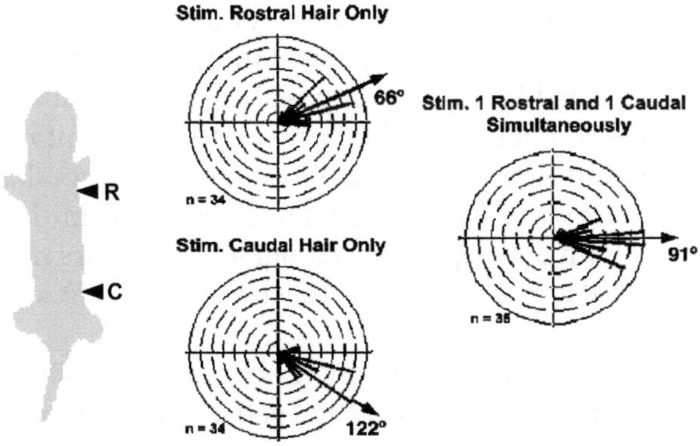

Fig. 13.5. Orientation responses when two body hairs were stimulated either individually or simultaneously. The schematic shows the location of the hairs tested (*arrowheads*). *Circular histograms* display the distribution of turns for each stimulus condition (*bars*) and the mean turn vector (*arrow*). Number of trials in each histogram is given by the number just below the horizontal axis of each. R = rostral hair, C = caudal hair. Reprinted from Neuroscience, Vol. 139 number 4, Crish et al., Population coding strategies and involvement of the superior colliculus in the tactile orienting behavior of naked mole-rats, Pages 1461–1466, Copyright 2006, with permission from Elsevier

N=3 animals). Hence, two very different motor patterns could be initiated to accomplish the same goal of bringing the snout to bear on the point of stimulation. In their tunnels, naked mole-rats, like other mole-rats, spend a considerable amount of time walking and running backwards, and their orientation behavior is well adapted in topography to their "tubular world".

The results presented above demonstrate that naked mole-rats can use their sparse array of guard hairs to accurately orient toward discrete points of contact. It is yet to be determined if the degree of accuracy and reliability they display are typical of other mammals (certainly humans can be keenly aware of even a small deflection of a hair on the dorsum of the hand). Perhaps the most remarkable thing about the naked mole-rats' orientation behavior to body hair deflection is its utility as a model system for studying non-visual sensory-motor integration since the number of sensors (hairs) is tractable, they are spaced widely apart, and the behavioral output is robust. For example, Crish et al. (2006b) recently used the naked mole-rat model to demonstrate that simultaneous stimulation of two hairs generates an averaged orientation response (Fig. 13.5) similar to what is seen for visually and acoustically guided orientation behavior in other mammals (e.g. Chou et al. 1999; Tollin and Yin 2003). These data represent the first demonstration of this phenomenon for the somatosensory system.

The body hairs of naked mole-rats may function in a variety of behaviors in nature. Taken together, the experiments on orienting through the tunnel maze and orientation to hair deflection suggest that the body hairs are more critically involved in immediate sensory guided orientation when the hairs are actively contacted by an outside force as opposed to passive contact with the tunnel walls. Active contact would likely mean the proximity of a conspecific or a predator, and the "backing-up" response suggests that orientation to hair deflections is mainly a reflex to allow an animal to initiate or maintain further contact with a conspecific. Of course, body hairs may also play a role in functions that are yet to be tested such as sensing changes in air flow.

Most mammals use their sense of touch not only to orientate and localize stimuli, but also to communicate, and to identify and gather food. The star-nosed mole is a remarkable example of a subterranean mammal that uses its somatosensory system in feeding. This animal is a master at locating and identifying the small invertebrates it eats at a tremendously fast pace with its nose appendages (Catania and Remple 2004). Communication via seismic signaling has been reported for a variety of subterranean mammals to varying degrees (Narins et al. 1997; Nevo 1999 and references therein). There is evidence that the blind mole-rat uses its somatosensory system for long distance seismic communication (Nevo et al. 1991); however, the extent to which the signals are processed by the somatosensory versus the auditory systems is not yet clear (Rado et al. 1998). Interestingly, a more recent study (Kimchi et al. 2005) presents evidence that blind mole-rats use self-generated seismic signals to identify and localize obstacles (e.g. a stone or an open ditch) through the soil. These authors showed that the mole-rats were able to accurately localize the source direction of seismic waves using only their paws. The authors also suggest that blind mole-rats may perceive seismic information through both somatosensory and auditory pathways depending on the context. Since a blind mole-rat generates seismic signals by head-drumming, it seems unlikely that the animal would be able to use its head to first drum and then listen (via bone conduction) given the short latencies of returning signals. On the other hand, it seems that the auditory system may play an important role in intraspecific seismic signaling, where individuals duet by alternately drumming and then resting their jaws against the tunnel wall (jaw hearing). Another mammal that integrates mechanical signals across somatosensory and auditory channels is the mountain beaver which appears to have a unique auditory brainstem region that can respond to changes in air pressure within its burrows (Merzenich et al. 1973).

13.4 Somatosensory Specializations in the Central Nervous System of Naked Mole-rats

Given the importance of touch to animals that live in dark underground tunnels we might predict that areas in the brain processing touch information would be uniquely specialized. This turns out to be true for naked mole-rats in some surprising ways. However, before describing how mole-rat brains differ from those of other mammals, it is useful to discuss brain organization in a more typical rodent to get a perspective on the usual size and location of sensory areas. For mammals, the highest processing station for sensory information is the neocortex. This six-layered sheet of neural tissue receives input from the retina, the cochlea, and the mechanoreceptors of the body and generally contains separate areas for processing information for each of these sensory modalities. Figure 13.6A shows these different cortical areas for the laboratory rat. Visual cortex is located caudally, auditory cortex is located in more lateral cortex, and a more rostral area contains somatosensory cortex where touch information is processed.

A fundamental organizing principle for cortical areas is their organization into topographic maps of the sensory receptors they represent. For example the visual cortex (V1) is organized as a map of the retina (corresponding to a map of visual space) the auditory cortex is organized as a map of the cochlea (corresponding to a map of tones) and the somatosensory cortex (S1) is organized as a map of the body surface. Note that only half the body is represented in the somatosensory cortex of each hemisphere.

With this organizational scheme in mind, we can consider how mole-rats may differ in brain organization from the typical mammalian plan. Figure 13.6B shows the somatosensory areas in a naked mole-rat (see Henry et al. 2006 for details). Two obvious specializations are apparent from this comparison. First, the somatosensory areas are much larger as a percentage of the cortex in mole-rats than in laboratory rats. Second, this larger somatosensory cortex extends into caudal cortex to occupy much of the area that would normally be devoted to vision. In fact, no visual responses have been recorded from the cortex of naked mole-rats, although it is possible a small visual area exists in far caudal cortex. We would predict that auditory cortex exists in caudal-lateral cortex, but this has not yet been identified in naked mole-rats and it would appear to occupy a relatively small region.

Clearly much of the neocortex that would normally be devoted to other sensory modalities processes touch information in naked mole-rats. However, this finding by itself is not very surprising given that mole-rats have little use for vision and many sounds are rapidly attenuated underground

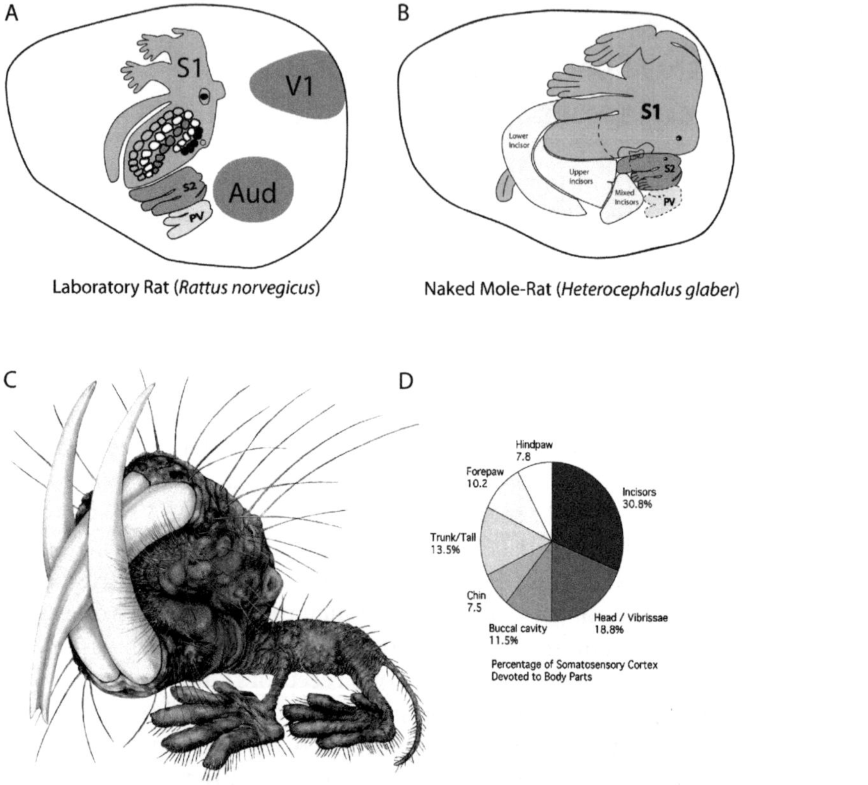

Fig. 13.6. Schematic illustrations of the organization of cortex in laboratory rats and naked mole-rats: **A** the left hemisphere of a rat showing primary somatosensory cortex (S1), primary visual cortex (V1) and auditory cortex (aud). The *circles* on the face area show the locations of the whisker representations; **B** the cortex of the naked mole-rat (from Henry et al. 2006). Note the relatively much larger area of cortex taken up by S1 for processing touch. S2 and PV are additional touch areas. Rostral is to the left and medial is up; **C** a mole-ratunculus and details of the somatosensory cortex in naked mole-rats. The schematic illustrates the proportions of cortex dedicated to different body parts. Note the exceptionally large incisors; **D** the percentages of somatosensory cortex devoted to different body parts

(Heffner and Heffner 1993). The most unusual result from investigating naked mole-rat cortex comes from analyzing how different body parts are represented within the map of touch receptors in the somatosensory area. This is best appreciated by examining a so-called mole-ratunculus that illustrates the body parts sized to match their relative proportions as represented in somatosensory cortex (Fig. 13.6C,D).

Amazingly, the somatosensory cortex of the naked mole-rat is dominated by the representation of the teeth – almost one-third of the area is devoted to inputs from the large, front incisors. This is unprecedented among mammals, and raises the question of why so much neural processing space is required for the teeth in mole-rats? The explanation for this adaptation lies in the unusual lifestyle of mole-rats.

A glance at the naked mole-rat's face suggests the teeth may play a special role in mole-rat behavior. Mole-rats use their teeth to dig tunnels and, as an adaptation to keep dirt out of their mouths, the incisors are not enclosed in the oral cavity as occurs for most other mammals. Digging tunnels requires sensory feedback to modulate bite force in response to changes in soil characteristics. In addition, objects embedded in the soil must be detected to avoid breaking the incisors. On a larger scale, the density and composition of soil in different areas may signal the presence of different food types (tubers) and such variations may also indicate the most energetically economical directions for digging tunnels.

Naked mole-rats, like other mole-rats, also use their teeth to move objects, carry young, for social interactions (incisor fencing) and, of course, for eating. In addition, our observations suggest that naked mole-rats touch their teeth to objects as they explore their environment to gather tactile information, and Poduschka (1978) has suggested that the common mole-rat uses its teeth to sense vibrations.

In the cortex, somatosensory maps are typically investigated by recording the responses of neurons in the cortex with microelectrodes while the body surface is stimulated. However some species have anatomical specializations in the cortex that reflect the organization of the map, and this can be revealed by processing cortical tissue for different histological markers. The most efficient approach is to first flatten out the cortical hemisphere before sectioning it parallel to the cortical surface. The middle layers of cortex (layer 4 in particular - which receives input from the thalamus) are then examined. Figure 13.7 shows examples of flattened cortical sections processed to reveal the metabolic enzyme cytochrome oxidase in a laboratory mouse and a naked mole-rat. In mice, there is a remarkably clear set of circles and subdivisions that each represent a different body part, or hair, from the contralateral side of the animal. The distinct black circles have been termed barrels (Woolsey and Van der Loos 1970) and each represents a single whisker from the contralateral side of the face. In naked mole-rats, the pattern is less distinct, yet it clearly resembles the pattern in mice, and some barrels representing the largest facial whiskers are apparent (Fig. 13.7B; Henry et al. 2006).

Cortical barrels are useful markers for investigating brain organization and emphasize the highly organized nature of the projection pathways that relay touch information from receptor arrays on the body to the trigeminal complex, thalamus, and finally to the cortex. The organizational details here are usually diagnostic of behavioral specializations (Adrian 1941). So note that there appears to be a more balanced representation of body

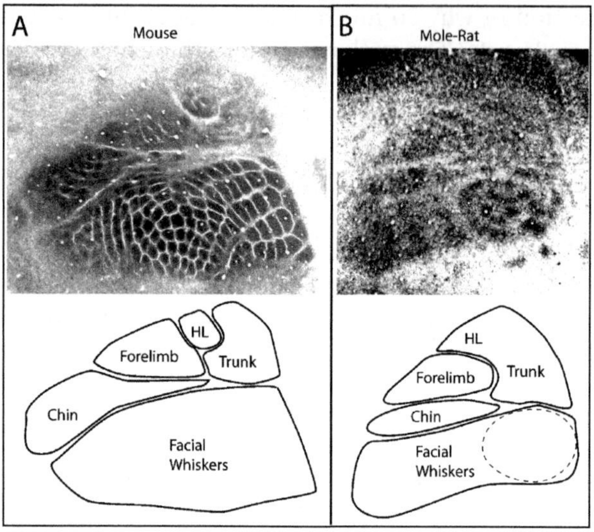

Fig. 13.7. Flattened sections of cortex processed for cytochrome oxidase in a mouse and naked mole-rat: **A** the flattened cortex of a mouse showing the distinctive anatomy that reflects the locations of body-part representations in S1. The *lower panel* shows roughly how the body parts are represented in relationship to the subdivisions visible above; **B** the flattened cortex of a mole-rat showing the distinctive anatomy that reflects the locations of body-part representations in S1. The *lower panel* shows roughly how the body parts are represented in relationship to the subdivisions visible above. The *dashed line* indicated the representation of the large whiskers on the face, where some barrels are apparent (see text)

and facial whiskers in naked mole-rats, rather that the strong bias toward over-representation of facial whiskers that is seen in mice. This would be consistent with the behavioral importance of the body hairs in naked mole-rat orientation described above.

13.5 Comparisons to Star-Nosed Moles

Star-nosed moles (*Condylura cristata*) are insectivorous moles, members of the order Erinaceomorpha (formerly Insectivora), and are thus only distantly related to rodents. However, it is hard to describe sensory specializations in subterranean mammals without considering some of the other touch specialists that have adapted to life underground. Star-nosed moles live in North America and make their living eating invertebrates in wetlands. For this purpose, the star-nosed moles have evolved one of the more unusual sensory specializations among mammals, consisting of 22 appendages that ring the nose and are used by the moles to feel their

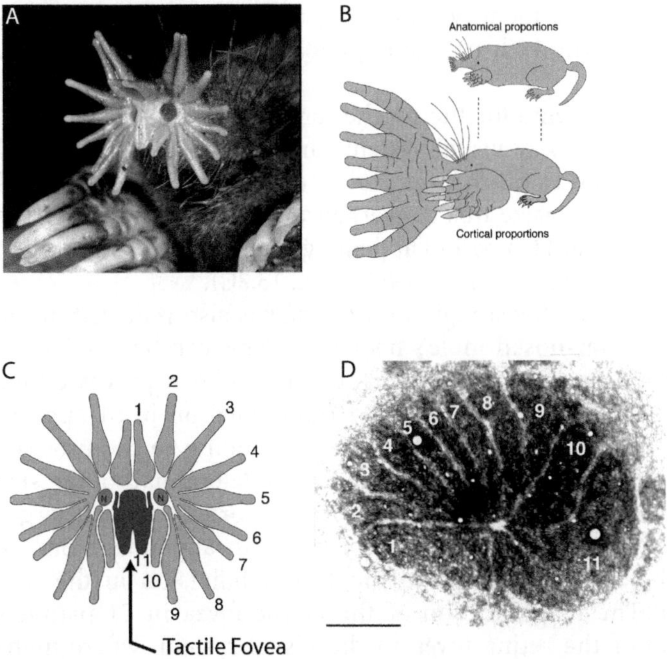

Fig. 13.8. The unusual sensory system of the star-nosed mole: **A** the nose and forelimbs of a star-nosed mole showing the large claws used for digging tunnels and the 22 appendages that ring the nostrils; **B** cortical magnification in star-nosed moles. Whereas mole-rat cortex is dominated by the representation of teeth, star-nosed moles over-represent the star; **C** the location of the tactile fovea in star-nosed moles. This region of the star is used for detailed investigations of object of interest, much like the retina is used for detailed examination of a visual scene in many other mammals; **D** the specialized cortex of a star-nosed mole. In the somatosensory cortex the 11 appendages of the star can be visualized in flattened sections processed to reveal the metabolic enzyme cytochrome oxidase. Note the greatly enlarged representation of the tactile fovea (appendage 11), emphasizing its behavioral importance rather than its size of innervation density (see Catania and Kaas 1997). Scale in D is 1 mm (*top*) and 500 μm (*bottom*)

way around dark tunnels (Fig. 13.8A). This specialized nose is commonly referred to as the "star".

As occurs for the incisors in mole-rats, the important touch receptors of the star are greatly magnified in the somatosensory cortex (Fig. 13.8B). However, when details of star-nosed mole behavior are examined, it becomes clear that not all parts of the star play an equal role in explorations (Catania and Remple 2004). The central, lowest appendages (the 11th pair in our numbering scheme) are used for detailed explorations of objects of interest, whereas the other 10 pairs of appendages are used for lower resolution touch to identify potentially important objects for further exploration with the 11th pair of appendages (Catania and Remple 2004).

This organization scheme for the mechanoreceptors on the star is remarkably similar to the way visual systems are organized, with a small high-resolution fovea and a much larger, lower resolution set of peripheral photoreceptors. To emphasize this convergence in organization with the visual system, the central 11th pair of appendages on the star is referred to as the tactile fovea (Catania and Kaas 1997) (Fig. 13.8C).

The importance of the fovea region on the star is also reflected in the organization of the star-nosed mole's neocortex. This can be readily appreciated by examining the flattened cortex that has been processed for cytochrome oxidase (Fig. 13.8D). The star representation in S1 appears as a set of dark stripes, with each stripe corresponding to the processing area for one of the appendages from the contralateral half of the star (Catania and Kaas 1995). The 11th stripe in the cortex is much larger than would be predicted from the size of the 11th appendage on the star (compare the 11th appendage to the 11th subdivision in the cortex in Fig. 13.8). The magnification of the tactile fovea in S1 parallels the magnification of the retina fovea in the visual system, where high-resolution photoreceptors take up the most space in primary visual cortex.

A similar organization scheme has been described for the auditory system of the common mole-rat (Müller et al. 1992) and some species of bats (Suga and Jen 1976; Suga 1989) which have an auditory fovea for processing behaviorally important auditory signals. These observations illustrate the important relationship between behavior and brain organization and highlight the flexibility of the central nervous system in adapting to the challenges of different animal's sensory environments. The convergent development of foveas in the visual, auditory, and somatosensory systems also suggests this is a particularly efficient way for sensory systems to be organized.

13.6 New Directions: Unique Somatic Organization for Processing Painful Stimuli

During the course of our studies, we made two interesting observations about the nerve fibers in the skin of naked mole-rats (Park et al. 2003). The first was not surprising – that the follicles of the body hair were substantially larger and more heavily innervated than those of the largest body hairs on rats and common mole-rats (a furred species of African mole-rat). This is

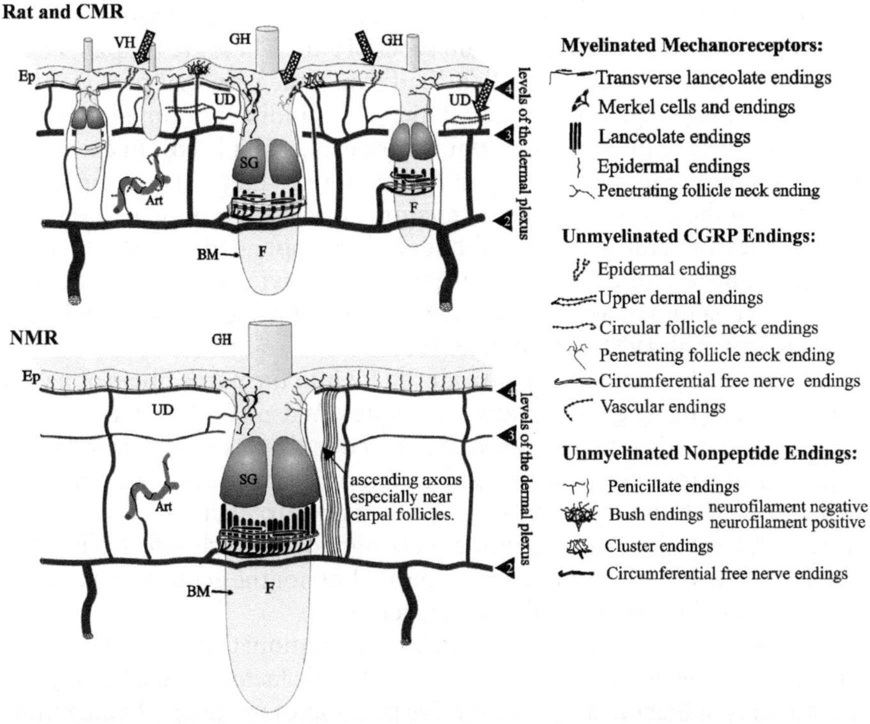

Fig. 13.9. Schematic showing innervation of a body hair and surrounding skin in naked mole-rat compared to that of other mammals, represented here by laboratory rats and common mole-rats (CMR). Note that naked mole-rats completely lack small guard hairs and vellus hairs. Also, the naked mole-rat's guard hairs (body hairs) have larger follicles and are more heavily innervated than those of the comparison species. *Arrows* indicate nerve fibers that labeled positively for CGRP and Substance P. Note the lack of these neuropeptides in the tissue from naked mole-rat. Ep, epidermis; F, follicle; GH, guard hair; UD, upper dermis; VH, vellus hair; Art, arteriole; BM, basement membrane; SG, sebaceous gland. For more detail see Park et al. (2003), re-published with permission

shown in the graphic representations in Fig. 13.9. However, it should be kept in mind that, functionally, many smaller structures may lead to the same result as few larger ones.

The second finding was a surprise. We found that nerve fibers on hair follicles, and in the skin and respiratory tract in general, lack the neuropeptides, Substance P (SP) and calcitonin gene related peptide (CGRP). This result came from an immunolabeling study where we used antibody labels for these, as well as other chemical signalers, to help identify specific subpopulations of nerve fibers. The neuropeptides that we found to be lacking in naked mole-rat skin are associated with two subpopulations of nerve fibers: (1) parasympathetic inputs to the arterioles of the skin which are usually involved in thermoregulation, and (2) unmyelinated sensory fibers that carry information about certain types of pain. These are the pain fibers thought to be involved in so-called "bad pain", the pain that persists after an initial injury (e.g. inflammatory pain). This finding prompted us to test several models of inflammatory pain (Park et al., submitted). In each case, the naked mole-rats showed no pain behavior (e.g., they did not respond to injection of the paw skin with the inflammatory agent capsaicin, the active ingredient from chili peppers), although they responded robustly to acute pinch and heat. These data are consistent with previous reports that naked mole-rats respond to acute heat and injection of formalin (Kanui and Hole 1990; Kanui et al. 1993; Towett et al. 1993).

In further experiments, we introduced SP into the nerve fibers of naked mole-rat skin using gene therapy techniques (Wilson and Yeomans 2000). Under these conditions, we found that naked mole-rats now respond to an injection of capsaicin with robust licking, as do mice and rats (humans rub and scratch the injection site). Thus, it appears that naked mole-rats have adapted to lose neuropeptide pain signalers, which effectively "disconnects" the inflammatory pain pathway (but not the acute pain pathway which normally does not utilize neuropeptides).

What would be the advantage of such an adaptation? Our current working hypothesis is related to a combination of lifestyle factors. Naked mole-rats are not only subterranean, but they live in colonies made up of many individuals and they are extremely social. Hence these animals crowd together in poorly ventilated spaces where they experience chronically high levels of CO_2. High levels of CO_2 are known to induce tissue acidosis, and life-threatening lung edema, both of which are mediated through peptidergic pain fibers (referred to in the lungs as irritant receptors). It seems likely that naked mole-rats have lost SP and CGRP as an adaptation to surviving in what would otherwise be a toxic atmosphere. Critical experiments to help support or refute this hypothesis will include measuring CO_2 levels in the nesting chambers of naked mole-rat burrows, and testing for a lack of neuropeptides in other mammals that experience chronically high levels of CO_2.

13.7 Conclusion

Subterranean rodents, and similarly challenged non-rodent mammals (e.g. Catania 2005a, b; Reep et al. 2002) are gaining recognition as touch specialists. This is important because sensory specialists of any modality are invaluable in advancing our understanding of sensory processing from receptors, to brain circuits and organization, to complex behaviors. In the somatosensory domain, the naked mole-rat and the star-nosed mole are emerging as excellent model systems for studying the neural organization and mechanisms underlying somatosensory processing. Furthermore, the naked mole-rat's adaptive loss of specific neuropeptides provides us with a new and powerful model system for asking questions about the normal role of these neuropeptides in sensory signaling. "And so hurrah for underground!" - Fyodor Dostoyevsky (Dostoyevsky 1993).

Acknowledgements. We thank Sam Crish, Christine Crish, Erin Henry, Hynek Burda, and John Zook for their thoughtful comments on the manuscript. Grant support to the Park, Comer and Catania laboratories was provided by NSF Grant 0238364, NIH grant R29 DC02850, and a UIC Campus Research Board grant.

References

Adrian ED (1941) Afferent discharges to the cerebral cortex from peripheral sense organs. J Physiol (Lond) 100:159–191

Artwohl J, Hill T, Comer C, Park T (2002) Naked mole-rats: unique opportunities and husbandry challenges. Lab Anim 31:32–36

Brett RA (1991) The ecology of naked mole-rat colonies: burrowing, food and limiting factors. In: Sherman PW, Jarvis JUM, Alexander RD (eds) The biology of the naked mole-rat. Princeton Univ Press, Princeton, NJ, pp 137–184

Burda H (2003) Adaptations for subterranean life. In: Kleiman DG, Geist V, Hutchins M, McDade MC (eds.) Mammals I. Volume 12 of Grzimek's animal life encyclopedia. Gale Group, Farmington Hills, Mich, pp 69–78

Burda H, Bruns V, Müller M (1990) Sensory adaptations in subterranean mammals. In: Nevo E, Reig OA (eds) Evolution of subterranean mammals at the organismal and molecular levels. Wiley-Liss, New York, pp 269–293

Catania KC (2005a) Star-nosed moles. Curr Biol 15:R863–864

Catania KC (2005b) Evolution of sensory specializations in insectivores. Anat Rec A 287:1038–1050

Catania KC, Kaas JH (1995) Organization of the somatosensory cortex of the star-nosed mole. J Comp Neurol 351:549–567

Catania KC, Kaas JH (1997) Somatosensory fovea in the star-nosed mole: behavioral use of the star in relation to innervation patterns and cortical representation. J Comp Neurol 387:215–233

Catania KC, Remple FE (2004) Tactile foveation in the star-nosed mole. Brain Behav Evol 63:1–12

Chou IH, Sommer MA, Schiller PH (1999) Express averaging saccades in monkeys. Vision Res 39:4200–4216

Crish SD, Rice FL, Park TJ, Comer CM (2003) Somatosensory organization and behavior in naked mole-rats I: vibrissa-like body hairs comprise a sensory array that mediates orientation to tactile stimuli. Brain Behav Evol 62:141–151

Crish SD, Dengler-Crish CM, Catania KC (2006a) Central visual system of the naked mole-rat (*Heterocephalus glaber*). Anat Rec A 288:205–212

Crish SD, Dengler-Crish CM, Comer CM (2006b) Population coding strategies and involvement of the superior colliculus in the tactile orienting behavior of naked mole-rats. Neuroscience 139:1461–1466

Dale RH, Innis NK (1986) Interactions between response stereotypy and memory strategies on the eight-arm radial maze. Behav Brain Res 19:17–25

Daly TJ, Buffenstein R (1999) Skin morphology and its role in thermoregulation in mole-rats, Heterocephalus glaber and *Cryptomys hottentotus*. J Anat 193:495–502

Dostoyevsky F (1993) Notes from underground (vintage classics). Translated by R. Pevear and L. Volokhonsky. Vintage Books, New York. Translation from the Russian text of the Soviet Academy of Sciences edition, volume 5 (Leningrad, 1973)

Heffner RS, Heffner HE (1993) Degenerate hearing and sound localization in naked mole rats (*Heterocephalus glaber*), with an overview of central auditory structures. J Comp Neurol 331:418–433

Henry EC, Remple MS, O'Riain MJ, Catania KC (2006) Organization of somatosensory cortical areas in the naked mole-rat (*Heterocephalus glaber*). J Comp Neurol 495:434–452

Hetling JR, Baig-Silva MS, Comer CM, Pardue MT, Samaan DY, Qtaishat NM, Pepperberg DR, Park TJ (2005) Features of visual function in the naked mole-rat *Heterocephalus glaber*. J Comp Physiol A 191:317–330

Kanui TI, Hole K (1990) Morphine induces aggression but not analgesia in the naked mole-rat (*Heterocephalus glaber*). Comp Biochem Physiol C 96:131–133

Kanui TI, Karim F, Towett PK (1993) The formalin test in the naked mole-rat (*Heterocephalus glaber*): analgesic effects of morphine, nefopam and paracetamol. Brain Res 600:123–126

Kimchi T, Terkel J (2004) Comparison of the role of somatosensory stimuli in maze learning in a blind subterranean rodent and a sighted surface-dwelling rodent. Behav Brain Res 153:389–395

Kimchi T, Reshef M, Terkel J (2005) Evidence for the use of reflected self-generated seismic waves for spatial orientation in a blind subterranean mammal. J Exp Biol 208:647–659

Klauer G, Burda H, Nevo E (1997) Adaptive differentiations of the skin of the head in a subterranean rodent, *Spalax ehrenbergi*. J Morphol 233:53–66

Lackner JR, DiZio P (2005) Vestibular, proprioceptive, and haptic contributions to spatial orientation. Ann Rev Psychol 56:115–147

Merzenich MM, Kitzes L, Aitkin L (1973) Anatomical and physiological evidence for auditory specialization in the mountain beaver (*Aplodontia rufa*). Brain Res 58:331–344

Mills SL, Catania KC (2004) Identification of retinal neurons in a regressive rodent eye (the naked mole-rat). Vis Neurosci 21:107–117

Müller M, Laube B, Burda H, Bruns V (1992) Structure and function of the cochlea in the African mole rat (*Cryptomys hottentotus*): evidence for a low frequency acoustic fovea. J Comp Physiol A 171:469–476

Narins PM, Lewis ER, Jarvis JJ, O'Riain J (1997) The use of seismic signals by fossorial southern African mammals: a neuroethological gold mine. Brain Res Bull 44:641–646

Nevo E (1999) Physiological adaptive strategies. In: Nevo E (ed) Mosaic evolution of subterranean mammals: regression, progression and global convergence. Oxford Univ Press, Oxford pp 101–143

Nevo E, Heth G, Pratt H (1991) Seismic communication in a blind subterranean mammal: a major somatosensory mechanism in adaptive evolution underground. Proc Natl Acad Sci USA 88:1256–1260

Park TJ, Comer C, Carol A, Lu Y, Hong HS, Rice FL (2003) Somatosensory organization and behavior in naked mole-rats: II. Peripheral structures, innervation, and selective lack of neuropeptides associated with thermoregulation and pain. J Comp Neuro 465:104–120

Park TJ, Lu Y, Hu J, Brand A, Anirudhan G, Heppenstall PA, Milenkovic N, Wetzel C, Laurito CE, Wilson SP, Lewin GR (submitted) Inflammatory pain insensitivity in the African naked mole rat (*Heterocephalus glaber*)

Poduschka W (1978) Abwehrreaktion der Mullratte, *Cryptomys hottentotus* (Lesson, 1826). Säugetierk Mitt 26:260–268

Rado R, Terkel J, Wollberg Z (1998) Seismic communication signals in the blind mole-rat (*Spalax ehrenbergi*): electrophysiological and behavioral evidence for their processing by the auditory system. J Comp Physiol A 183:503–511

Reep RL, Marshall CD, Stoll ML (2002) Tactile hairs on the postcranial body in Florida manatees: a mammalian lateral line? Brain Behav Evol 59:141–154

Save E, Cressant A, Thinus-Blanc C, Poucet B (1998) Spatial firing of hippocampal place cells in blind rats. J Neurosci 18:1818–1826

Suga N (1989) Principles of auditory information-processing derived from neuroethology. J Exp Biol 146:277–286

Suga N, Jen PH (1976) Disproportionate tonotopic representation for processing CF-FM sonar signals in the mustache bat auditory cortex. Science 194:542–544

Thigpen LW (1940) Histology of the skin of a normally hairless rodent. J Mammal 21:449–456

Toldi J, Farkas T, Volgyi B (1994) Neonatal enucleation induces cross-modal changes in the barrel cortex of rat. A behavioural and electrophysiological study. Neurosci Lett 167:1–4

Tollin DJ, Yin TC (2003) Psychophysical investigation of an auditory spatial illusion in cats: the precedence effect. J Neurophysiol 90:2149–2162

Towett PK, Kanui TI, Titus I (1993) Effects of pethidine, acetylsalicylic acid, and indomethacin on pain and behavior in the mole-rat. Pharmacol Biochem Behav 45:153–159

Tucker R (1981) The digging behavior and skin differentiations in *Heterocephalus glaber*. J Morphol 168:51–71

Wilson SP, Yeomans DC (2000) Genetic therapy for pain management. Curr Rev Pain 4:445–450

Woolsey TA, Van der Loos H (1970) The structural organization of layer IV in the somatosensory region (SI) of mouse cerebral cortex: the description of a cortical field composed of discrete cytoarchitectonic units. Brain Res 17:205–242

Part III
Life Histories, Behavioural Ecology, Demography

14 From Natural Histories to Life Histories – A Homage to a Comparative Approach

Hynek Burda

"Natural history", a term which has successfully ousted the earlier "bionomy", describes the way of living of animals of a given species. Knowledge of the natural history is necessary for practical reasons (e.g. pest management or successful animal husbandry) as well as for scientific understanding. Natural history is also a window into "life history", one of the most frequent terms in behavioural and population ecology, meaning indeed the history, i.e. the whole course of life of an (average) individual. Given the specificity and narrow confinement of the underground ecotope resulting in its uniformity or at least similarity in many aspects across diverse habitats (reviewed, e.g. in Nevo 1979, 1999; Burda et al., this volume), we may expect that not only morphologies or physiologies, but also the life histories of diverse species of subterranean rodents have been convergently moulded. However, whereas morphology and physiology are mostly punctual snapshots of single individuals, insight into life histories requires long-term (field) studies involving many individuals. No wonder that for most species of subterranean rodents life history patterns are still unknown.

Time, energy, and nutrients are limited resources for organisms, and adaptive morphological or physiological modifications may enable their more effective exploitation. However, every modification represents a certain trade-off: Increasing one parameter means decreasing another one. Students of life histories are specifically interested in trade-offs included in the "reproductive plan" of individuals and pose the question which rules and choice strategies enhance their evolutionary fitness. Correspondingly, the main components of life history are maturity (the age at first breeding), parity (number of breeding events), fecundity (number of offspring per breeding event), age at death, and age at other events influencing individual's lifetime reproductive traits. Each of these components influence other life history parameters.

As Nevo (1979, 1999) stressed in his comprehensive overview of the evolution of subterranean mammals, the relatively constant, climatically predictable and safe subterranean ecotope leads to convergent K-selection. Correspondingly, subterranean mammals are selected for delayed repro-

Hynek Burda: Department of General Zoology, University of Duisburg-Essen, 45117 Essen, Germany, E-mail: hynek.burda@uni-due.de

Subterranean Rodents: News from Underground
S. Begall, H. Burda, C.E. Schleich (Eds.)
© Springer-Verlag Berlin Heidelberg 2007

duction, lower fecundity, and slower pre- and postnatal development with longer lifespan. Populations in subterranean mammals are at equilibrium, with population sizes near carrying capacity. Indeed, none of the hitherto studies of life history patterns in diverse species of subterranean rodents has falsified this prediction. Furthermore, as Nevo (1979, 1999) argued, low productivity of the subterranean ecotope leads to resource competition resulting usually in solitariness accompanied by strong territoriality and competition within species. Indeed, most species of subterranean mammals are of solitary habits (Nevo 1999; Burda et al. 2000).

Therefore, reporting 25 years ago that the naked mole-rat (*Heterocephalus glaber*) is eusocial (i.e. living in large family groups with only a single female reproducing) and thus standing at the other end of the range of social systems (Jarvis 1981), triggered an enormous interest in this species. In spite of this, our knowledge of the naked mole-rat's life history is based largely on the study of captive animals. No wonder that results of more recent thorough long-term field studies (Braude 2000; Ciszek 2000) which disturb the established attractive popular story (which is extraordinary, yet simple and well understandable) have been largely ignored or only partly and marginally cited and so far not really taken into account. Apart from these two papers, no significant news on behavioural ecology of the naked mole-rat have been reported since 2000. (However, see the below paragraph addressing the phenomenon of aging and longevity.) In a single ethological study relevant to life-history of the naked mole-rat published within the last five years, Clarke and Faulkes (2001) revisited the problem of intracolonial aggression. Apparently, new insights into life and natural histories of the naked mole-rat, and new stimuli into the research in this field will come from population genetic studies reported in this book by Braude, though in a different section.

Some of the naked mole-rats' hairy counterparts, namely genera of African mole-rats *Cryptomys* and *Fukomys* can also be regarded as social or eusocial, and in the 1990s a debate arose on the evolution of eusociality in bathyergid mole-rats (for contrary views see Bennett and Faulkes 2000 and Burda et al. 2000; also briefly reviewed by Šumbera et al., this volume). Only a few other reports relevant to life histories have been published on these otherwise well studied species since 2000. Hazell et al. (2000), Spinks et al. (2000), and Scantlebury et al. (2006) described adult dispersal in mole-rats and discussed it from the point of view of evolution of eusociality. Two important studies dealt with patterns of genetic relatedness in families of (eu)social mole-rats. Burland et al. (2002) showed that, consistently with the previous experience from laboratories, most breeding pairs in wild colonies of the Damaraland mole-rat are indeed unrelated. Furthermore, the finding confirmed the previous theoretical argument by Burda (1999) and Burda et al. (2000) that eusociality represents an extended monogamy and normal familial levels of relatedness are sufficient for its occurrence in mammals. On the other hand, high levels of extra pair paternity were found in social *Cryptomys hottentotus* (Bishop et al. 2004) challenging thus the role of kin selection and genetically based altruism in maintaining cooperative societies.

The point at issue in the debate on the evolution of eusociality is basically the following: the proponents of the so-called aridity food-distribution hypothesis (AFDH) emphasize ecological factors and consider eusociality in mole-rats an evolutionary endpoint (cf. Bennett and Faulkes 2000), whereas the opposite view stresses the role of phylogenetic constraints, considering solitariness a derived trait (cf. Burda et al. 2000). It is apparent that to solve the question new data are required which must involve phylogenetically related taxa living in different habitats and unrelated taxa occurring in comparable habitats. Recently, the scientific community of "subterraneologists" has started to grow, and new species have been introduced into research programmes. The new findings, some of which are reviewed in this section, verify or falsify hitherto interpretations and represent a real enrichment to our knowledge.

1) The recent finding that Zambezian eusocial mole-rats represent a lineage which should be separated from South African *Cryptomys* (and which was designated by the generic name *Fukomys*) (Ingram et al. 2004; Kock et al. 2006; see also Van Daele et al., this volume), also challenged the interpretation of diverse previous studies, in which various parameters of sociality in, e.g. *Cryptomys hottentotus* and *C. damarensis* (now considered *Fukomys damarensis*) were correlated with diverse parameters of habitats. Taxonomic revision shows that in such studies not only ecologies were different but at the same time also two different genera were compared, without taking phylogeny into account. On the other hand, finding that giant mole-rats (*F. mechowii*) actually represent a basal lineage of the genus *Fukomys*, yet in spite of occupying mesic habitats are highly social (cf. Kawalika and Burda, this volume), similar to *F. damarensis* representing a more recent divergence and occupying arid regions, is also a strong argument for intrinsic rather than extrinsic causes of sociality.

2) The studies of the Zambian giant mole-rats *Fukomys mechowii*, reviewed in this section by Kawalika and Burda (Chap. 15) particularly demonstrate how any inclusion of a new species into our repertoire of knowledge can change our interpretations of observed facts. When Mathias Kawalika triggered the research of giant mole-rats 15 years ago, their biology was virtually unknown, and the animals were considered to be of solitary habits, thus supporting the previously dominating hypothesis about evolution of eusociality. As we have shown, the assumption about solitariness of giant mole-rats was completely wrong. Moreover, the study demonstrates the wealth of vernacular knowledge of indigenous people

(for whom giant mole-rats have represented some of the most familiar wild mammals) – knowledge, which has been ignored and untapped for a long time.

3) Real progress has been made in the study of the silvery mole-rats (*Heliophobius argenteocinereus*): almost unknown ten years ago, *Heliophobius* has become the best known solitary bathyergid mole-rat by now. Šumbera et al. summarize and discuss in Chap. 16 published as well as unpublished results of recent research on the biology of this interesting species. Contrary to solitary species of bathyergids living mainly in sandy soils in South Africa, *Heliophobius* is a generalist surviving in a variety of habitats with different ecological conditions, including sites with seasonally very hard soil and a very low food supply, falsifying (or at least questioning) thus some of the premises of the AFDH. Sociality of bathyergids was also related to parameters of reproductive biology. Similarly, reproductive and developmental traits in *Heliophobius* suggest their ancestral character connecting bathyergid mole-rats with other hystricognath rodents, rather than adaptive traits. Relative shortening of prenatal and postnatal development in *Bathyergus* and *Georychus* (cf. Bennett and Faulkes 2000) has to be considered a derived trait, consistent with the fact that these genera represent later divergence events.

4) Recently, predictions of the AFDH were tested in another group of subterranean rodents involving solitary and social members (even) within a single genus. Lacey and Wieczorek (2003) found social *Ctenomys sociabilis* and solitary *C. haigi* living sympatrically in a mesic habitat, whereas in an arid steppe with a low food supply only the solitary *C. haigi* was living. No difference in distribution of clumped food resources was found between both sites. Thus, two of the three key ecological parameters contradicted expectations derived from the AFDH. Further relevant studies on ctenomyids are reviewed by Lacey and Cutrera in this volume.

5) Zhang reviews in Chap. 17 the biology and ecology of plateau zokors (*Eospalax fontanierii*) – myosplacine solitary rodents from the Tibetan Plateau of China. The author presents much basic data on morphological and physiological adaptations as well as on the ecology of this subterranean rodent living under extreme climatic high altitude conditions. Again, the fact that zokors thrive under such conditions – and can even afford to spend 85–90% of their lifetime in their underground nests, devoting only about 2–3 h a day to digging, maintaining their tunnels and foraging – demonstrates clearly that our views of what is considered an extreme and harsh environment and how the expected solution of the problem looks like, are very simplified and subjective.

6) As stated above, one of the important components, and indeed an integrative trait, of life histories is age at death. Already earlier reports (cf. Sherman et al. 1991) reported extraordinary longevity in the naked mole-

rats. Recently, the maximum lifespan in this species was settled at more than 28 years (Buffenstein and Jarvis 2002; Sherman and Jarvis 2002). It was recognized that this species represents an interesting experimental model for gerontological research (Buffenstein 2005) and several studies of proximate factors potentially effecting longevity followed (see references in Dammann and Burda, this volume). Most recently Dammann and Burda (2006) demonstrated that Ansell's mole-rats show similar longevity records and – more importantly and interestingly – show differential life expectancy patterns between breeding and non-breeding animals. The authors summarize in Chap. 18 what is known about aging patterns in subterranean rodents and argue that the Bathyergidae provide an interesting substrate to examine the influence of diverse factors on senescence. At the same time the authors show that our knowledge is still strongly biased in favour of very few mole-rat species.

How and Where to Go from Here?

To address questions about ultimate mechanisms of evolution of life history patterns we still need: (a) much more comparative data, involving diverse taxa and habitats; (b) control for phylogeny being a substantial part of ecological studies; (c) efforts to falsify rather than only to verify existing hypotheses; (d) long term demographic studies involving monitoring of individual life histories in the field – such studies (using, e.g. genetic and capture-mark-recapture methods) are in progress into naked mole-rats (Braude et al., this volume), in the Damaraland mole-rats (Bennett et al., this volume) or in tuco-tucos (cf. Lacey and Cutrera, this volume). Results of thorough long term field studies on Siberian mole-voles (*Ellobius talpinus*) published by Yevdokimov (2001) show interesting parallels in many aspects of behavioural ecology and life histories of these arvicoline rodents to eusocial African *Fukomys* mole-rats. These findings, so far published only in Russian, are, unfortunately still unknown to Western scientists, and mole-voles, which also provide an example par excellence for morphological convergence with mole-rats, are seldom, if ever, considered in models of evolution of sociality. Further study of Siberian social mole-voles and their counterparts, sympatrically living solitary zokors, will surely bring new insights into evolution of life histories in mammals.

Furthermore, although proximate and ultimate explanations represent two different levels of analysis, they are not mutually independent. Particularly considering the proximate aspects may forward our understanding of evolution of the respective trait in ultimate terms.

Last but not least, we should always keep in mind that the objects of our interest, however unique they appear, are mammals. Their solutions to given problems are basically mammalian ones. To understand many behavioural or life history traits, it may be quite inspiring to look around at other mammals – in the case of bathyergid mole-rats particularly at other hystricognaths or arvicoline voles (cf. Burda 1999; Burda et al. 2000).

References

Bennett NC, Faulkes CG (2000) African mole-rats: ecology and eusociality. Cambridge Univ Press, Cambridge

Bishop JM, Jarvis JUM, Spinks AC, Bennett NC, O'Ryan C (2004) Molecular insight into patterns of colony composition and paternity in the common mole-rat *Cryptomys hottentotus hottentotus*. Mol Ecol 13:1217–1229

Braude S (2000) Dispersal and new colony formation in wild naked mole-rats: evidence against inbreeding as the system of mating. Behav Ecol 11:7–12

Buffenstein R (2005) The naked mole-rat. A new long-living model for human aging research? J Gerontol A 60:1369–1377

Buffenstein R, Jarvis JUM (2002) The naked mole rat – a new record for the oldest living rodent. Sci Aging Knowl Environ 21:7

Burda H (1999) Syndrome of eusociality in African subterranean mole-rats (Bathyergidae, Rodentia), its diagnosis and aetiology. In: Wasser SP (ed) Evolutionary theory and processes: modern perspectives. NL-Dordrecht: Kluwer Academic Publ, pp 385–418

Burda H, Honeycutt RL, Begall S, Locker-Grütjen O, Scharff A (2000) Are naked and common mole-rats eusocial and if so, why? Behav Ecol Sociobiol 47:293–303

Burland TM, Bennett NC, Jarvis JUM, Faulkes CG (2002) Eusociality in African mole-rats: new insights from patterns of genetic relatedness in the Damaraland mole-rat (*Cryptomys damarensis*). Proc R Soc Lond B 269:1025–1030

Ciszek D (2000) New colony formation in the "highly inbred" eusocial naked mole-rat: outbreeding is preferred. Behav Ecol 11:1–6

Clarke FM, Faulkes CG (2001) Intracolony aggression in the eusocial naked mole-rat, *Heterocephalus glaber*. Anim Behav 61:311–324

Dammann P, Burda H (2006) Sexual activity and reproduction delays aging in a mammal. Curr Biol 16:R117–R118

Hazell RWA, Bennett NC, Jarvis JUM, Griffin M (2000) Adult dispersal in the co-operatively breeding damaraland mole-rat (*Cryptomys damarensis*): a case study from the Waterberg region of Namibia. J Zool Lond 250:19–25

Ingram CM, Burda H, Honeycutt RL (2004) Molecular phylogenetics and taxonomy of the African mole-rats, genus *Cryptomys* and the new genus *Coetomys* Gray, 1864. Mol Phyl Evol 31:997–1014

Jarvis JUM (1981) Eusociality in a mammal: cooperative breeding in naked mole-rat colonies. Science 212:571–573

Kock D, Ingram CM, Frabotta LJ, Burda H, Honeycutt RL (2006) On the nomenclature of Bathyergidae and *Fukomys* n. gen. (Mammalia: Rodentia). Zootaxa 1142:51–55

Lacey EA, Wieczorek JR (2003) Ecology of sociality in rodents: a ctenomyid perspective. J Mammal 84:1198–1211

Nevo E (1979) Adaptive convergence and divergence of subterranean mammals. Ann Rev Ecol Syst 10:269–308

Nevo E (1999) Mosaic evolution of subterranean mammals: regression, progression and global convergence. Oxford Univ Press, Oxford

Scantlebury M, Speakman JR, Oosthuizen MK, Roper TJ, Bennett NC (2006): Energetics reveals physiologically distinct castes in a eusocial mammal. Nature 440:795–797

Sherman PW, Jarvis JUM (2002) Extraordinary life spans of naked mole-rats (*Heterocephalus glaber*). J Zool Lond 258:307–311

Sherman PW, Jarvis JUM, Alexander RD (1991) The biology of the naked mole-rat. Princeton Univ Press, Princeton, NJ

Spinks AC, Jarvis JUM, Bennett NC (2000) Comparative patterns of philopatry and dispersal in two common mole-rat populations: implications for the evolution of mole-rat sociality. J Anim Ecol 69:224–234

Yevdokimov NG (2001) Population ecology of the common mole-vole (in Russian). Russ Acad Sci (Ural Sect) Press, Yekaterinburg

15 Giant Mole-rats, *Fukomys mechowii*, 13 Years on the Stage

Mathias Kawalika†, Hynek Burda

15.1 Introduction

Fukomys – a new genus of bathyergid mole-rats recently emancipated from the genus *Cryptomys* – comprises African mole-rat species from out of South Africa (Ingram et al. 2004; Kock et al. 2006). *Fukomys* mole-rats are remarkably polymorphic (Rosevear 1969; Ansell 1978; Honeycutt et al. 1991; Burda et al. 1999), so that it is not possible to provide unambiguous diagnostic morphological traits or measurements. Whereas most of the *Fukomys* and *Cryptomys* mole-rat species show only subtle morphological differences, and can be diagnosed on their karyological, genetic or allozymic base, one species is markedly different: the giant mole-rat, *Fukomys mechowii*. In fact, indigenous people in various localities of Zambia distinguish between giant and "common" mole-rats even by name (*kakoko* and *shilufukwe* or *mfuko*, respectively).

In the present chapter we summarise the studies on the giant mole-rat performed during the last 15 years. Until the publication of our first report (Burda and Kawalika 1993) the biology of giant mole-rats was virtually unknown, and Kingdon (1974) encompassed all the then available knowledge in one sentence: "*Cryptomys mechowi* is a little known form of reportedly solitary habits". As we have shown the assumption about solitariness of giant mole-rats was fully wrong.

The giant mole-rat (*F. mechowii*) stands out from numerous *Fukomys* species due to its large body size, and distribution in mesic areas of the subequatorial central Africa (Democratic Republic of Congo, Angola and Zambia; Honeycutt et al. 1991; Burda 2001). This species may help to elucidate the evolution and adaptive significance of eusociality in African mole-rats (Burda and Kawalika 1993), a question, which is not yet solved (Burda et al. 2000).

Mathias Kawalika†: Ndola, Zambia

Hynek Burda: Dept. General Zoology, Institute for Biology, University of Duisburg-Essen, Universitätsstr. 5, D-45117 Essen, Germany, E-mail: hynek.burda@uni-due.de

Subterranean Rodents: News from Underground
S. Begall, H. Burda, C.E. Schleich (Eds.)
© Springer-Verlag Berlin Heidelberg 2007

Whereas the taxonomy (Faulkes et al. 1997, 2004; Filippucci et al. 1997; Macholan et al. 1993; Faulkes et al. 2004; Ingram et al. 2004), parasitology (Scharff et al. 1996, 1997), reproduction (Bennett and Aguilar 1995; Bennett et al. 2000; Burda and Kawalika 1993; Scharff et al. 1999), behaviour of captive animals (Credner et al. 1997; Wallace and Bennett, 1998; Heth et al. 2002), neuroanatomy (Lindenlaub and Burda 1994; Peichl et al. 2004) and physiology (Bennett et al. 1994) of *F. mechowii*, have been studied, information on free living colonies is still meagre and anecdotal (Burda and Kawalika 1993; Hill 1941, Scharff et al. 2001).

15.2
Taxonomy, Nomenclature and Phylogeny

The giant mole-rat was first described under the name *Georychus mechowii* by Peters (1881) from Malange, North Angola, reference grid 0916C. Subsequent authors recognized that the species belongs to the genus *Cryptomys*, but did not replicate the spelling of the species name accurately, so that for many years "*mechowi*" has become familiar (cf. Allen 1939; Ellerman et al. 1953). The correct spelling is, however, "*mechowii*". Thomas (1906) described *Cryptomys mellandi* from Mpika, reference grid 1131C, which was later considered a subspecies or synonym of the giant mole-rat, *C. mechowii* (cf. Ansell 1978). Our knowledge of ecology and biology of the giant mole-rat and its systematic relationships to other members of the genus is based on the study of animals from the Copperbelt Province in Zambia. Molecular studies revealed that, despite large geographic distance, these mole-rats belong to the same species as giant mole-rats from Kinshasa, Dem. Rep. Congo, ref. grid 0415A (Faulkes et al. 1997, 2004). Nevertheless, until mole-rats from Malange and Mpika, i.e. type localities, were examined, calling the giant mole-rats from the Copperbelt Province (and those from Kinshasa) as "*mechowii*" has to be considered as preliminary. Molecular phylogenies identified *F. mechowii* together with its sister taxon, *F. bocagei* from Angola, as the most basal lineage of a distinct clade of mole-rats (Faulkes et al. 2004; Ingram et al. 2004). Ingram et al. (2004) argued that this clade should be given status of a separate genus and called it, *in errore*, *Coetomys*. The nomenclature was thoroughly revised and a new genus name *Fukomys*, was coined (Kock et al. 2006).

The low chromosome number ($2n = 40$), lowest among all bathyergids studied thus far, and a high proportion of biarmed chromosomes in the karyotype (Macholan et al. 1993) seem to be related to the basal position (suggesting earlier radiation) in the phylogenetic tree (cf. Faulkes et al. 2004; Ingram et al. 2004). Thanks to the availability of molecular cladograms, our previous interpretation of chromosomal speciation in *Cryptomys-Fukomys* (Kawalika et al. 2001) has to be reconsidered. We assume that chromosomal speciation in *Fukomys* was characterized by Robertsonian fission rather than fusion. Interestingly, allozymatic traits (Filippucci et al. 1997) and olfactory mating preferences (Heth et al. 2002) were less reliable to reveal phylogenetic relationships. We suggest that the differentiation rate of these traits is adaptive and, at least in the case of mating preferences, reflects also the speciation history and the role of isolation barriers.

15.3 Morphology of the Giant Mole-rats

15.3.1 Body size

Both *Fukomys mechowii* and its sister taxon *F. bocagei* are of larger body sizes and represent a basal lineage of the genus (Faulkes et al. 2004; Ingram et al. 2004). In fact, the larger body size has been considered an ancestral trait in bathyergids (Jarvis and Bennett 1991).

Wild-captured adult males (380 ± 95 g, range 250–600 g, $n = 20$) are significantly heavier than adult non-pregnant females (260 ± 35 g, range 200–355 g, $n = 20$; Scharff et al. 2001). Additionally, M.K. captured a male wandering aboveground, in Ndola, weighing 700 g. On the other hand, adult giant mole-rats from Chibale and Ndola-Chichele which (though being of the typical giant mole-rat's karyotype (see above) were rather small (about 200 g) and even after five years in captivity have not grown up.

Reproductive males have broad heads due to massive chewing muscles. The sexual dimorphism in body mass in *F. mechowii* may be an evidence of intrasexual competition, specifically behavioural suppression and hierarchy among males. Although the incest avoidance, as in smaller *F. anselli* (cf. Burda 1995), seems to determine the maintenance of the family structure and "monopolization" of reproduction in giant mole-rats, (Bappert, unpublished) we suggest that incest avoidance between the mother and sons is weaker than that between father and daughters or sisters and brothers, and some sort of behavioural reproductive suppression in the male line may exist (unpublished). It may be also of significance that the hitherto, although still relatively rare, findings of the surface wandering mole-rats were all males (see below).

15.3.2 Pelage Colour

Pups and juvenile giant mole-rats are dark grey or dark brown, and with ageing they become subsequently paler and more brownish. Old adults are golden ochre. The same colour age-related polymorphism was reported also in *F. anselli* (cf. Burda 1989). Since mole-rats are blind and live in darkness of their burrows it is doubtful that the colour polymorphism could be of adaptive significance. Also it should be noted that mole-rats occur in soils of different type and colours, yet there is no apparent difference in their pelage colour which could be correlated with soil coloration (contrary to the situation in blind mole-rats, *Spalax ehrenbergi*, from Israel, cf. Heth et al. 1988). We suggest that the predictable colour age-polymorphism is a plesiomorph trait in the *Fukomys* genus. The fact that, e.g. *F. damarensis* or *Fukomys* from Southern Malawi (Šumbera, unpublished) do not display these changes could reflect their recent origin (a kind of "neoteny"). Noteworthy, adaptive coloration has been suggested in the naked mole-rat where countershaded animals may be better camouflaged during their dispersal run. Colour is lost in older reproductives who have already made their aboveground dispersal and will never leave the burrow again (Braude et al. 2001).

The white head-spot is missing in most animals. In some few individuals (about one per cent) a whitish patch is apparent on the head. A white head-spot is typical of *F. anselli*, *F. kafuensis*, *F. damarensis*, *F. micklemi* and other *Fukomys* species from southern and western Zambia (southwards of the 15th latitude) (cf. Burda et al. 1999; Van Daele et al. 2004) but is missing in most *F. mechowii* and in the examined "Kasama" mole-rats, *F. amatus*, and *F. whytei* (cf. Macholan et al. 1998; Kawalika et al. 2001; Burda et al. 2005). It may be of significance that the head-spots were small or missing also in "common" mole-rats collected in the North-Western province (Van Daele et al. 2004). Even in species where the head-spot is present it is variable in size and shape. Lovegrove et al. (1993) speculated about the meaning of the white head-spot and suggested that, being actually an unpigmented part of skin, it may facilitate penetration of light to the pineal organ and thus be involved in photoperception and control of photoperiodicity. However, chronobiological data for *Cryptomys* and *Fukomys* are very ambivalent and large polymorphism in the size and shape of the white head-spot, even in its very presence, weaken the idea of its functional significance. Nevertheless, based on preliminary survey, we suggest that there is a latitude gradient of the average size of the head-spot, with a trend to increase southwards and correlating thus with a prolonging photoperiod. Interestingly, the white head-spot occurs also in *Heliophobius* and *Cryptomys* (and indeed in diverse mammalian species) and is thus a trait, which may have arisen independently, and parallelly, in different lineages – indicating its adaptive value. The patchy distribution of the white head spot and the variation in the character suggests active evolution of this character.

15.3.3 Skull

Skulls ($n = 10$) display large, thin-walled, elliptical infraorbital foramina. The finding contradicts the assumption by Honeycutt et al. (1991) that thick-walled outer foramina characterize the *Fukomys* (*F. damarensis*, *F. mechowii* and *F. bocagei*) group, while thin-walled foramina should characterise the *Cryptomys* group. In any case, it seems that the size and shape of the infraorbital foramen can be used as a species-diagnostic trait, while it is probably not useful as a correlate of phylogenetic relatedness.

15.4 Distribution

15.4.1 Geographic Distribution

Fukomys mechowii have been collected in Zambia from diverse localities in the Copperbelt Province, Central Province (near Chibale, cf. Scharff et al. 2001, Kawalika 2004 and unpublished), and from the North-Western Province in Salujinga and Watopa (Van Daele et al. 2004). On the other hand, the occurrence of the giant mole-rat in Malawi and in the Southern Province of Zambia as reported by Ansell (1978) could not be confirmed. It is unclear whether the populations in Kinshasa in the Democratic Republic of Congo (Faulkes et al. 1997, 2004; Kisasa et al. 2004; Palata-Kabudi et al. 2005; Palata-Kabudi, personal communication) and Malange in Angola (Peters 1881) represent isolates or whether there is a continuous distribution.

15.4.2 Habitat Characteristics

In Zambia, this species is found in a variety of habitats in cultivated fields, bushland, marshland (the so-called dambos) and forests (pine forest as well as dense acacia forest), and in a variety of soil types. Soil characteristics vary vertically as well as horizontally. In most soils (but in the lateritic ferralsol in the acacia forest) digging was rather easy throughout the year in upper horizons (Scharff et al. 2001 and unpublished own observations). In the D.R. Congo, *F. mechowii* occupies mesic savannas (cultivated fields, tree orchards, and bare land) with sandy soils of the acolian type (Kisasa et al. 2004; Palata-Kabudi et al. 2005; Palata-Kabudi, personal communication). Regions of occurrence are generally of mesic character (annual precipitation 940–1,435 mm) yet differing in climatic regimes: whereas there are seven virtually rainless months (precipitation below 25 mm/month) in Mpika or Kabwe, there are only three rainless months in Kinshasa.

15.4.3 Distributional Borders and Sympatry with Other *Fukomys* Species

Besides the distributional boundary corresponding roughly to the precipitation isoline of 900 mm annual rainfall and 80 annual rainy days, there seems to be a clear geo-morphological barrier, separating both distinct forms of mole-rats: the Muchinga Escarpment, the Mulungushi River, the Lukanga swamps and the Kafue River. In the South-West of Zambia, giant mole-rats seem to be restricted in their distribution by Kalahari sands and higher temperatures.

In Zambia, at the western edge of distributional area, in North-Western Province, in Watopa, the giant mole-rat occurs in sympatry with "common" small mole-rats ($2n = 56$), in Salujinga with those having a karyotype $2n = 44$ (Van Daele et al. 2004). In the Central Province, in the South, the giant and common mole-rats (probably *F. anselli*) live sympatrically around Kabwe; to the East, in Chibale, *F. mechowii* occurs sympatrically with *F. amatus* ($2n = 50$) (Scharff 1998; Macholan et al. 1998).

Existence of smaller forms of this species can be surmised also in Kitwe, where the tunnel diameter was 7.5–8.5 cm. In Kasama, local hunters assured M.K. that there are four types of mole-rats in the area: common (small) mole-rats; giant mole-rats; multi-coloured (white belly, brown upper body) giant mole-rats; and white super giant mole-rats. Further studies are needed to verify or falsify the claims of local hunters.

15.5 Burrow Systems

The smallest burrow systems (encompassing 0.2 ha) occurred at dambos. Burrow systems in acacia forests were slightly larger while those in pine forests and open fields encompassed about 2–3 ha. The number of mounds in areas occupied by mole-rats amounts from 1–2 to 18 per 100 m^2 (up to 225 mounds in an area of 25 × 50 m). Building of mounds (but not digging per se) seems to be a seasonal affair restricted during the dry season.

The main structure of burrow systems of *F. mechowii* was similar to other members of the genus as well as for other Bathyergidae (Hickman 1979, 1990; Jarvis and Bennett 1991; Nevo 1999; Bennett and Faulkes 2000; Šumbera et al. 2003b) and consisted of nests, food chambers, defecation chambers, superficial foraging burrows and deeper connecting tunnels (see Scharff et al. 2001 for details). All burrow systems are sealed with mounds and additional plugs. Diameter of tunnels is on average 8 cm (range 6.5–11.5 cm). The main tunnel ("runway") in one examined burrow system has been 200 m long, 30 cm deep, and mounds being 2–3 m apart.

The level from the surface to the roof of the tunnel was minimum 5 cm, maximum about 200 cm. There was no correlation between the diameter (range 8.3–11.5 cm) and the depth of the tunnels (15–60 cm) (n = 61, corr. coeff. 0.155) studied in two burrow systems, indicating that also adult grown up animals are burrowing and foraging near the surface.

Eight nests (diameter: 20–40 cm; height: 10–20 cm) from four colonies were excavated in an average depth of 90 ± 40 (range 50–160) cm. Consistently, Kisasa et al. (2004) reported a nest in a depth of 80 cm in the D.R. Congo. In some areas (Chichele), nests seem to be located always in termite hills, which may be a strategy to avoid the nest being flooded during the rainy season. Grass and root fibres constitute common litter material for nests of mole-rats. In one nest excavated in D.R. Congo, *Aframomum stipilatus* was identified as nest material (Kisasa et al. 2004). Use of plastic materials as nest materials has also been reported in other subterranean rodents (cf. Burda et al., this volume). In general, food and defecation chambers are located near the nest (cf. also Kisasa et al. 2004), a strategy encountered in other bathyergids and subterranean rodents (Nevo 1999).

15.6 Food

Giant mole-rats from cultivated fields feed on sweet potatoes, Irish potatoes, cassava, groundnuts, ginger, and corn while the plant diet in non-cultivated areas involve grass rhizomes (particularly in forests), roots, bulbs, and tubers of diverse weeds, shrubs, and trees. Other food plants of *F. mechowii* in the Copperbelt province in Zambia are *Albizia* sp., *Aframomomum biauriculatum*, *Crossopteryx febrifuga*, *Lannea discolor*, *Mucuna* sp., *Pristachya* sp., *Pseudolachnostylis* sp., *Rhynchosia resinosa*, *Sissus* sp., *Steganotaenia* sp., *Syzygium guineense*, *Tacca* sp., and *Tephrosia* sp. In D.R. Congo, food chambers contained 10–25 kg of tubers, bulbs and roots of wild plants, mainly

Aframomum stipilatus, Anisophyllea pongei, Aspargus africanum, Brachycorythis pleistophyllea, Crinum jagus, Euphilopus sp., *Pteridium aquilinum* (Kisasa et al. 2004; Palata-Kabudi et al. 2005). We have also found three separate small storage chambers which contained animals prey (about 20 items in each: earthworms and larvae of a scarabeid beetle, probably *Anomala* sp.) (Burda and Kawalika 1993). Carnivorous habits have been evidenced by the examination of stomach contents and by recording feeding habits of captive animals (own observations) as well as by the parasite spectrum (Scharff et al. 1997).

Similar to other species of *Fukomys*, giant mole-rats their food in special chambers, a feature prevalent in subterranean mammals (Nevo 1999). Dependent on location of the burrow system (i.e. cultivated fields or uncultivated landscape), these animals construct food chambers either filled with mixed contents or filled only with one available sort of food. We have not found diversified food caches in field colonies, perhaps due to the uniformity and prevalence of food (e.g. sweet potatoes) in some sites or a large diversity of food types at other sites. However, (separate) storing of animal prey should be considered an unusual feature and indeed this has not been recorded in any other subterranean rodent so far (Burda and Kawalika 1993).

15.7 Colony Size and Structure

Thus far, no complete families originating with certainty from a single burrow system have been captured. Nevertheless, family size of more than 20 members could be confirmed (Scharff et al. 2001). According to experienced mole-rat hunters, colonies of 40 individuals are not exceptional and they may consist of up to sixty or more members (Burda and Kawalika 1993). Mole-rat hunters and farmers have confirmed that a single family may persist at one place for at least ten years.

Three age-classes are distinguished according to body weight: juvenile (< 50 g), subadult (50–200 g in females and 50–250 g in males) and adult (> 200 g in females, and > 250 g in males). This division is based on reproductive and growth data in captive animals (Scharff et al. 1999) with average weight at weaning (being reached at about three months) and the weight at maturity (first successful breeding). In a sample of 72 mole-rats captured within two months (July–August) at the peak of the dry season adults amounted to 53%, subadults to 39% and juveniles to 8%. The estimated age of the juveniles and subadults suggests that there is no distinct breeding period for Zambian giant mole-rats. This is also supported by reports of interviewed hunters, by a long term laboratory study (Scharff et al. 1999) and observations of Ansell (1978), who caught new-born animals throughout the year. Similarly, *Fukomys anselli* (Burda 1989, 1990), *F. damarensis*, and *H. glaber* (Jarvis and Bennett 1991) are continuous non-seasonal breeders.

Juveniles make a small proportion of the sample (8%), suggesting that recruitment, dispersal and turnover of giant mole-rats are very low. This contrasts with naked mole-rats (Braude 2000) with very high recruitment and high losses due to predation or dispersal. Apparently, large colony sizes in these eusocial species are attained in different ways (Burda et al. 2000). Although a low proportion of juveniles is common in subterranean rodents (Nevo 1979, 1999; Busch et al. 2000), the percentage of juvenile giant mole-rats is one of the lowest reported.

15.8 Sex Ratio

The sex ratio among wild captured young (< 1 year) giant mole-rats is male-biased (1.22, $n = 40$) but among adults (> 1 year), the proportion of males decreases (0.96, $n = 45$) (Table 15.1). Taking all wild-captured animals together, the overall sex ratio is equal (1.07). Sampling was roughly equally distributed throughout space and time as well as with respect to capture methods so that there was probably no bias due to the sampling method. Data on captive colonies are inconsistent to the field data due to a strong female biased neonate sex ratio of 0.54 among laboratory-born animals (Scharff et al. 1999) and could be explained as follows:

1. The skew towards neonate females is an artefact of captivity breeding and does not reflect the natural situation. A reverse (i.e., male biased) sex ratio should be expected among neonates in the field to explain a higher proportion of males among young wild-captured mole-rats.

Table 15.1. Sex ratio in wild-captured and captive-born giant mole-rats

	Wild captured				Captivity
		> One year	< One year		
	Total	Adult > 250 g (M) > 200 g (F)	Subadult 50–250 g (M) 50–200 g (F)	Juvenile 25–50 g	Neonate < 20 g
Total	85	45	31	9	13
Male	44	22	17	5	36
Female	41	23	14	4	67
Sex ratio	1.07	0.96	1.21	1.35	0.54
			1.22		

M = males, F = females

2. With growing older, males disappear from the population. Indeed, there are slight indications for higher dispersal of adult males. Thus far we have recorded three giant mole-rats wandering on the surface (see below) – in all three cases very heavy, adult males were involved. On the other hand, there is no indication for higher intrinsic mortality in captive male giant mole-rats (see Dammann and Burda, this volume).

3. There is a sex-linked age-polyethism, expressed in higher activity and/or trapability of subadult males. With increasing age, males may become more cautious or less active. Indeed, the hitherto observations on captive animals provide support for this interpretation (Dammann 2006).

Finding answers to the questions, why sex ratios in newborn captive-bred giant mole-rats deviate so markedly from equality, may prove to be significant for getting insight into ecology of mole-rats in particular, and for understanding phenomena affecting sex-ratios in mammals in general. The exciting hypothesis, suggested here, could explain the phenomenon and should be tested: In *F. mechowii*, two types of females occur: homogametic females: XX and heterogametic females: XY*. Due to early loss of YY* zygotes, XY* females would produce twice as many phenotypical females as males. XY* females have been described in the South American field mice (*Akodon*) where XY* females persist at high frequencies (up to 30%) and are better breeders than XX females (Bianchi 2002). An unusual sex determination system has been found also in an unrelated subterranean rodent: the mole-lemming, *Ellobius* (Just et al. 2002). Existence of heterogametic (XY*) female giant mole-rats at different frequencies in the populations and in our captive breeding stock and higher mortality of younger males (observed also in breeding colonies) could explain all the observed phenomena.

15.9 Surface Activity

Several independent mole-rat hunters have reported to us that (giant) mole-rats used to be observed and picked up on the open ground, during daylight, away from their burrow system. It appears that most cases of finding a mole-rat on the ground occur between March and May which in Zambia coincides with the end of rains, as has also been found in naked mole-rats (Braude 2000), and with the harvesting season. M. Kawalika himself has evidenced surface activity of adult males on three occasions (February, April, May). Diverse items found in nests of mole-rats (plastic bag remains, nylon stockings, green leaves, nuts) seem to be collected aboveground (although the alternative explanation is that they have been partly buried during farming activities and were pulled down from below).

Interestingly, the period when mole-rats found aboveground are mostly recorded coincides with the period when breeding dispersal and mating takes place also in the solitary silvery mole-rats (*Heliophobius argenteocinereus*) in neighbouring Malawi (Šumbera et al. 2003a). However, breeding dispersal in mole-rats is apparently provoked by an incidental encounter of two unfamiliar animals of opposite sex (own laboratory observations). It should be noted that families of mole-rats are very xenophobic and any intruder will be immediately attacked and killed by residents of the same sex. Also, breeding in mole-rats is seasonally not limited.

In summary, although abundant food and soft soil may promote establishing of new territories (burrow systems), other facts do not support the idea of a synchronised seasonal breeding dispersal (searching for mate) in mole-rats. Also, at the end of the rainy period, rains become more sporadic and flooding of burrow systems cannot explain more observations of aboveground-moving mole-rats. The higher agricultural activity and associated disturbances can be excluded as well: burrow systems of mole-rats are extensive and deep; farmers use simple tools (hoes) and mole-rats, if disturbed, escape in the depth and not on the surface. Last but not least, the observations have been made in the non-cultivated land as well. There is no obvious reason why aboveground foraging should be concentrated in time when also underground food is abundant and soil is well workable.

Noteworthy, the analysis of owl pellets of two barn owls (nesting and hunting in an area where *F. mechowii* is abundant) collected over four years (Kawalika 2004, and unpublished) has not revealed any remains of mole-rats. However, adult giant mole-rats (weight above 250 g) may be too big a prey item for this bird. In any case, it can be concluded that either the younger mole-rats (less than one year old) do not dare to go aboveground or do not do it at night when owls are hunting. Noteworthy, skulls of young *Heliophobius* were found in a much smaller sample of barn owl pellets from Malawi (Šumbera et al., this volume).

15.10 Economic Importance

Though for a long time virtually unknown to scholars (see Introduction), giant mole-rats are well known to the local population: they are serious pests to many crops (like cassava, sweet potatoes, groundnuts), their meat is considered delicious and highly appreciated, and nest materials and parts of animals are used as (medicinal and magical) charms. This information refers not only to the situation in Zambia (cf., also Burda and Kawalika

Fig. 15.1. Meeting the street vendors of giant mole-rats outside Ndola City. Ladies on the photo are, *from the left*: Doyen Kawalika and Simone Lange – the dominating smiling person on the right is the late Dr. Mathias Kawalika. Photo by courtesy of Simone Lange

1993) but also to that in D.R. Congo (Kisasa et al. 2004; Palata-Kabudi, personal communication). Some people have specialized in hunting these animals. An experienced hunter captures yearly 600–800 giant mole-rats, visiting as many burrow systems as possible. Captured giant mole-rats in the Copperbelt Province are sold at markets and by vendors along the main roads (Fig. 15.1). Giant mole-rats are serious agricultural and horticultural pests, both in Zambia and in the D.R. Congo. There are reported cases at the Ministry of Agriculture of Zambia that fields have been abandoned due to damage to the crop which can reach as high as over 95%. In spite of that there have been no systematic studies done to assess the amount of damage and to suggest control or management strategies.

Acknowledgements. The field research of the late Dr. Mathias Kawalika was enabled to a great extent thanks to the generous support by the Zoological Garden - Freunde Verein e.V. in Basel, Switzerland.

References

Allen GM (1939) A checklist of African mammals. Bull Mus Comp Zool Harvard Coll 83:1–763

Ansell WFH (1978) The mammals of Zambia. National Parks and Wildlife Service, Chilanga, Zambia: 1–237

Bennett NC, Aguilar GH (1995) The reproductive biology of the Zambian mole-rat, *Cryptomys mechowi* (Rodentia: Bathyergidae). S Afr J Zool 30:1–4

Bennett NC, Faulkes CG (2000) African mole-rats: ecology and eusociality. Cambridge Univ Press, Cambridge, NY

Bennett NC, Aguilar GH, Jarvis JUM, Faulkes CG (1994) Thermoregulation in three species of Afrotropical subterranean mole-rats (Rodentia: Bathyergidae) from Zambia and Angola and scaling within the genus *Cryptomys*. Oecologia 97:222–227

Bennett NC, Molteno AJ, Spinks AC (2000) Pituitary sensitivity to exogenous GnRH in giant Zambian mole-rats, *Cryptomys mechowi* (Rodentia: Bathyergidae): support for the 'socially induced infertility continuum'. J Zool Lond 252:447–452

Bianchi NO (2002) *Akodon* sex reversed females: the never ending story. Cytogen Genome Res 96:60–65

Braude SH (2000) Dispersal and new colony formation in wild naked mole-rats: evidence against inbreeding as the system of mating. Behav Ecol 11:7–12

Braude S, Ciszek D, Berg N, Shefferley N (2001) The ontogeny and distribution of countershading in colonies of the naked mole-rat, *Heterocephalus glaber*. J Zool Lond 253: 351–357

Burda H (1989) Reproductive biology (behaviour, breeding, and postnatal development) in subterranean mole-rats, *Cryptomys hottentotus* (Bathyergidae). Z Säugetierk 54: 360–376

Burda H (1990) Constraints of pregnancy and evolution of sociality in mole-rats. Z Zool Syst Evol-forsch 28:26–39

Burda H (1995) Individual recognition and incest avoidance in eusocial common mole-rats rather than reproductive suppression by parents. Experientia 51:411–413

Burda H (2001) Determinants of the distribution and radiation of the African mole-rats (Bathyergidae, Rodentia). Ecology or geography? In: Denys C, Granjon L, Poulet A (eds) African small mammals. Collection colloques et séminaires, Institut de Recherche pour le Developpement Editions, Paris, France, pp 263–277

Burda H, Kawalika M (1993) Evolution of eusociality in the Bathyergidae: The case of the giant mole-rat (*Cryptomys mechowi*). Naturwiss 80:235–237

Burda H, Zima J, Scharff A, Macholan M, Kawalika M (1999) The karyotypes of *Cryptomys anselli* sp. nova and *Cryptomys kafuensis* sp. nova: new species of the common mole-rat from Zambia (Rodentia, Bathyergidae). Z Säugetierk 64:36–50

Burda H, Honeycutt RL, Begall S, Locker-Grütjen O, Scharff A (2000): Are naked and common mole-rats eusocial and if so, why? Behav Ecol Sociobiol 47:293–303

Burda H, Šumbera R, Chitaukali WN, Dryden GL (2005) Taxonomic status and remarks on ecology of the Malawian mole-rat, *Cryptomys whytei* (Thomas, 1897) (Rodentia, Bathyergidae). Acta Theriol 50:529–536

Busch C, Antinuchi CD, Del Valle JC, Kittlein MJ, Malizia AI, Vassallo AI, Zenuto R (2000) Population ecology of subterranean rodents. In: Lacey EA, Patton JL, Cameron GN (eds) Life underground: the biology of subterranean rodents. Univ Chicago Press, Chicago, pp 179–226

Credner S, Burda H, Ludescher F (1997) Acoustic communication underground: vocalization characteristics in subterranean social mole-rats (*Cryptomys* sp., Bathyergidae). J Comp Physiol A 180:245–255

Dammann P (2006) Sexualität und Seneszenz bei eusozialen Graumullen (*Fukomys*, Bathyergidae, Rodentia), PhD Thesis, Univ of Duisburg-Essen, Essen

Ellerman JR, Morrison-Scott TCS, Hayman RW (1953) Southern African mammals 1758–1951: a reclassification, Trustees of the British Museum (Natural History), London

Faulkes CG, Bennett NC, Bruford MW, O'Brien HP, Aguilar GH, Jarvis JUM (1997) Ecological constraints drive social evolution in the African mole-rats. Proc R Soc Lond B 264: 1619–1627

Faulkes CG, Verheyen E, Verheyen W, Jarvis JUM, Bennett NC (2004) Phylogeographic patterns of speciation and genetic divergence in African mole-rats (Family Bathyergidae). Mol Ecol 13:613–629

Filippucci MG, Kawalika M, Macholan M, Scharff A, Burda H (1997) Allozyme differentiation and taxonomic status of Zambian giant mole-rats, *Cryptomys mechowi* (Bathyergidae, Rodentia. Z Säugetierk 62:172–178

Heth G, Beiles A, Nevo E (1988) Adaptive variation of pelage color within and between species of the subterranean mole rat (*Spalax ehrenbergi*) in Israel. Oecologia 74: 617–622

Heth G, Todrank J, Burda H (2002) Individual odours and social recognition: similarity in the qualities of individual odours within colonies and across species of African eusocial mole rats (*Cryptomys* sp.). J Mammal 83:569–575

Hickman GC (1979) Burrow system structure of the Bathyergid *Cryptomys hottentotus* in Natal, South Africa. Z Säugetierk 44:153–162

Hickman GC (1990) Adaptiveness of tunnel system features in subterranean mammal burrows. In: Nevo E, Reig OA (eds) Evolution of subterranean mammals at the organismal and molecular levels. Wiley-Liss, New York, pp 185–210

Hill JE (1941) A collection of mammals from Dondi, Angola. J Mammal 22:81–85

Honeycutt RL, Allard MW, Edwards SV, Schlitter DA (1991) Systematics and evolution of the family Bathyergidae. In: Sherman PW, Jarvis JUM, Alexander, RD (eds) The biology of the naked mole-rat. Princeton Univ Press, Princeton, NJ, pp 46–65

Ingram CM, Burda H, Honeycutt RL (2004) Molecular phylogenetics and taxonomy of the African mole-rats, genus *Cryptomys* and the new genus *Coetomys* Gray, 1864. Mol Phylogeny Evol 31:997–1014

Jarvis JUM, Bennett NC (1991) Ecology and behavior of the family Bathyergidae. In: Sherman PW, Jarvis JUM, Alexander, RD (eds) The biology of the naked mole-rat. Princeton Univ Press, Princeton, NJ, pp 66–96

Just W, Baumstark A, Hameister H, Schreiner B, Reisert I, Hakhverdyan M, Vogel W (2002) The sex determination in *Ellobius lutescens* remains bizarre. Cytogen Genome Res 96:146–153

Kawalika M (2004) Rodents of Ndola (Copperbelt Province, Zambia). PhD Thesis. Univ Duisburg-Essen, Essen

Kawalika M, Burda H, Brüggert D (2001) Was Zambia a cradle of the genus *Cryptomys* (Bathyergidae, Rodentia)? A further new ancestral (?) species of *Cryptomys* from Zambia. In: Denys C, Granjon L, Poulet A (eds) African small mammals IRD Editions, Collection Colloques et seminaires, Paris, pp 253–261

Kingdon J (1974) East African mammals. An atlas of evolution in Africa. Vol. 2. Hares and rodents. Academic Press, London New York

Kisasa R, Palata JC, Pwema V (2004) Premiers essais d'élevage du rat-taupe (*Cryptomys mechowi*, Rongeur: Bathyergidae) à Kinshasa, République démocratique du Congo. Tropicultura 22:180–184

Kock D, Ingram CM, Frabotta LJ, Burda H, Honeycutt RL (2006) On the nomenclature of Bathyergidae and *Fukomys* n. gen. (Mammalia: Rodentia). Zootaxa 1142:51–55

Lindenlaub T, Burda H (1994) Functional allometry of the semicircular ducts in subterranean mole-rats *Cryptomys* (Bathyergidae, Rodentia). Anat Rec 240:286–289

Lovegrove BG, Heldmaier G, Ruf T (1993) Circadian activity rhythms in colonies of 'blind' molerats, *Cryptomys damarensis* (Bathyergidae). S Afr J Zool 28:46–55

Macholan M, Burda H, Zima J, Misek I, Kawalika M (1993) Karyotype of the giant mole-rat, *Cryptomys mechowi* (Rodentia, Bathyergidae). Cytogenet Cell Genet 64:261–263

Macholan M, Scharff A, Burda H, Zima J, Grütjen O (1998) The karyotype and taxonomic status of *Cryptomys amatus* (Wroughton, 1907) from Zambia (Rodentia, Bathyergidae). Z Säugetierk 63:186–190

Nevo E (1979) Adaptive convergence and divergence of subterranean mammals. Ann Rev Ecol Syst 10:269–308

Nevo E (1999) Mosaic evolution of subterranean mammals: regression, progression, and global convergence. Oxford Univ Press, Oxford

Palata-Kabudi JC, Kisasa R, Poso O, Pwema V (2005) Regime alimentaire du rat-taupe geant, *Cryptomys mechowi*, sur le plateau de Bateke, Republique Democratique du Congo. Mammalia 69:445–450

Peichl L, Němec P, Burda H (2004) Unusual cone and rod properties in subterranean African mole-rats (Rodentia, Bathyergidae). Europ J Neurosci 19:1545–1558

Peters W (1881) (Species description - *Georychus mechowii*). Sitz Ber Ges Naturf Freunde Berlin 1881:133

Rosevear DR (1969) The rodents of West Africa. Trustees of the British Museum, London

Scharff A (1998) Systematik und Verhaltensökologie sambischer Sandgräber (Bathyergidae, Rodentia), PhD Thesis, Univ Essen, Essen

Scharff A, Tenora F, Kawalika M, Barus V, Burda H (1996) Helminths from Zambian molerats (*Cryptomys*, Bathyergidae, Rodentia). Helminthologia 33:105–110

Scharff A, Burda H, Tenora F, Kawalika M, Barus V (1997) Parasites in social subterranean Zambian mole-rats (*Cryptomys spp*., Bathyergidae, Rodentia). J Zool Lond 241: 571–577

Scharff A, Begall S, Grütjen O, Burda H (1999) Reproductive characteristics and growth of Zambian giant mole-rats, *Cryptomys mechowi* (Rodentia: Bathyergidae). Mammalia 63:217–230

Scharff A, Locker-Grütjen O, Kawalika M, Burda H (2001) Natural history of the giant mole-rat, *Cryptomys mechowi* (Rodentia: Bathyergidae) from Zambia. J Mammal 82: 1002–1015

Šumbera R, Burda H, Chitaukali WN (2003a) Reproductive biology of a solitary subterranean bathyergid rodent, the silvery mole-rat (*Heliophobius argenteocinereus*). J Mammal 84:278–287

Šumbera R, Burda H, Chitaukali WN, Kubová J (2003b) Silvery mole-rats (*Heliophobius argenteocinereus*, Bathyergidae) change their burrow architecture seasonally. Naturwiss 90:370–373

Thomas O (1906) *Georychus mellandi*. Ann Mag Nat Hist 7(XVIII):141

Van Daele PAAG, Dammann P, Kawalika M, Meier J-L, Van De Woestijne C, Burda H (2004) Chromosomal diversity in *Cryptomys* mole-rats (Rodentia: Bathyergidae) in Zambia; with the description of new karyotypes. J Zool Lond 264: 317–326

Wallace ED, Bennett NC (1998) The colony structure and social organization of the giant Zambian mole-rat, *Cryptomys mechowi*. J Zool Lond 244:51–61

16 Biology of the Silvery Mole-rat (*Heliophobius argenteocinereus*). Why Study a Neglected Subterranean Rodent Species?

Radim Šumbera, Wilbert N. Chitaukali, Hynek Burda

16.1 Introduction

African mole-rats (Bathyergidae, Rodentia) have become the most studied African rodents within the last two decades. Due to a number of studies covering a broad range of topics, many aspects of the biology of some bathyergids are currently better-known than those of laboratory or common Eurasian or American wild rodents. Within the family Bathyergidae, members of three genera (*Bathyergus*, *Georychus*, and *Heliophobius*) are solitary; members of three other genera (*Heterocephalus*, *Fukomys* and *Cryptomys*) are highly social or even eusocial (cf. Jarvis et al. 1994; Burda et al. 2000). Regarding the divergence of social systems and occurrence of eusociality, attention has focused almost exclusively on the social mole-rats. Among the solitary bathyergids, the silvery mole-rat *Heliophobius* (monotypic genus with one recognised species *H. argenteocinereus*) has remained the least known taxon, in spite of the fact that *H. argenteocinereus* is among the most widely distributed bathyergid species occurring in southern Kenya, Tanzania, southeast D.R. Congo, eastern Zambia, Malawi, and northern Mozambique (Bennett and Faulkes 2000; Burda 2001).

Initial studies of the biology of *Heliophobius* were published before mole-rats awoke the broad interest of scientists. However, they were performed on small samples of a single population from the northern part of the species' distribution (southern Kenya). For example, McNab (1966) analysed metabolism in two subadult animals, George (1979) described the karyotype of a single female, Jarvis and Sale (1971) analysed the burrow systems of one adult female and three subadults and Jarvis (1973) recorded activity of a single radio-tagged individual.

Radim Šumbera: Faculty of Biology Sciences, University of South Bohemia, Branišovská 31, 37005 České Budějovice, Czech Republic, E-mail: sumbera@tomcat.bf.jcu.cz

Wilbert N. Chitaukali: Biology Department, Chancellor College, P.O. Box 280, Zomba, Malawi

Hynek Burda: Dept. General Zoology, Institute for Biology, University of Duisburg-Essen, Universitätsstr. 5, D-45117 Essen, Germany

Subterranean Rodents: News from Underground
S. Begall, H. Burda, C.E. Schleich (Eds.)
© Springer-Verlag Berlin Heidelberg 2007

A few years ago Burda et al. (2000) pointed out that the knowledge of the biology of solitary bathyergids may help to elucidate one of the most interesting yet also most controversial problems (cf. Bennett and Faulkes 2000; Burda et al. 2000) – that of the evolutionary origin and adaptive meaning of eusociality in bathyergid mole-rats. The absence of satisfactory information on solitary genera was really surprising because *Bathyergus* and *Georychus* live in areas where their social relatives have been intensively studied for about two decades. Since 1999 we have been studying Malawian and Zambian silvery mole-rats in the field and in the laboratory. Currently, *Heliophobius* has become the best known solitary bathyergid mole-rat. Here we review published and some of unpublished results of our field research on the biology of this enigmatic species.

Fieldwork was carried out mainly in southern Malawi, in three different habitats: grassland in Blantyre-Limbe and farmland in Mulanje-Chitakali in 2000, and in miombo woodland (dominated by *Brachystegia* trees in this part of Africa) in Mpalanganga estate in Zomba district in 2005 (Table 16.1). Short-term data and animal collections were done in alpine meadows at Nganda slopes in the Malawian Nyika Plateau in northern Malawi in April-May 1997 (10°26'S, 33°50'E, altitude 2,200–2,300 m) and in miombo mixed with agricultural spots in the the Lubalashi Area in the Central Province of Zambia (14°80'S; 29°85'E, altitude about 850 m) about 160 km east of Lusaka in August 1996.

16.2 Taxonomy and Body Size

Two recent independent studies of the molecular phylogeny of bathyergids (Faulkes et al. 2004; Ingram et al. 2004) have identified *Heliophobius* as the second (to *Heterocephalus*) most basal lineage of bathyergid mole-rats. Whereas genetic differences within populations were low, genetic differences among different geographic regions (Kenya, Zambia and Malawi) were marked (12S mtDNA corrected pairwise differences = 7.3–13.3%).

Slight karyotypic differences have also been detected among localities. *Heliophobius* from Kenya exhibits a karyotype of $2n = 60$ chromosomes (George 1979). In contrast, *Heliophobius* in Zambia and Malawi have a karyotype of $2n = 62$ (Scharff et al. 2001; Šumbera, unpublished).

In general, the Malawian and Zambian silvery mole-rats correspond to the morphological descriptions of animals from Kenya (cf. Jarvis and Sale 1971; Kingdon 1974). In spite of the absence of any information on sexual dimorphism in *Heliophobius* from the northern part of the distribution, there is sexual dimorphism in size, with males from southern Malawi being heav-

Table 16.1. Characteristics of three localities in southern Malawi where studies were carried out in 2000 and 2005. I–XII = months of the year (January–December). Mean monthly rainfall (mm) is given in the first row for each locality. The values below are standard deviations. The climate in Malawi is characterized by a rainy season (October/November–March/April), a cold dry season (April–July), and a hot dry season (August–October)

Locality	Geographic position	Altitude (m a.s.l.)	Habitat	I	II	III	IV	V	VI	VII	VIII	IX	X	XI	XII	Total
Blantyre-Limbe	15°47'S, 35°04'E	1120	Grassland	259	199	191	**72**	**15**	14	13	4	**2**	19	102	220	1107
–	–	–	–	107	75	120	42	8	6	11	5	3	15	56	114	258
Mulanje-Chitakali	16°02'S, 35°30'E,	620	Farmland	271	253	269	161	**58**	**60**	41	23	24	41	141	271	1610
–	–	–	–	114	108	139	105	35	33	35	20	31	40	75	113	311
Mpalanganga Est.	15°27'S, 35°16'E,	1070	Woodland	303	254	197	85	**24**	**14**	**8**	**6**	**12**	29	124	280	1342
–	–	–	–	118	85	112	64	22	14	7	6	22	28	80	136	129

Precipitation data from closest stations for the particular locality were calculated from the Global Historical Climatological Network Database (http://www.ncdc.noaa.gov/ol/climate/research/ghcn/ghcn.html). The monthly means are based on measurements of 13 years (1965–1978) for Blantyre-Chichiri, 25 years (1954–1978) for Mulanje-Chitakali, and 21 years for Zomba (1953–1973). The mean annual rainfall is given.
Shading marks periods of the study in particular localities

ier (190 ± 58 g, range 63–331 g) than females (162 ± 47 g, range = 51–271 g) (Šumbera et al. 2003a). Wild-captured silvery mole-rats from southern Malawi thus tend to be larger than animals from Kenya weighing 160 g (range 142–168 g) (Kingdon 1974) or Zambia where one male weighed 200 g and females 146 g (range 118–170 g) (Scharff et al. 2001). However, the sex difference in mass is not as pronounced as in the dimorphic solitary *Bathyergus suillus* and *B. janetta* (Davies and Jarvis 1986) or in the eusocial *Fukomys mechowii* (cf. Kawalika and Burda, this volume). Interestingly, large specimens of silvery mole-rats were also captured in northern Malawi; a male captured at Nganda, Nyika Plateau weighed 360 g, and a female from this locality weighed 230 g.

Since sampling was limited to a few localities within the wide area of occurrence (cf. Burda 2001), additional sampling across the entire distribution of the species is required prior to any formal taxonomic decisions on species status of different populations of silvery mole-rats.

16.3 Population Density and Structure

In the three localities studied in southern Malawi, the silvery mole-rats are territorial animals living solitarily in separated, unconnected burrow systems. The mean minimum distance between systems in grassland habitat at Blantyre-Limbe was 8.4 ± 4.4 m (range 2.6–18.7 m) (Šumbera et al. 2003b). Contrary to the density of 58 animals/ha estimated in arid grassland in Kenya (Davies and Jarvis 1986), mole-rats in natural habitats in Limbe and Mpalanganga in Malawi occur in much lower densities of about 5 mole-rats/ha (Šumbera et al. 2003b; Šumbera and Šklíba, in preparation). However, the estimation of high population density of silvery mole-rats in Kenya was deduced from home ranges of only four, mainly subadult, mole-rats, and this value is likely an overestimate. Indeed, one conspicuous feature of burrow systems of *Heliophobius* in Nyika and in eastern Zambia was that they were widely separated from each other, raising the question of how these solitary subterranean rodents find reproductive partners.

Sex ratios were not different from equality at Blantyre and Mulanje (Šumbera et al. 2003a), although the sex ratio for adults at Mpalangaga was extremely female-biased (Šumbera and Šklíba, in preparation).

There is a little information on age structure and lifespan in silvery mole-rats. In captivity, body mass of *Heliophobius* is correlated with age during at least the first year of life, but this relation is probably weaker at later age (Šumbera et al. 2003a). Some large animals captured as adults (probably at least two years old) have lived in captivity for an additional five years and are still alive (Šumbera, unpublished data). In Mpalanganga, adults make up about 50% of the population; solitary living subadults and pups sharing female territory each make about 25% of the population (Šumbera and Šklíba, in preparation).

16.4 Food and Habitat Preferences

The preference of silvery mole-rats for warmer sandy areas with low rainfall at lower altitudes (250–600 mm per year; Jarvis and Sale (1971), Kingdon (1974)) may be typical for *Heliophobius* from the northern margin of its distribution but not representative for the whole range. Generally, *Heliophobius* occurs in more mesic areas (> 900 mm annual rainfall) including mountain plateaus at altitudes above 2,000 m a.s.l. (Burda 2001). Across its distribution, which is characterised primarily by the *Brachystegia* woodland, the silvery mole-rat inhabits a wide range of habitats, including those with seasonally very hard soil (e.g., grasslands in Malawi or Kenya), to those with light farmland soils, where several harvests per year are accompanied by extensive mechanical disturbance of burrows due to digging and spading (Jarvis and Sale 1971; Šumbera 2003).

Similar to other subterranean rodents, the silvery mole-rat is a food generalist, feeding on underground organs of a variety of plants. In the grassland at Blantyre-Limbe, the bulbs following species were found to be eaten: *Vigna vexillata, Dolichos kilimandscharicus, Gladiolus dallenii, Hypoxis sp.* and several unidentified species (Šumbera 2003). Here, the mole-rats preferred sites with higher food supply and larger geophytes; biomass of geophytes at sites where mole-rats occurred were three times higher than at sites where mole-rats were absent, although geophyte abundance did not differ. A large dominant *Vigna* seemed to be a main factor positively influencing occurrence of mole-rats. In food caches in alpine meadows of the Nyika Plateau, Burda and Chitaukali found green leaves, roots, tubers and bulbs of following plant taxa: *Carex* sp., *Chlorophyton* sp., *Dissa* sp., *Dolichos kilimandscharicus, Eriosema psoralea, E. shirensa, Euphorbia* sp., *Gladiolus* sp., *Heteromorpha arborescens, Polygala vivigata, Satyrium* sp., *Steganotaenia* sp. About 30 plant, mostly geophyte species were apparently consumed in Mpalanganga.

16.5 Burrow Systems and Burrowing

A mesic climate, a layer of soil above the tunnels and a vegetation cover contribute to daily and seasonally stable microclimates in silvery mole-rat burrows (cf. Šumbera et al. 2004; Burda et al., this volume). Architecture of burrow systems in *Heliophobius* is similar to that in other subterranean rodents. Burrow systems were rather simple at Blantyre and Mulanje (Fig. 16.1), but more complicated at Mpalanganga (Fig. 16.2). Within their territories, mole-rats concentrate on systematic exploration of particular areas (Fig. 16.2). Burrowing activity did not cease even when the soil was getting very hard, but there was remarkable seasonality in patterns of burrowing and food storage during hot dry season. Systems mapped during the hot dry season were more reticulated with a higher proportion of refilled old burrows (Fig. 16.1) (Šumbera et al. 2003b). When the soil is hard, animals probably search for food digging short tunnels within the framework of the original burrow system. A similar seasonal switch in burrowing was reported for the social *Fukomys damarensis* (Jarvis et al. 1998).

In cultivated areas, deeper burrows (Šumbera et al. 2004) and longer systems, encompassing several fields with different crops could be a strategy to avoid mechanical disturbance of burrows and seasonal depletion of the food supply. Burrow systems in the farmland in Mulanje were longer (105 ± 47 m) than those (47 ± 28 m) examined at almost the same time in the grassland in Limbe (Šumbera 2003; Šumbera et al. 2003b, Fig. 16.1). However, long and deep burrow systems were also found in pristine alpine meadows (Burda and Chitaukali unpublished data).

16.6 Reproduction

Apart from a few anecdotal reports regarding the breeding period and litter size (Copley 1950) or minimal length of pregnancy (> 87 days) (Jarvis 1969), there was little information on reproductive biology in *Heliophobius* available until recently.

16.6.1 Mating Behaviour

Both sexes show complex courtship, with individuals alternately lying down on their backs – a behaviour which can be considered a ritualised form of agonistic behaviour (Šumbera 2001). Aggression was low both in inter-

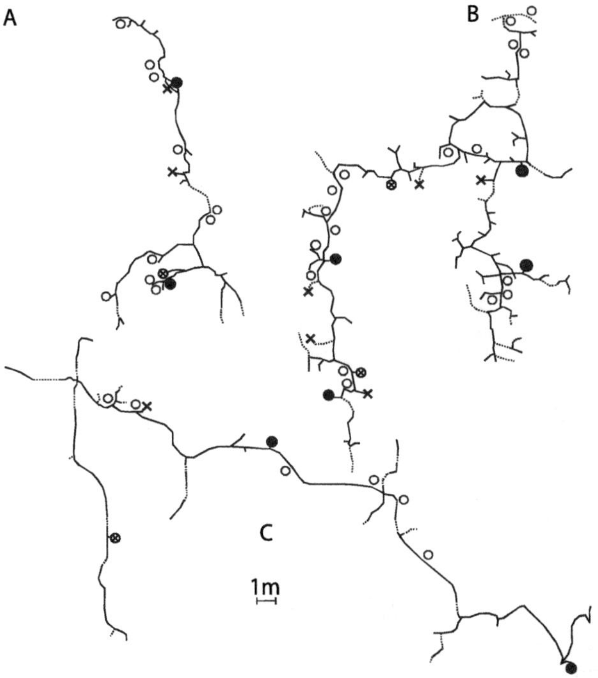

Fig. 16.1A–C. Representative burrow patterns of systems excavated in cold dry season (i.e. 1–3 months after end of rains) in: **A** Blantyre-Limbe; **C** Mulanje-Chitakali; **B** during hot dry season (i.e. 5 months after end of rains) in Blantyre-Limbe. *Open circle* = mound, *filled circle* = nest, *crossed circle* = food chamber, *cross* = sanitary area. *Dashed lines* indicate backfilled parts of systems. Fractal dimension describing complexity of systems was in Limbe 1.19 ± 0.03 (n = 7) and 1.2 ± 0.1 (n = 7) in Mulanje during cold dry season, 1.33 ± 0.1 (n = 7) in hot dry season in Limbe (Fractal dimension ranges from 1 to 2, values approaching 2 indicate a more reticulated structure)

and intrasexual encounters during breeding season (Šumbera and Burda 2003). A remarkable absence of drumming as a means of long-distance communication, low population densities, together with the already mentioned separation and spacing of burrow systems, suggest that searching for mates in *Heliophobius* occurs above ground.

16.6.2 Seasonality of Breeding

Like all other solitary subterranean rodents, the silvery mole-rat is a seasonal breeder (Copley 1950; Scharff et al. 2001; Šumbera et al. 2003a). In southern Malawi and eastern Zambia, pups are born in August and September. Considering a pregnancy to be about three months, we assume that

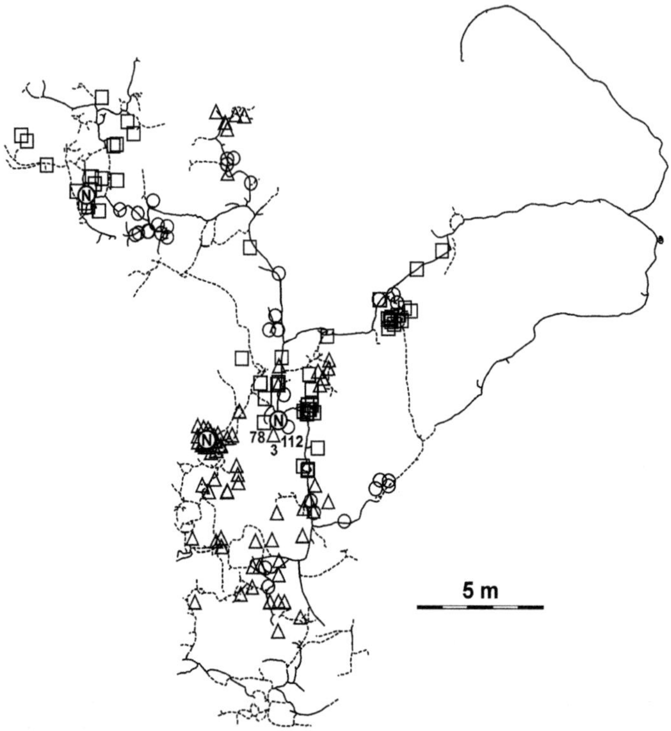

Fig. 16.2. Pattern of spatial activity of an adult female in *Brachystegia* woodland measured by radio-tracking during three months of the dry season 2005. *Squares* represent recordings done in June (08.06–17.06), *triangles* in July (10.07–19.07), and *circles* in August (10.08–19.08). August recordings were done a few days after pups were born. N = nest chamber (a *number* next to each symbol gives the number of radio-fixes inside the nest within each particular period). *Dashed lines* symbolise backfilled tunnels. A non-branched part of the system on the right side probably represents main tunnels of the burrow system used before the rainy season. Side tunnels of this older part of the burrow system were probably refilled with soil. They may be maintained to allow a shift of territory in case of food shortage or in some other unpredictable events

mating takes place in May and June, i.e., one to two months after the end of the rains (Scharff et al. 2001; Šumbera et al. 2003a). Contrary to most other seasonal breeders, *Heliophobius* lives in areas with a minimal annual variability in the day length. The nature of the zeitgeber for breeding activity remains unclear. Factors triggering breeding in tropical mammals with long gestation remain largely unknown (Goodman 1998). The main periodic environmental factor in Malawi is a regular alternation of rains and drought. Breeding in *Heliophobius* is probably not synchronised with the onset of rains but with a complex of ecological factors connected with the end of the rains. The role of the rains in triggering seasonal reproduction may be indirect, such as via increased access to geophytes (Bennett et al. 2000). Interestingly, an association between breeding and the end of the rains was supported in 2005 when the rainy season in Malawi ended one to two months earlier than usual, and, accordingly, births occurred earlier than usual.

16.6.3
Pregnancy, Litter Size and Postnatal Development

Pregnancy in the silvery mole-rat was estimated to last at least three months (Jarvis 1969; Šumbera et al. 2003a). This finding falsifies the previous assumption that, in contrast to social bathyergids and octodontids, short gestation is common in solitary bathyergids, geomyids and spalacids (Bennett et al. 2000). Litter size for *Heliophobius* from Malawi was 3.2 ± 0.9, similar to the one reported from Kenya (Copley 1950; Kingdon 1974). The prenatal and postnatal development of pups is slow and in many ways comparable to that in (eu)social species of bathyergids (cf. Šumbera et al. 2003a). Given the pattern of long pregnancy, slow postnatal development and a distinct breeding season, females apparently deliver only a single litter per year. Nothing is known about the mating system in the silvery mole-rat. Nevertheless, low sexual dimorphism in size (less than 15%), an equal sex ratio (at least in some sites) and low population densities suggest a monogamous or a promiscuous (rather than a polygynous) mating system with low competition for females.

16.6.4
Dispersal

Natal dispersal in solitary species is probably initiated by aggressive interactions between mother and offspring. In captive silvery mole-rats the onset of serious sparring between mother and pups was correlated with attaining a certain body mass rather than with age, so competition for space in burrows could be a simple proximate mechanism triggering natal dispersal in growing young (Šumbera et al. 2003a). Silvery mole-rats disperse mainly aboveground as suggested by wide separation of burrow systems and presence of skulls of young *Heliophobius* in barn owl pellets (Šumbera and Šklíba, unpublished).

16.7 Economic and Epidemiological Importance

The influence of subterranean rodents on community ecology has been studied for several species (reviewed in Cameron 2000; Reichman, this volume). Less information is available on their biology in cultural landscapes. In rural areas, bathyergids could act as serious agricultural and horticultural pests (De Graaff 1981; Kawalika and Palata, personal communication) and they also constitute a source of animal proteins (De Graaff 1981; Burda and Kawalika 1993; own observations from Malawi). Concordantly, the silvery mole-rat was reported as a serious pest (Copley 1950). In Mulanje, it feeds on all available crops with large belowground organs: cassava, sweet potato, yam, koko or ground nuts (Šumbera 2003). Frequently, a mole-rat eats only a small part of a tuber and then attacks a new plant. In light arable soil mole-rats are able to dig many meters of tunnels per day, thus reaching several plants. Fully grown tubers of one cassava plant may reach several kg together (mean weight of tubers of 10 randomly selected cassava plants was 1.8 ± 1.3 kg; Šumbera 2003). Damage is high not only because of yield reduction, but also because of decreasing quality of bulbs. Moreover, mole-rats also consume or destroy roots and rhizomes of crops not cultivated for subterranean parts (i.e. maize, pigeon peas, fruit trees).

Due to their longevity, subterranean life and consumption of perennial geophytes, subterranean rodents could be a suitable model indicating environmental quality or even health risks for human populations. For example, analysis of four heavy metals and their bioaccumulation potential in the tissues of silvery mole-rats (Šumbera et al. 2003c) suggests only very slight environmental pollution in a rural area in Mulanje, as well as around the industrial centre of Blantyre-Limbe. Therefore there is no health risk connected with consumption of mole-rats with respect to tested heavy metals. Similarly, subterranean rodents could be an indicator for occurrence of emmonsiosis caused by soil fungi *Emmonsia* affecting lungs of infected animals including man (Hubálek et al. 2005).

16.8 Parasites and Sociality in African Mole-rats

Compared to many other rodent species, African mole-rats have been rather understudied from the point of view of parasitology. Scharff and Burda (1998) pointed out that in spite of ecological factors favouring parasitic infections (high and stable humidity and temperature) in the subterranean environment as well as communal life and coprophagy, parasitic infection in eusocial mole-rats of two *Fukomys* species was relatively low (Scharff et al.

1997). In another study Scharff et al. (2001) found neither ectoparasites nor helminths among eleven silvery mole-rats captured in Zambia – a finding supporting the idea that solitary style of life reduces the risk of infestation. In contrast, Kundu and Faulkes (2004), who studied patterns of MHC selection in some bathyergids (including 12 silvery mole-rats from four localities in Kenya and Tanzania), concluded that the overall risk of parasite transmission in social African mole-rats is lower than in the solitary silvery mole-rat. Subsequent parasitological examination involving a larger sample from Malawi and Zambia, revealed that *Heliophobius* hosts a diversity of parasites, such as mites (Šumbera et al. 2004) and helminths, a nematode (*Protospirura muricola*) and a cestode (*Inermicapsifer arvicathidis*) (Tenora et al. 2003). In addition, four new species of coccidians of the genus *Eimeria* were described (Koudela et al. 2000; Modrý et al. 2005).

These findings challenge the previous conclusion that the silvery mole-rat is not (or seldom) parasitized and that a solitary lifestyle leads to lower infestation. Certainly the question of a relationship between parasite load and social and mating systems remains open and deserves further examination. Further parasitological studies of mole-rats of diverse species are needed that take into account possible temporal, geographical, ecological (climatic), age and sexual factors. It is possible that horizontal parasite transfer occurs within colonies of (eu)social species, yet, due to effective isolation between colonies, the risk of inter-colonial infection is limited. In fact, solitary mole-rats could be more subject to infections due to periodic contacts with mates. For the purpose of comparative studies, members of a particular family group of social mole-rats cannot be considered as independent units.

16.9 Evolution of Sociality in the African Mole-rats

Understanding the evolution and adaptive value of social behaviour is one of the main goals of evolutionary and sociobiological studies. The African mole-rats are one of the most suitable models, because of different degrees of sociality across a relatively small number of species sharing the same, strictly subterranean lifestyle. Here, we focus on the evolution of social life in bathyergids reviewing the two most relevant hypotheses regarding sociality and we show how recent findings from silvery mole-rats help us to understand this aspect of behaviour.

16.9.1
Hypotheses on the Evolution of Sociality in Bathyergids

The ecological Aridity Food Distribution Hypothesis (AFDH) considers eusociality as a derived trait and an endpoint of social evolution (Jarvis et al. 1994; Bennett and Faulkes 2000). It is based on the fact that in arid areas with sparse and unpredictable rainfall, food resources are limited and clumped in distribution. Random "blind" foraging would be risky and there should be selective pressure on offspring in solitary animals to delay dispersal, stay in the nest and cooperate in searching for food. Dispersal is enabled only when environmental conditions (rains) allow the animals to build their own burrow systems and establish new colonies. The alternative hypothesis of phylogenetic constraints simply considers sociality as an ancestral trait of hystricognath rodents (80% genera being social) and eusociality as the endpoint of monogamy (cf. Burda et al. 2000). In this scenario, the solitary way of life in some genera of bathyergids is considered a derived trait. These scenarios do not differ significantly with regard to the adaptive meaning of eusociality. Instead, the main point of the conflict is a simple question: What was the beginning? Is bathyergid sociality a derived or a plesiomorphic trait? Unfortunately, an objective test – a phylogenetic reconstruction – is not helpful here. Tested against phylogenetic trees for bathyergids (Faulkes et al. 2004; Ingram et al. 2004), social life could have arisen – or been lost – twice. Several further arguments in support of the solitary nature of the bathyergids' ancestor were reviewed in Bennett and Faulkes (2000) but disputed by Burda et al. (2000). Moreover, since groups in (eu)social mole-rats originate as extended families by accumulation of non-dispersing offspring and not as colonies through aggregation of unrelated individuals, pair bonding lasting beyond the mating season, i.e. monogamy, is probably at the beginning of each group.

16.9.2
Silvery Mole-rat and Evolution of Sociality

Contrary to solitary species living mainly in sandy soils in South Africa, *Heliophobius* is a generalist surviving in a variety of habitats with different ecological conditions, including sites with seasonally very hard soil and a very low food supply. A density of 2.5 geophytes/m^2 found at Limbe (Šumbera et al. 2003b) is even lower than that set up by mathematic models (Lovegrove 1991) as a lower limit ($<$ 4 geophytes/m^2) excluding solitary life. Moreover, food resources in natural habitats at Limbe and Mpalanganga were clumped (Šumbera and Šklíba, in preparation), thus falsifying the prediction of AFDH that solitary species would avoid such areas.

Sociality of bathyergids was also related to parameters of reproductive biology. Long pregnancy and slow postnatal development – previously considered as both a cause for sociality (combined with inability to store sufficient body fat) (Burda 1990) or a consequence (Bennett et al. 1991) – were also found in the silvery mole-rat (growth curves for different bathyergids are compared in Šumbera et al. 2003a). Thus, reproductive and developmental traits in *Heliophobius* suggest their ancestral character connecting bathyergid mole-rats with other hystricognath rodents. Relative shortening of pre- and postnatal development in *Bathyergus* and *Georychus* (cf. Bennett and Faulkes 2000) has to be considered a derived trait, consistent with the fact that these genera represent later divergence events. Furthermore, previous studies suggested that metabolic rate in bathyergid mole-rats is independent of body mass (Lovegrove and Wissel 1988). Accordingly, it was postulated that mole-rats would benefit from a large workforce of small foragers, with no or only little increase in metabolic costs. Nevertheless, recent studies of the metabolism of *Heliophobius* and a re-assessment of the allometry of metabolism in a larger number of species clearly demonstrated that resting metabolic rates in African mole-rats follow typical mammalian body mass dependence (Zelová et al., in press). This finding thus falsifies one of the important premises of AFDH.

To conclude, there is no reason to reject the scenario of a secondary loss of sociality in African mole-rats. Sociality in African mole-rats could have been a preadaptation to occupy the subterranean ecotope, particularly in unpredictable habitats, similar to the scenario suggested for the highly cooperative vole *Lasiopodomys mandarinus* (Smorkatcheva 1999). The AFDH may explain why cooperative foraging is useful for survival in arid habitats, but cannot discriminate between group living as a consequence of harsh environments and sociality as a precondition for occupying it. If the scenario of ancestral sociality in bathyergids is true, we should reformulate the problem and ask what has led to (and what has enabled) independent loss of sociality in three lineages of African mole-rats. To answer the question, we need to know more about the biology of the remaining solitary bathyergids – *Bathyergus suillus*, *B. janetta* and *Georychus capensis*.

16.10 Further Perspectives

The silvery mole-rat is certainly a species deserving the attention of scientists. It differs from the two solitary bathyergid genera from South Africa in many aspects of its biology, ecology and reproduction. As the only solitary, strictly subterranean rodent from the tropics, *Heliophobius* is an interesting model to study factors influencing circadian and circannual activity and reproduction. The ecological flexibility of *Heliophobius* and its ability to survive in various habitats, as well as its presumed aboveground activity and the related orientation and communication problems (absence of seismic communication) should be studied more specifically and thoroughly. Other aspects of the biology of this species such as the mating system, timing of dispersal or behaviour during the rainy season are still almost unknown. There are still some other intriguing aspects of the biology of silvery mole-rats (like pacifism and tameness frequently recorded in captured animals, genetic structure of different populations, mating system) which deserve thorough study.

Acknowledgements. We are grateful to the National Research Council of Malawi (NRCM) for permission to do field work and export animals, the company "Fargo Ltd" in Blantyre-Limbe, many farmers in Mulanje-Chitakali (Chipoka village) and family Christie from Mpalanganga estate for permitting us to work on their private lands and for creating excellent conditions for our work. We are obliged to Markéta Elichová, Jana Kubová and Jan Šklíba for cooperation in our field research on biology of the silvery mole-rats and many assistants who helped us in the field and laboratory. We thank Jan Šklíba for helpful comments on the text and for preparing Fig. 16.2 and Eileen A. Lacey and Cristian E. Schleich for their valuable comments and suggestions on the manuscript. Our research on Malawian mole-rats is continuously supported by following institutions: University of South Bohemia, České Budějovice, Czech Republic, Museums of Malawi, Blantyre and Chancellor College, Zomba, both from Malawi and University of Duisburg-Essen, Germany. Research of Radim Šumbera is generously supported mainly by Grant Agency of the Czech Republic no. 206/00/1699 and 206/04/P116 and a grant of the Ministry of Education of the Czech Republic (MSMT 6007665801).

References

Bennett NC, Faulkes CG (2000) African mole-rats: ecology and eusociality. Cambridge Univ Press, Cambridge

Bennett NC, Jarvis JUM, Aguilar GH, McDaid EJ (1991) Growth and development in six species of African mole-rats (Rodentia: Bathyergidae). J Zool Lond 225:13–26

Bennett NC, Faulkes CG, Molteno AJ (2000) Reproduction in subterranean rodents. In: Lacey EA, Patton JL, Cameron GN (eds) Life underground: the biology of subterranean rodents. Univ Chicago Press, Chicago, pp 145–177

Burda H (1990) Constraints of pregnancy and evolution of sociality in mole-rats. With special reference to reproductive and social patterns in *Cryptomys hottentotus* (Bathyergidae, Rodentia). Z Zool Syst Evolut-forsch 28:26–39

Burda H (2001) Determinants of the distribution and radiation of the African mole-rats (Bathyergidae, Rodentia). Ecology or geography? In: Denys C, Granjon L, Poulet A (eds) African small mammals. Collection colloques et séminaires, Inst de Recherche pour le Developpement Editions, Paris, France, pp 263–277

Burda H, Kawalika M (1993) Evolution of eusociality in the Bathyergidae. The case of the giant mole-rats (*Cryptomys mechowi*). Naturwiss 80:235–237

Burda H, Honeycutt RL, Begall S, Locker-Grütjen O, Scharff A (2000) Are naked and common mole-rats eusocial and if so, why? Behav Ecol Sociobiol 47:293–303

Cameron GN (2000) Community ecology of subterranean rodents. In: Lacey EA, Patton JL, Cameron GN (eds) Life underground: the biology of subterranean rodents. Univ Chicago Press, Chicago, pp 227–256

Copley H (1950) Small mammals of Kenya. Highway Press. Nairobi, Kenya

Davies KC, Jarvis JUM (1986) The burrow systems and burrowing dynamics of the mole-rats *Bathyergus suillus* and *Cryptomys hottentotus* in the fynbos of the south-western Cape, South Africa. J Zool Lond 209:125–147

De Graaff D (1981) The rodents of Southern Africa. Butterworth, Johannesburg

Faulkes GC, Verheyen E, Verheyen W, Jarvis JUM, Bennett NC (2004) Phylogeographical patterns of genetic divergence and speciation in African mole-rats (Family: Bathyergidae). Mol Ecol 13:613–629

George W (1979) Conservatism in the karyotypes of two African mole rats (Rodentia, Bathyergidae). Z Säugetierk 44:278–285

Goodman RL (1998) Seasonal reproduction, mammals. In Knobil E, Neil JD (eds) Encyclopaedia of reproduction, Academic Press, San Diego, pp 341–351

Hubálek Z, Burda H, Scharff A, Heth G, Nevo E, Šumbera R, Peško J, Zima J (2005) Emmonsiosis of subterranean rodents (Bathyergidae, Spalacidae) in Africa and Israel. Med Mycol 43:691–697

Ingram CM, Burda H, Honeycutt RL (2004) Molecular phylogenetics and taxonomy of the African mole-rats, genus *Cryptomys* and the new genus *Coetomys* Gray, 1864. Mol Phyl Evol 31:997–1014

Jarvis JUM (1969) The breeding season and litter size of African mole-rats. J Reprod Fert 6:237–248

Jarvis JUM (1973) Activity patterns in the mole-rats *Tachyoryctes splendens* and *Heliophobius argenteocinereus.* Zool Afr 8:101–119

Jarvis JUM, Sale JB (1971) Burrowing and burrow patterns of East African mole-rats *Tachyoryctes, Heliophobius* and *Heterocephalus.* J Zool Lond 163:451–479

Jarvis JUM, O'Riain MJ, Bennett NC, Sherman PW (1994) Mammalian eusociality: a family affair. Trends Ecol Evol 9:47–51

Jarvis JUM, Bennett NC, Spinks AC (1998) Food availability and foraging by wild colonies of Damaraland mole-rats (*Cryptomys damarensis*): implications for sociality. Oecologia 113:290–298

Kingdon J (1974) East African mammals. An atlas of evolution in Africa. Voll II, Pt B, Hares and rodents. Academic Press, New York

Koudela B, Šumbera R, Sedláček F (2000) *Eimeria burdai* sp.n. (Apicomplexa: Eimeriidae), a new parasite species from subterranean African silvery mole-rat, *Heliophobius argenteocinereus.* Folia Parasitol 47:97–99

Kundu S, Faulkes GC (2004) Patterns of MHC selection in African mole-rats, family Bathyergidae: the effects of sociality and habitat. Proc R Soc Lond B 271:273–278

Lovegrove BG (1991) The evolution of eusociality in molerats (Bathyergidae): a question of risks, numbers, and costs. Behav Ecol Sociobiol 28:37–45

Lovegrove BG, Wissel C (1988) Sociality in molerats – metabolic scaling and the role of risk sensitivity. Oecologia 74:600–606

McNab BK (1966) The metabolism of fossorial rodents: a study of convergence. Ecology 47:712–733

Modrý D, Jirků M, Šumbera R (2005) Three new species of *Eimeria* (*Apicomplexa: Eimeriidae*) from the silvery mole- rat *Heliophobius argenteocinereus* Peters, 1846 (*Rodentia: Bathyergidae*) from Malawi. J Parasitol 91:1200–1203

Scharff A, Burda H (1998) Eusociality, life underground and parasites. Trends Ecol Evol 13:199

Scharff A, Burda H, Tenora F, Kawalika M, Baruš V (1997) Parasites in social subterranean Zambian mole-rats (*Cryptomys* spp., Bathyergidae, Rodentia). J Zool Lond 241: 1–6

Scharff A, Macholán M, Zima J, Burda H (2001) A new karyotype of *Heliophobius argenteocinereus* (Bathyergidae, Rodentia) from Zambia with field notes on the species. Mamm Biol 66:376–378

Smorkatcheva AV (1999) The social organisation of the Mandarine vole, *Lasiopodomys mandarinus*, during reproductive period. Z Säugetierk 64:344–345

Šumbera R (2001) The mating behaviour of the silvery mole-rat. In: Denys C, Granjon L, Poulet A (eds) African small mammals. Collection colloques et séminaires, Inst de Recherche pour le Developpement Editions, Paris, France, pp 353–364

Šumbera R (2003) Ecology and reproduction of the Silvery mole-rat (*Heliophobius argenteocinereus*). Can it help us to understand evolution of sociality in African mole-rats? PhD. Dissertation, Univ South Bohemia, České Budějovice

Šumbera R, Burda H (2003) Pacifistic behaviour in a solitary subterranean rodent, the silvery mole-rat (*Heliophobius argenteocinereus*, Bathyergidae), Contributions to 28th Int Ethol Conf, Florianópolis, Brazil 2003, Rev de Etol Suppl 5:130

Šumbera R, Burda H, Chitaukali WN (2003a) Reproductive biology of a solitary subterranean bathyergid rodent, the Silvery mole-rat (*Heliophobius argenteocinereus*). J Mammal 84:278–287

Šumbera R, Burda H, Chitaukali WN, Kubová J (2003b) Silvery mole-rat (*Heliophobius argenteocinereus*, Bathyergidae) change their burrow architecture seasonally. Naturwiss 90:170–173

Šumbera R, Baruš V, Tenora F (2003c) Heavy metals in the silvery mole-rat, *Heliophobius argenteocinereus* (Bathyergidae, Rodentia) from Malawi. Folia Zool 52:149–153

Šumbera R, Chitaukali WN, Elichová M, Kubová J, Burda H (2004) Microclimatic stability in burrows of an Afrotropical solitary bathyergid rodent, the silvery mole-rat (*Heliophobius argenteocinereus*). J Zool Lond 263:409–416

Tenora F, Baruš V, Prokeš M, Šumbera R, Koubková B (2003) Helminths parasitising the silvery mole-rat (*Heliophobius argenteocinereus*) from Malawi. Helminthologia 40: 227–232

Zelová J, Šumbera R, Sedláček F, Burda H (in press) The energetics in the solitary subterranean rodent, *Heliophobius argenteocinereus*, and the allometry of RMR in African mole-rats (Bathyergidae). Comp Biochem Physiol A

17 The Biology and Ecology of Plateau Zokors (*Eospalax fontanierii*)

Yanming Zhang

17.1 Introduction

The Tibetan plateau is located in the Southwest of the People's Republic of China (27°–40° N, 75°–105° E) with an average elevation of more than 4,000 m. It occupies 2.5 million km^2, approximately 25% of the area of China. An estimated 70% is high altitude grassland, and Tibetan pastoralism constitutes the primary use of this rangeland habitat (Xia 1988; Smith and Foggin 1999; Zhang Y et al. 2003). Daily temperature varies greatly from −15°C to 0°C in January and 0°C to 16°C in July, with an annual mean temperature < 0°C (Xia 1988). There is no frostless season and extensive areas of thick permafrost can be found even in August in mountains and grasslands (Smith et al. 1986). The major vegetation ecotypes are classified as alpine meadow, alpine shrub, alpine prairie and alpine steppe meadow (Xia 1988). Here, plateau pikas (*Ochotona curzoniae*), plateau voles (*Microtus* sp.), and plateau zokors (*Eospalax fontanierii*) are the most abundant small mammals.

Plateau zokors belong to the rodent family Spalacidae (according to Norris et al. 2004 to the subfamily Myospalacinae of the family Muridae) and are small (females: ca. 220 g for females, males: ca. 270 g) rodents spending most of their life below ground (Zhang Y et al. 2003; Norris et al. 2004). They show year-round activity in excavating their burrow systems and installing caches to store food for periods when fresh vegetation is not readily available. Like other subterranean rodents, plateau zokors exhibit morphological, physiological and behavioral adaptations for digging and living in the underground environment (Reichman and Smith 1990; Zhang Y and Liu 2003).

Yanming Zhang: Northwest Plateau Institute of Biology, Chinese Academy of Sciences, Xining, Qinghai, 810001, People's Republic of China, E-mail: zhangym@nwipb.ac.cn

Subterranean Rodents: News from Underground
S. Begall, H. Burda, C.E. Schleich (Eds.)
© Springer-Verlag Berlin Heidelberg 2007

17.2 Morphological Characteristics

Plateau zokors use their extremities to excavate underground tunnels and to acquire food. They possess stocky and strong forelimbs and laterally flattened claws frequently supported by a bony phalanx extending into them. The forelimbs are always used for digging, while the hind limbs stabilize the body during the forelimbs' movements and often serve to kick loosened soil backward (Wang Z et al. 1979; Zeng et al. 1984; Wang Q et al. 2000b).

Plateau zokors also use their heads, exhibiting hardened pads, for digging and for transporting the loosened soil (Su and Wang 1992). The specialized large, strong incisors mostly handle taproots, pulling the whole plant into the tunnel system and then dislodging the dirt (Zhang Y et al. 2003).

17.3 Physiological Characteristics

The plateau zokor's basal metabolic rates (*BMR*) vary in different seasons (Table 17.1). These values are significantly higher either than the expected values based on Kleiber's equation ($BMR = 3.42W^{-0.25}$, Kleiber 1961) or those of other fossorial rodents such as *Geomys pinetis* (McNab 1966) and *Spalax leucodon* (Gorecki and Christov 1969), but lower than those of the aboveground small mammals inhabiting the same area, such as plateau pikas and root voles (*Microtus oeconomus*, Wang Z et al. 1979; Wang D et al. 1991). The maximal metabolic response after norepinephrine injection in different months (May, July, September) is on average $1.60\,\mathrm{ml\,O_2\,g^{-1}\,h^{-1}}$, equivalent to 122% of their *BMR* (Su et al. 1991). The regulatory nonshivering thermogenesis values in the given months amount to $0.35\,\mathrm{ml\,O_2\,g^{-1}\,h^{-1}}$. The burrowing metabolic rate of a plateau zokor is $4.63\,\mathrm{ml\,O_2\,g^{-1}\,h^{-1}}$ (Su and Wang 1992).

Their daily energy consumption per gram of body mass is greater than that of a Tibetan sheep by a factor of 4.1 (Wang Z et al. 1980; Pi 1982). Because of these high energetic demands (667.6 KJ/day) and their year-round activity, plateau zokors play a major consumer role despite their relatively small body size.

The body temperature of plateau zokors averages 34.72°C and varies only slightly when the ambient temperature of air is in the range 0–27°C. However, it increases rapidly at ambient temperatures above 30°C (Zeng et al. 1984). The heart frequency of zokors (296.6 beats/min) is lower than that of plateau pikas (332.7 beats/min), and their heart (1.65 g), which ac-

Table 17.1. Comparison of the body weight and *BMR* between *Eospalax fontanierii* and some other subterranean[a] or epigeic sympatric[b] small mammals

Species	Month or Season	Average body weight (g)	*BMR* (ml O_2 g^{-1} h^{-1})	Me (%)	Distribution and altitude (m a.s.l.)	Reference
Eospalax fontanierii	Spring (May)	241.2	1.32	153.9	Qinghai-Tibetan plateau, 3200	Wang Z et al. (1979)
	Summer (July)	196.0	1.35	148.7		
	Autumn (September)	250.2	1.19	139.0		
Spalax leucodon[a]		178.0	0.63	67.0	Europe	Gorecki and Christov (1969)
Geomys bursarius[a]		197.0	0.70	77.0		Bradley and Yousef (1975)
Geomys pinetis[a]		200.0	0.71	78.0	North America	McNab (1966)
Thomomys talpoides[a]	Summer	106.0	1.03	97.0	North America, 2300	Bradley et al. (1974)
	Winter	111.0	1.33	124.0		Gettinger (1975)
Spalacopus cyanus[a]		136.0	0.75	74.0	Chile, 2500	Contreras (1986)
Microtus oeconomus[b]	Spring (May)	21.9	3.23	205.0	Qinghai-Tibetan plateau, 3200	Wang D et al. (1991)
	Summer (July)	25.9	3.56	235.0		
	Autumn (September)	23.0	3.42	220.0		
Ochotona curzoniae[b]	Spring (May)	138.0	1.67	168.1	Qinghai-Tibetan plateau, 3200	Wang Z et al. (1979)
	Summer (July)	117.5	2.15	208.4		
	Autumn (September)	128.7	1.46	144.2		
Ochotona cansus[b]	Spring (May)	61.9	2.82	223.9	Qinghai-Tibetan plateau, 3200	Wang D et al. (1991)
	Summer (July)	56.2	2.94	237.0		
	Autumn (September)	58.2	2.65	225.1		

Me (%) = $BMR \cdot 100/3.4W^{-0.25}$

counts for 0.57% of total body weight, is heavier than that of the plateau pika (0.84 g, 0.38%, respectively). The diameter of erythrocytes in plateau zokors (6.35 μm) is significantly larger than in plateau pikas (5.99 μm).

17.4 Habitat Selection and Distribution

Plateau zokors are widely distributed in the alpine meadows, steppe meadows, alpine shrubs, farmlands, banks and wasteland of the provinces Qinghai, southern Gansu and western Sichuan. The distribution and population density of plateau zokors are limited by elevation, vegetation, precipitation and anthropogenic disturbance (Zhang Y et al. 1999, 2003). Their habitat is characterized by moist soils and degraded grassland dominated by forbs (Shi et al. 1994; Zhang Y et al. 1999). They tend to avoid the Gobi desert, irrigated lands, and boulder fields.

It has been estimated that these animals occupy a range of approximately $3.8 \times 10^4\,\mathrm{km}^2$ at an average density of 15 animals/ha, with density estimates in the 1980s in Qinghai ranging from 5 to more than 70 per ha (Wang Q and Fan 1987; Zhang Y et al. 1999). During the 1990s, annual eradication programs were carried out by local officials and farmers with the result of reducing zokor populations to less than 31.6% of their former abundance (Hall et al. 1999; Zhang Y et al. 1999; Zhang Y 2000).

17.5 Burrow System and Environment

Plateau zokors construct very complicated burrow systems that consist of one or two main nests, foraging and transportation tunnels, food store caches, and blind endings. Females occupy deeper nests (approximately 2–2.5 m) than males (approximately 0.8–1.5 m) (Zhou and Dou 1990).

When constructing tunnels, plateau zokors move the loosened soil to the surface and deposit it in mounds. On average one mound with a mean volume of ca. $0.007\,\mathrm{m}^3$ is formed by a single zokor everyday (Wang Q and Fan 1987). Tunnelling is often extensive; the average tunnel length of the whole burrow system is about 100 m and mounds may cover as much as 15–20% of the surface (Wang Q and Fan 1987; Zhang Y 1999; Zhang Y et al. 2003).

The soil is an excellent insulator, buffering microclimate from the extreme temperature variations in the aboveground environment (cf. Burda et al., this volume). In mid-summer (July), Zeng et al. (1984) and Li et al. (2004) recorded ambient temperatures in an alpine meadow ranging between 4 and 24°C, while the temperatures of foraging burrows (approximately 12–18 cm deep) varied only between 3.5 and 12°C. In winter (January), temperatures measured in an empty nest ranged from 0 to 5°C while outside temperatures went from −4 to −38°C. The sealed burrow systems also maintain a relatively high moisture content. Zeng et al. (1984) measured the relative humidity in burrows of zokors as 79.0–87.3% while the aboveground humidity was a relatively low level (31.7–53.3%). The high humidity contributes to economizing water balance by zokors (Frank 1988). The evaporative water loss of zokors measured in laboratory (0.79 mg/g) is lower than that of plateau pikas (1.38 mg/g), which use open burrows (Zeng et al. 1984).

The gas concentration in the zokor's burrows also differs from that in ambient atmospheric conditions. Zeng et al. (1984) recorded reduced oxygen levels inside burrows (17.04–18.43%) and a significantly increased concentration of carbon dioxide (0.22–1.46%) compared with local atmospheric values of 20.45% (O_2) and 0.03% (CO_2).

17.6 Foraging and Diet

Plateau zokors in alpine meadows prefer areas with soft soils and high primary productivity. In alpine meadows they forage mainly at a depth of 3–20 cm (Zhang Y 1999). They have broad diets, consuming both roots and shoots of annual and perennial grasses, forbs, and a few shrubs, (Wang Q et al. 2000a; Zhang Y 2000). Also our data indicate that plateau zokors mainly forage the roots, rhizomes and other underground parts of weeds, that forbs occurring in alpine meadows were preferred over grasses and that succulent belowground storage organs, such as *Notopterygium forbesiide*, *Potentilla anserina*, and *Morina chinensis*, were consumed preferentially (Wang Q et al. 2000a; Zhang Y 2000).

17.7 Activity Rhythms

Like many other completely subterranean species, plateau zokors stay active throughout the year and their activity patterns are also affected by the surrounding conditions, such as temperature and weather (Wang Q and Fan 1987). They are active in spring and fall, exhibit a somewhat reduced level of activity in the relative heat of summer, and remain in their nests longer during winter when the soil is frozen and only loosened with difficulty.

Generally, strictly subterranean animals are less influenced by a circadian cycle and frequently show activity patterns not clearly associated with day/night cycles (cf. Reichman and Smith 1990). Contrary to this assumption, plateau zokors do exhibit circadian rhythms. They spend 85–90% of their lifetime in their underground nests, and about 2–3 h a day feeding and using the tunnel systems that they actively excavate and maintain. There are two daily peaks of activity during spring, summer and autumn: one occurs between 17:00 and 21:00 h, and another between 01:00 and 06:00 h (Zhang Y 2000). In winter, a single peak of activity near the nest occurs mostly between 12:00 and 22:00 h (Zhou and Dou 1990).

17.8 Reproduction

A radio-telemetry study at Haibei Research Station indicated that adult males and females never live together, even when females are in oestrus, but that mating occurs from the end of March to June at the intersection of two tunnels of a male and a female (Zhou and Dou 1990). An analysis of 20 burrow systems of zokors showed that, in the mating period, the male digs a few long tunnels to intercept the burrow system of a female; these tunnels usually have two branches to increase the chance of meeting a female. The adult sex ratio bias towards females (1:1.67) and the interspersion of male and female home ranges probably increases the chance of encounters for mating (Zhou and Dou 1990). A situation with two rival males meeting each other has never been reported (Zhang Y et al. 2003). A male may mate with several females during one reproduction cycle, and a female may also mate with a few males in this time. Therefore, the zokor mating system can be classed as promiscuous.

Adult females of plateau zokors give birth once a year between April to July. Litter size ranges between 1 and 5 embryos (average 2.91 ± 1.08, $n = 131$), and the neonate sex ratio is strongly female biased (only 38.4% of neonates being males) (Zheng 1980). The lactation period lasts about 50 days (Zhang D et al. 1995).

17.9 Dispersal

The dispersal mechanisms of plateau zokors are not yet clear. The energetic costs of a potential dispersal through the soil appear to be impossibly high. However, aboveground dispersal would result in a high risk of predation. A few studies reported that zokor skulls were found

Table 17.2. Capture and recapture numbers and recapture rates of plateau zokors from May 1987 to October 1989 in an alpine meadow (Zhang Y et al. 1993)

Month/year	Captures total		First captures		Recaptures		Recapture rate (%)	
	♂	♀	♂	♀	♂	♀	♂	♀
May 87	12	24	12	24				
Oct 87	22	26	15	13	7	13	36.8	35.1
May 88	9	15	7	6	2	9	8.3	25.7
Oct 88	15	17	11	10	4	7	30.8	31.2
May 89	9	16	7	9	2	7	11.8	29.2
Oct 89	9	11	6	6	3	5	25.0	25.0

Recapture rate (%) = 100 (no. of recaptures)/(no. of recaptures + no. of previous captures)

in the pellets of many raptors (Cui et al. 2003), revealing that plateau zokors probably disperse across the surface. Capture-mark-recapture data collected over three years (Table 17.2) indicate that rates of male recapture are lower than female, especially in the spring period. These results indicate that plateau zokors have male-biased dispersal (Zhang Y et al. 1993).

17.10 Territoriality and Home Range

In general, plateau zokors are highly aggressive towards each other, irrespective of gender (Zhang Y et al. 1993). The populations tend to be spatially clumped (Jing et al. 1991; Zong et al. 1991; Zhang Y 1999). Home range varies between seasons. The average home range of males ($1790 \pm 720\,m^2$, $n = 5$) is larger than that of females ($260 \pm 112\,m^2$, $n = 6$) in spring. In other seasons, there were no significant differences between the home ranges of males ($560 \pm 101\,m^2$, $n = 8$) and females ($640 \pm 130\,m^2$, $n = 8$) (Zhang Y et al. 1993).

17.11 Population Dynamics

Investigations of the population dynamics of plateau zokors over five years (from 1984 to 1989) in an alpine meadow (Fig. 17.1) showed that population densities differed between spring (May) and autumn (October), but that there was no significant difference between these respective months over the different years (Zhang Y et al. 1991). During the reproductive season the population density increases until the size of the population in October

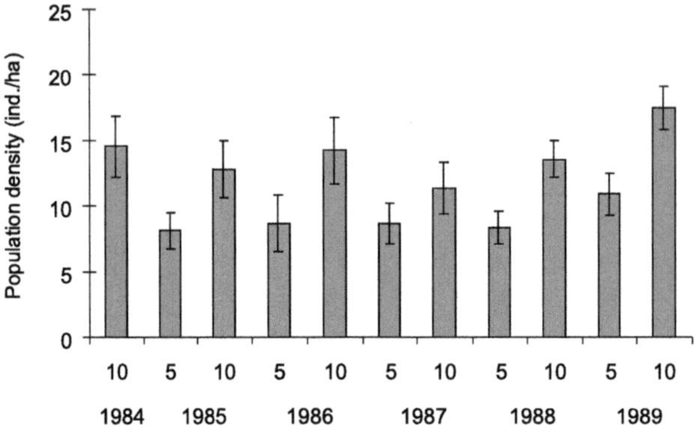

Fig. 17.1. Population densities (± SE) of plateau zokors in an alpine meadow from 1984 to 1989 (Zhang Y et al. 1991). Population estimates were obtained in May (month 5) and October (month 10) each year

is 1.56 times that in the previous May. The population then decreases from October to the following May, as the reproduction activity stops and population numbers are reduced by the severe winters (Zhang Y et al. 1991).

17.12 Burrow Cohabitants

Burrows dug by plateau zokors are used as breeding habitat by native amphibians, reptiles, birds and small mammals other than the zokor itself (Fan et al. 1990; Smith and Foggin 1999). There is also intense competition for burrow sites between plateau zokors and surface-living plateau pikas (Zong et al. 1986; Fan et al. 1990; Zhang Y et al. 1998). Where these two species occur sympatrically, this competitive interaction limits each species from extreme increases in population density (Zong et al. 1986; Zhang Y et al. 1998). Abandoned plateau zokor burrows are often colonized by root voles (*Microtus oeconomus*) and Gansu pikas (*O. cansus*) as a refuge for breeding and for protection from predators. Some birds depend on plateau zokor burrows for nesting and shelter. Hume's ground jay (*Pseudopodoces humilis*) and several species of the snow finch (*Montifringilla adamsi, M. blanfordi, M. davidiana, M. ruficollis, M. tacazanowski*) nest primarily in pika or zokor burrows (Zhang X 1982; Smith et al. 1990; Smith and Foggin 1999).

17.13 Plateau Zokors as Prey

Across the Tibetan plateau, both the plateau pika and the plateau zokor are the dominant small mammal herbivores. Other species, such as woolly hares (*Lepus oiostolus*), Himalayan marmots (*Marmota himalayana*), the

Gansu pika, the Daurian pika (*Ochotona daurica*), root voles, or the Qinghai vole (*Microtus fuscus*) are scarce and patchily distributed. Most predatory animals living on the Tibetan plateau including Eurasian ferrets (*Mustela nigripes*), foxes (*Vulpes ferrilata* and *V. vulpes*), steppe polecats (*Mustela eversmanni*), Chinese mountain cats (*Felis bieti*), the Pallas's cat (*Otocolobus manul*) and the Eurasian lynx (*Lynx lynx*) rely mainly on the abundant plateau pikas for their diet (Smith et al. 1990; Zhou and Dou 1990; Zhou et al. 1994a, b, 1995; Schaller 1998; Smith and Foggin 1999). However, populations of the surface-living plateau pikas can be affected strongly by the climate and their density plummets regionally when atypical heavy snowstorms occur in winter. During such periods, plateau zokors become the primary food resource for many predators. Steppe polecats, Chinese mountain cats, the Pallas's cat and the Eurasian lynx then are forced to change their diet to plateau zokors (Li 1989). Eurasian polecats, foxes and many raptor species also depend at least partly on plateau zokors as prey. Thus, plateau zokors act as an alternative food source for many carnivores in the Tibetan plateau ecosystem (Li 1989; Zhou et al. 1994a, b, 1995; Wei et al. 1996).

Some large mammalian predators such as wolves (*Canis lupus*) and brown bears (*Ursus arctos*) prey on small animals and include plateau pikas and plateau zokors in their diet (Li 1989; Schaller 1998).

It has been generally assumed (Li 1989) that most of the large predatory birds of the Tibetan plateau, such as golden eagles (*Aquila chrysaetos*), upland buzzards (*Buteo hemilasius*), saker falcons (*Falco cherrug*), goshawks (*Accipiter gentilis*), and black kites (*Milvus migrans*), depend primarily upon plateau pikas as a food resource. Recently however, Cui et al. (2003) found that a significant amount of pellets under nests of upland buzzards, and saker falcons contained remains of plateau zokors, showing that the subterranean zokors represent an important component in the diet of avian predators of the plateau.

17.14 Relationship of Plateau Zokors and Livestock

The community of small herbivores in alpine meadows is determined predominantly by the structural attributes of the habitat, which in turn is largely a function of the intensity of livestock grazing. Species diversity and abundance of small herbivorous mammals show significant positive correlations with the grazing intensity of livestock, although there is a non-significant positive correlation between rodent species evenness and grazing intensity (Liu et al. 1991). The combined grazing of yaks (*Bos grunniens*), sheep (*Ovis aries*) and horses (*Equus sp.*) lowers the degree of cover and the height of vegetation, enhancing plateau zokors' occurrence compared to population densities in natural meadows due to a change in the plant community structure (plants with taproots become dominant, which cause an increase in underground food availability) (Liu et al. 1991; Bian et al. 1994; Zhang Y et al. 1998).

In most cases, degraded vegetation resulting from overgrazing by livestock provides advantageous ecological conditions for the formation and spread of high-density populations of plateau zokors and plateau pikas (Liu et al. 1991; Bian et al. 1994). Plateau zokors, along with plateau pikas, are more likely to accelerate the deterioration of rangelands that are already overgrazed (Shi 1983; Zhang Y et al. 1998). Thus, the most effective way to manage population densities of small mammals on the high alpine grasslands of the Tibetan plateau would be to improve the management of the grazing system.

17.15 Conclusion

Plateau zokors are the only subterranean herbivores across the Tibetan plateau of China, exhibiting elaborate anatomical, physiological, and behavioral adaptations associated with a subterranean existence. The morphological traits of plateau zokors reveal a remarkable convergence with other fossorial mammals in their body plan, including claws, incisors, forelimbs, head, and nose, which can be directly associated with digging. Their physiological adaptations enable them to deal with low oxygen and high carbon dioxide levels in this region where thick permafrost is widespread. The daily energy consumption of a plateau zokor per gram of body weight is much higher than surface-living herbivores due to the high energetic demands of excavating and foraging within the hard substrate underground. Like many other subterranean rodents, plateau zokors are highly territorial, solitary and aggressive. Females of plateau plateau zokors give birth only once a year to a litter of three pups. Population densities fluctuate seasonally but are relatively stable between years. Abandoned burrows of plateau zokors are important shelter resources for native amphibians, reptiles, birds and other small mammals during their respective breeding seasons. Additionally plateau zokors provide a significant food source for many avian and mammalian predators. Plateau zokors thus play a vital ecological role in alpine meadows on the Tibet plateau.

Acknowledgements. This work was supported by the National Science Foundation of China (No.30470311, and 30670357), The Chinese Key Technological Research and Development Program (No.2005BA529A05), and The Western Light Projects of China.

References

Bian J, Fan N, Jing Z, Shi Y (1994) Studies on the successive relation between small mammal community and plant community in alpine meadow. Acta Theriol Sin 14:209–216 [In Chinese with English summary]

Bradley WG, Yousef MK (1975) Thermoregulatory responses in the plains pocket gopher, *Geomys bursarius*. Comp Biochem Physiol A 52:35–38

Bradley WG, Miller JS, Yousef MK (1974) Thermoregulatory patterns in pocket gopher: desert and mountain. Physiol Zool 47:172–179

Contreras LC (1986) Bioenergetics and distribution of fossorial *Spalacopus cyanus* (Rodentia): Thermal stress, or cost of burrowing. Physiol Zool 59:20–28

Cui Q, Lian X, Zhang T, Su J (2003) Food habits comparison between *Buteo hemilasius* and *Bubo bubo*. Chin J Zool 38:57–63 [In Chinese with English summary]

Fan N, Jing Z, Zhou W (1990) On the invading behavior of plateau zokor and a new approach to its control. Acta Theriol Sin 10:114–120 [In Chinese with English summary]

Frank C (1988) The effects of moldiness level on seed selection by *Dipodomys spectabilis*. J Mammal 69:358–362

Gettinger RD (1975) Metabolism and thermoregulation of a fossorial rodent, the northern pocket gopher (*Thomomys talpoides*). Physiol Zool 48:311–322

Gorecki A, Christov L (1969) Metabolic rate of the lesser mole rat. Acta Theriol 14: 441–448

Hall K, Boelhouwers J, Driscoll K (1999) Animals as erosion agents in the alpine zone: Some data and observations from Canada, Lesotho, and Tibet. Arct, Antarct, Alp Res 31:436–446

Jing Z, Fan N, Zhou W, Bian J (1991) Integrated management of grassland rodent pest in Panpo area. Chin J Appl Ecol 2:32–38 [In Chinese with English summary]

Kleiber M (1961) The fire of life. Wiley, New York.

Li D (1989) Economically Important Vertebrates of Qinghai Province. Qinghai People's Publishing House, Xining China [In Chinese with English summary]

Li Y, Zhao X, Cao G, Zhao L, Wang Q (2004) Analyses on climates and vegetation productivity background at Haibei Alpine Meadow Ecosystem Research Station. Plat Meteorol (China) 23:558–567 [In Chinese with English summary]

Liu J, Wang X, Liu W, Nie H (1991) Effect of experimental grazing level of Tibetan sheep on rodent communities I. Analyses of structure and function for rodent communities. In: Liu J, Wang Z (eds) Alpine meadow ecosystem fascicle 3. Science Press, Beijing, China, pp 21–27 [In Chinese with English summary]

McNab BK (1966) The metabolism of fossorial rodents: a study of convergence. Ecol, 47:712–733

Norris RW, Zhou K, Zhou C, Yang G, Kilpatrick CW, Honeycutt RL (2004) The phylogenetic position of the zokors (Myospalacinae) and comments on the families of muroids (Rodentia). Mol Phylogenet Evol 31:972–978

Pi N (1982) Energy dynamics of the population of sheep in alpine meadow ecosystem: The preliminary estimation of energy flow through the population of Tibetan sheep in alpine meadow ecosystem. In: Xia W (ed) Alpine meadow ecosystem fascicle 1. Science Press, Beijing, China, pp 67–84 [In Chinese with English summary]

Reichman OJ, Smith SC (1990) Burrows and burrowing behavior by mammals. Curr Mammal 66:197–244

Schaller GB (1998) Wildlife of the Tibetan Steppe. Univ Chicago Press, Chicago

Shi Y (1983) On the influence of rangeland vegetation to the density of plateau pikas (*Ochotona curzoniae*). Acta Theriol Sin 3:181–187 [In Chinese with English summary]

Shi Y, Zhang D, Bian J (1994) Population densities and distribution of the plateau pikas and plateau zokors in the Haibei state and Tianjun county. Grassland of China 6:51–52 [In Chinese with English summary]

Smith AT, Foggin JM (1999) The plateau pika (*Ochotona curzoniae*) is a keystone species for biodiversity on the Tibetan plateau. Anim Conserv 2:235–240

Smith AT, Smith HJ, Wang X, Yin X, Liang J (1986) Social behavior of the steppedwelling blacklipped pika. Natl Geo Res 2:57–74

Smith AT, Formozov NA, Hoffmann RS, Zheng C, Erbajeva MA (1990) The pikas. In: Chapman JA, Flux JAC (eds) Rabbits, hares and pikas: status survey and conservation action plan. IUCN, Gland, Switzerland, pp 14–60

Su J, Wang Z (1992) Studies on the population energetics of plateau zokor I. Average daily metabolic rate and burrowing metabolic rate. Acta Theriol Sin 12:200–206 [In Chinese with English summary]

Su J, Wang Z, Hu X (1991) The physiological basis of cold adaptation in the fossorial rodent (*Myospalax baileyi*): Basal metabolism (*BMR*) and nonshivering thermogenesis (NST). In: Liu J, Wang Z (eds) Alpine meadow ecosystem fascicle 3. Science Press, Beijing, China, pp 139–147 [In Chinese with English summary]

Wang D, Wang Z, Feng Y (1991) Strategies for survival of small mammals in a cold alpine environment III. Thermoregulation of *Ochotona cansus* and adaptive convergence of small mammals to cold and high altitude. In: (Liu J, Wang Z (eds) Alpine meadow ecosystem fascicle 3. Science Press, Beijing, China, pp 125–137 [In Chinese with English summary]

Wang Q, Fan N (1987) Studies on the digging activities and exploration about the method of number estimation of plateau zokor. Acta Theriol Sin 7:283–290 [In Chinese with English summary]

Wang Q, Zhang Y, Wei W, Bian J (2000a) Food habit of the plateau zokor. Acta Theriol Sin 20:198–210 [In Chinese with English summary]

Wang Q, Zhou W, Wei W, Zhang Y, Fan N (2000b) The burrowing behavior of *Myospalax baileyi* and its relation to soil hardness. Acta Theriol Sin 20:277–283 [In Chinese with English summary]

Wang Z, Zeng J, Han Y (1979) Studies on the metabolism of the mouse hare (*Ochotona curzoniae*) and the mole rat (*Myospalax fontanierii*). Acta Zool Sin 25:75–85 [In Chinese with English summary]

Wang Z, Zeng J, Han Y, Zhang X (1980) Energy dynamics of some small mammals in an alpine meadow ecosystem. Digestibility and assimilation of natural foods on the mouse hare and the mole rat in an alpine meadow ecosystem. Acta Zool Sin 26:184–195 [In Chinese with English summary]

Wei W, Zhou W, Fan N, Biggins DE (1996) Activity rhythm and home range of alpine weasel. Acta Theriol Sin 16:35–42

Xia W (1988) A brief introduction to the fundamental characteristics and the work in Haibei Research Station of Alpine Meadow Ecosystem. In: Xia W (ed) The proceedings of the international symposium of alpine meadow ecosystem. Science Press, Beijing, China, pp 1–10

Zeng J, Wang Z, Shi Z (1984) Metabolic characteristics and some physiological parameters of mole rat (*Myospalax baileyi*) in alpine area. Acta Biol Pla Sin 3:163–171 [In Chinese with English summary]

Zhang D, Zhou W, Wei W, Wang Q (1995) Study on the reproduction and reproductive behaviour of the plateau zokor. In: Cheng Y (ed) Studies on mammal biology in China. Forestry Publishing House, Beijing, China, pp 174–179 [In Chinese with English summary]

Zhang X (1982) Studies on breeding biology of 10 species of passerine in alpine meadow. Acta Zool Sin 28:190–199 [In Chinese with English summary]
Zhang Y (1999) Effect of plateau zokor on characters and succession of plant communities in alpine meadow. Zool Res (China) 20:435–440 [In Chinese with English summary]
Zhang Y (2000) Studies on the pattern of animal-plant interaction: The effects of plateau zokor on the biogeochemical cycling of alpine meadow ecosystem and its response to the chemical defense of plants. PhD. Dissertation, Chin Acad Sci, Beijing, China. [In Chinese with English summary]
Zhang Y, Liu J (2003) Effects of plateau zokors (*Myospalax fontanierii*) on plant community and soil in an alpine meadow. J Mammal 84:644–651
Zhang Y, Zhou W, Fan N, Zhang D (1991) Population dynamics and prediction of the plateau zokors. In: Liu J, Wang Z (eds) alpine meadow ecosystem fascicle 3. Science Press, Beijing, China, pp 175–179 [In Chinese with English summary]
Zhang Y, Fan N, Zhou W (1993) Study on the population ecology of the plateau zokors. Chin J Vector Biol Contr 4:359–361 [In Chinese with English summary]
Zhang Y, Fan N, Wang Q, Jing Z (1998) The changing ecological process of rodent communities during rodent pest managements on alpine meadow. Acta Theriol Sin 18:137–143 [In Chinese with English summary]
Zhang Y, Fan N, Wang Q (1999) The effects of some environmental factors on population abundance of plateau zokor. Acta Biol Pla Sin 14:101–109 [In Chinese with English summary]
Zhang Y, Zhang Z, Liu J (2003) Burrowing rodents as ecosystem engineers: the ecology and management of plateau zokor (*Myospalax fontanierii*) in alpine meadow ecosystems on the Tibetan Plateau. Mammal Rev 33:284–294
Zheng S (1980) Studies on the reproduction in the mole rat (*Myospalax fontanieri*). Acta Zool Sin 1:465–477 [In Chinese with English summary]
Zhou W, Dou F (1990) Studies on activity and home range of plateau zokor. Acta Theriol Sin 10:31–39 [In Chinese with English summary]
Zhou W, Wei W, Biggins DE (1994a) Daily active range and home range of polecat in Qing-Zang plateau. Acta Biol Pla Sin 12:173–179 [In Chinese with English summary]
Zhou W, Wei W, Biggins DE (1994b) Study on population dynamic of polecats and its effective factors. Acta Biol Pla Sin 12:161–171 [In Chinese with English summary]
Zhou W, Wei W, Biggins DE (1995) Activity rhythms and distribution of natal dens for red foxes. Acta Theriol Sin 15:267–272 [In Chinese with English summary]
Zong H, Xia W, Sun D (1986) The influence of a heavy snow on the population density of small mammals. Acta Biol Pla Sin 5:85–90 [In Chinese with English summary]
Zong H, Fan N, Yu F, Zhu J (1991) The research on the population spatial patterns of the plateau zokor (*Myospalax baileyi*) and the plateau pika (*Ochotona curzoniae*) in the alpine meadow ecosystem. Acta Ecol Sin 11:125–126 [In Chinese with English summary]

18 Senescence Patterns in African Mole-rats (Bathyergidae, Rodentia)

Philip Dammann, Hynek Burda

18.1 Introduction

From an evolutionary viewpoint, senescence is an intriguing phenomenon. Many theories have been developed to identify its ultimate and proximate causes, but the process is so complex that, to date, barely any question about how and why organisms age has been fully answered. Consequently, even well-established theories on the evolution and mechanisms of senescence are still subject to research and debate.

For several reasons, African mole-rats of the family Bathyergidae are especially suited for aging studies. First, all members of the family are strictly subterranean and thus presumably suffer a lower risk of dying from predators or climatic extremes than surface-dwelling rodents. Since all relevant evolutionary senescence theories are based on the assumption that the rate of internal decay (i.e. senescence) is ultimately shaped by the extrinsic mortality risk (cf. e.g. Medawar 1952; Williams 1957; Kirkwood 1977), aging in mole-rats is predicted to be generally slow by these theories, making bathyergids very suited to test them. Second, in contrast to most other groups of subterranean mammals which are usually solitary (Nevo 1999; Burda et al. 2000), this family exhibits an exceptional diversity of social and mating systems, ranging from solitary and polygamous to social and monogamous species. Social and reproductive strategies are fundamental life history components and hence might affect longevity (as an integrative life history trait), too. The Bathyergidae therefore provide an interesting model to examine the influence of these factors on senescence. Third, the Bathyergidae are the only mammalian family known to date to include species with a truly eusocial organization, namely the naked mole-rat, *Heterocephalus glaber*, and several species of the genus *Fukomys* (*Fukomys* is a genus recently emancipated from the genus *Cryptomys* com-

Philip Dammann: Dept. General Zoology, Institute for Biology, University of Duisburg-Essen, 45117 Essen, Germany and Central Animal Laboratory, University of Duisburg-Essen Medical School, 45117 Essen, Germany, E-mail: philip.dammann@uni-due.de

Hynek Burda: Dept. General Zoology, Institute for Biology, University of Duisburg-Essen, 45117 Essen, Germany

Subterranean Rodents: News from Underground
S. Begall, H. Burda, C.E. Schleich (Eds.)
© Springer-Verlag Berlin Heidelberg 2007

prising species of "*Cryptomys*" from out of South Africa; Ingram et al. 2004, Kock et al. 2006). Similar to hymenopterans and termites, colonies of these mole-rat species are characterized by monopolization of reproduction by few individuals (mostly a single breeding pair), reproductive altruism of their offspring (which involves non-breeding and cooperative alloparental brood care) and long-lasting philopatry resulting in an overlap of about six adult generations (Burda et al. 2000). Evolution of eusociality has led to a very peculiar aging-phenotype in eusocial insects, namely extreme longevity (some ant queens can live up to 30 years), caste-specific aging rates within species, and a complete reversal of the trade-off between reproductive investment and longevity (cf. Hölldobler and Wilson 1990) which is characteristic of most other organisms studied thus far (cf. Bell 1980; Stearns 1992). Since adoption of similar life styles often leads to convergence in life history traits (cf. Partridge 2001), the question arises whether eusocality is associated with similar aging characteristics in mole-rats, too.

Here we review what is known about senescence in bathyergid mole-rats, and add some new comparative aspects. Our goal is to integrate all these data in order to describe the aging phenotype of bathyergids comprehensively, i.e. on all taxonomic levels: interfamilial (comparing longevity of the family Bathyergidae to other rodent families), intrafamilial (comparing longevity of different genera and species within the Bathyergidae), and intraspecific (comparing aging rates of the reproductive and non-reproductive "castes" in eusocial species).

18.2 Longevity of Bathyergid Mole-rats Compared to Other Rodent Families

In mammals, longevity (expressed as maximum species life span, MLSP) is allometrically correlated with body size (Calder 1983; Promislow and Harvey 1990). We analyzed this relationship for rodents alone, both on the species and family level (see Fig. 18.1, where further details on data sources and methods are also provided). A simple rule of thumb to identify taxa with an extraordinarily high (or low) potential for longevity is provided by this allometry: the longer the distance from the regression line, the more exceptionally do the respective taxa age.

Amongst rodents, naked mole-rats *Heterocephalus glaber* have the highest MLSP for their body size: 28.3 years (Buffenstein and Jarvis 2002). As apparent from Fig. 18.1A, this remarkably high potential for long life span in relation to body size is characteristic not only for naked mole-rats, but also for other bathyergids. The Bathyergidae have the second highest MLSP for their body weight out of 26 analyzed rodent families, outranged

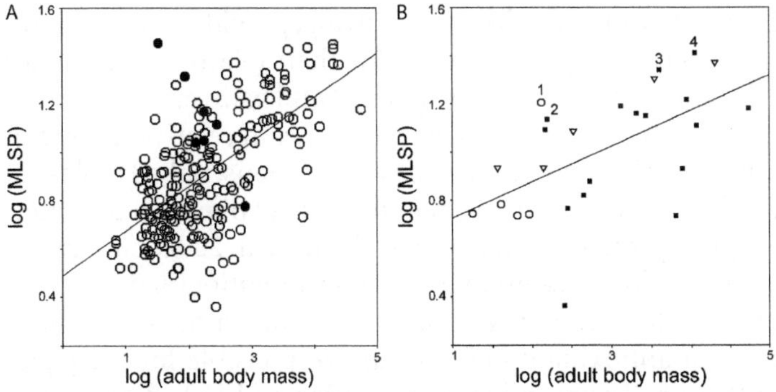

Fig. 18.1A–B. Maximum species life span (MLSP) and body mass in rodents. Data were obtained from our own recordings and Carey and Judge (2000), Weigl (2005), de Magalhaes et al. (2005), Nevo, personal communication, Bennett, personal communication, and Jarvis, personal communication. Data declared as scarce, unreliable, incomplete, etc. were not included: **A** across species. *Filled circles*: bathyergid species ($n = 7$), *open circles*: species from other rodent families ($n = 205$). Body mass data are means over both sexes. The *regression line* is shown (linear regression after Pearson, $F = 135.54$; $df = 211$; $r^2 = 0.392$; $P < 0.0001$); **B** across families. *Circles*: Myomorpha ($n = 5$); *reversed triangles*: Sciuromorpha ($n = 5$); *filled squares*: Caviomorpha ($n = 16$; classification following Corbet and Hill 1980). Data on body mass and MLSP are means over all representatives of their respective families. The *regression line* is shown (linear regression after Pearson, $F = 11.17$; $df = 25$; $r^2 = 0.318$; $P = 0.003$). The four families with the longest life span relative to their body mass are numbered consecutively according to their distance from the regression line (in [brackets]: number of analyzed species per family): 1: Spalacidae [1], 2: Bathyergidae [7], 3: Erethizontidae [3], 4: Hystricidae [5]

only by the equally subterranean Spalacidae (Fig. 18.1B). Interestingly, the next two families exhibiting relatively long life span are the New and Old World porcupines (Erethizontidae and Hystricidae). This is in accordance with evolutionary aging theories, because most members of these families have evolved effective predator avoidance mechanisms, either ecologically (underground shelter) or morphologically (spines). Effect of phylogeny has to be considered particularly in this case because all of the mentioned rodent families but the Spalacidae belong to the suborder Hystricognatha. However, the hystricognath families are equally distributed above and below the regression line (cf. Fig. 18.1B) so that a phylogenetical bias does not seem to be the reason of the observed high MLSP.

One proximate cause of the aging process can be accumulated damage of biomolecules, cell components and tissues caused by reactive oxygen species (ROS) which are inevitably produced in the mitochondria as a side product of aerobic metabolism ("oxidative stress", e.g. Finkel and Holbrook 2000). Therefore, in fossorial and subterranean species, it is interesting to speculate that long-term living in an atmosphere with low oxygen pressure might reduce the negative effects of oxidative stress, providing a mechanistic rather than an evolutionary explanation for their relatively slow aging rates. However, the longevity records for spalacid and bathyergid mole-rats in Fig. 18.1 have been obtained in the laboratory, i.e. in relative hyperoxia compared to the natural situation. Under these conditions, naked mole-rats exhibit higher levels of oxidative stress than mice in all biomolecules measured so far (i.e. lipids, proteins, and DNA; Andziak et al. 2004, Buffenstein 2005). Naked mole-rats do not seem to have superior antioxidant defences compared to mice neither (Andziak et al. 2005). The authors concluded that it is unlikely that antioxidant defences are responsible for the eight-fold longevity difference between laboratory mice and naked mole-rats. While it would definitely be of interest to test whether oxidative stress in mole-rats under natural conditions (i.e. in the hypoxic atmosphere of their burrows; see also Burda et al., this volume) is lower than in the laboratory (making thus higher antioxidant defence not necessary), it is nevertheless apparent that – at least in naked mole-rats – antioxidant activities are not limiting factors in aging and/or that the naked mole-rats are highly tolerant of oxidative damage (Buffenstein 2005). Comparable studies on other long-lived mole-rat species would be highly informative but are not yet available.

18.3 Intrafamilial Longevity Patterns

Within the family Bathyergidae, the typical positive relation between body size and longevity (see above) is absent. Instead, the pattern is literally reversed: the smallest species (*Heterocephalus glaber*) has the highest and the largest species (*Bathyergus suillus*) the shortest MLSP for its size, respectively, leading to a strong negative correlation between body size and maximum life span within this family whose computed statistical values are highly significant (Fig. 18.2). This is interesting, but probably incidental and not causal: there is little reason to believe that body size per se had the opposite effect on longevity in bathyergids than in all other mammals. Thus, other factors must be taken into account.

Figure 18.2 suggests that social bathyergids (genera *Heterocephalus*, *Cryptomys*, and *Fukomys*) live longer than the solitary species (genera *Bathyergus*, *Georychus*). So the question arises: do sociality and/or monogamy promote longer lives in this family? Theoretically, both traits have the potential to reduce individual extrinsic mortality under certain circumstances and thus might be causative sensu evolutionary senescence theories (see above). For example, all else being equal, sociality should reduce individual predation risk by buffering exposure to predators (Alexan-

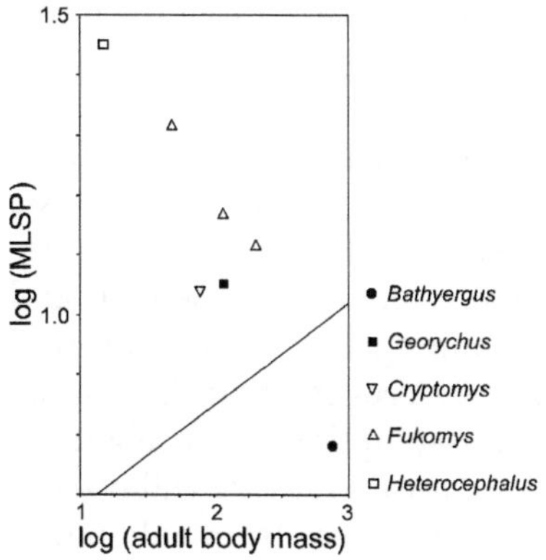

Fig. 18.2. Maximum species life span (MLSP) and body mass within the Bathyergidae. All genera that make up the family are represented except *Heliophobius*, since data on longevity are too preliminary for this genus (Šumbera, personal communication). *Filled symbols*: solitary/polygamous species; *open symbols*: social/monogamous species. The computed values of a linear regression analysis are: $F = 31.7$; $df = 6$; $r^2 = 0.318$; $P = 0.002$, but the corresponding regression line is not shown because the negative correlation between longevity and body mass is probably not causal (see text). Instead, the *regression line* for all rodent species (see Fig. 18.1A) is shown for orientation

der 1974; Ebensperger 2001). Among the few species which may prey on subterranean mammals, in sub-Saharan Africa the most common are the mole-snake *Pseudaspis cana* and (in some areas) the African striped weasel (*Poecilogale albinucha*). Both actively enter the burrow systems, and then probably attack the first burrow inhabitant they randomly encounter. Supposing that snakes and weasels usually kill only one prey individual per hunting event, it is evident that a mole-rat living in a group is less likely to be the particular individual met by the invader than a solitary burrow inhabitant who can hardly avoid this confrontation.

Another aspect is that the risks and costs of mate search must differ between polygamous and monogamous species that form stable pair bonds, because the former have to find a mate each year, whereas the latter do so only once in a lifetime. Reproductive dispersal and mate search are associated with substantial survival costs in many mammalian species (Stearns 1992) and are thought to be particularly costly for subterranean mammals, no matter whether they take place underground (high energetic expenditure of digging) or at the surface, where subterranean animals are surely more vulnerable to predators than surface-dwellers due to their limited locomotor and sensory abilities.

Furthermore, breeders in strictly monogamous *and* social species (*Fukomys*, *Heterocephalus*, *Cryptomys*; but only very few other mammals) should enjoy decreasing costs of mate competition as their family grows, presuming that their (adult) offspring behaves xenophobic against invading conspecifics and thus forms a "protective workforce" – not only against predators, but also against potential same-sex-rivals. This is the case for all social bathyergids (cf. Bennett and Faulkes 2000). Decreasing mortality risk with advancing age should therefore be characteristic of social mole-rats (because colony size increases), whereas in the solitary species there is no reason to assume such age-dependency of environmentally imposed costs and risks, at least not after the animals are fully grown. All this matches very well with the seemingly slower aging rate of the social bathyergids.

It has to be noted that empirical data and theoretical calculations (cf. Burda et al. 2000 and references cited therein) suggest higher extrinsic mortality and group-turnover in *Heterocephalus* than in *Fukomys* mole-rats, despite *Heterocephalus* lives in much larger groups (cf. Bennett and Faulkes 2000 and references therein). Thus, among the social bathyergid species, group size per se does not necessarily reduce mean individual mortality risk, which seems to contradict the assumption of a straightforward effect of sociality on internal aging rate. However, as we will discuss in more detail in the following section, there is evidence from wild mole-rat colonies that breeders and non-breeders ("workers") are exposed to differential risks due to "caste-specific" behavioural differences. Therefore, high colony turn-over and comparatively high predation risk for one subgroup (frequent workers) does not exclude that members of another subgroup can live extremely long once they have abandoned these risky tasks for one or the other reason. In accordance with that, in the social species the longevity record is regularly held by a breeding animal, not a worker (Buffenstein and Jarvis 2002; Dammann 2006; Dammann and Burda 2006).

However, one must keep in mind that there is a strong bias towards the social species in our knowledge and in availability of comparative data. Whereas thousands of individuals of naked mole-rats (and several hundreds in some *Cryptomys* and *Fukomys* species) have been maintained in various laboratories during the last 30 or 20 years, respectively, the interest in study and captive husbandry of their solitary counterparts has been triggered only recently. Thus, in a study of Sherman and Jarvis (2002), longevity data of 1140 naked mole-rats were analyzed. "Only" 86 (= 7.5%) of them lived longer than 15 years, 52 (= 4.6%) lived longer than 20 years. (However, note that Buffenstein 2005 stated that "if one excludes data from terminal experiments or human negligence, more than 80% of our original animals lived longer than 24 years in "captivity", without, however, providing further details on sample size.) For comparison, systematic maintenance of solitary silvery mole-rats (*Heliophobius argenteocinereus*) in the laboratory has started less than ten years ago and the total number of captivity-kept individuals has not exceeded a few dozen (cf. Šumbera et al., this volume). Thus, based just on statistical grounds and on length of research, even if the silvery mole-rats had a similar longevity potential as the naked mole-rats, we have had little chance to find it out so far.

In conclusion, there are good indications that social and mating characteristics might affect aging in bathyergids, but a final answer to this question has to be postponed until more data for the solitary, polygamous species became available.

18.4 Intraspecific Aging Rates in Eusocial Bathyergids

Using long-term breeding data of a eusocial bathyergid species, the Ansell's mole-rat *Fukomys anselli*, we could recently demonstrate that reproductive division of labour goes along with a "caste"-specific aging pattern in this mammal. In fact, Ansell's mole-rat breeders of both genders lived approximately twice as long as their non-breeding counterparts. No difference was found between males and females in either reproductive caste. The oldest female breeder had lived for about 15 years when data collection was stopped; this female was still alive and reproducing. The longest lived male breeders had reached ages of about 20 years. In contrast, all non-breeders have died before their eighth birthday (Fig. 18.3).

The differences in aging rates where not due to variation in social rank, intrinsic quality, or time budgets allocated to feeding, resting, or working (Dammann and Burda 2006). Obviously, pair-bonding and/or sexual activ-

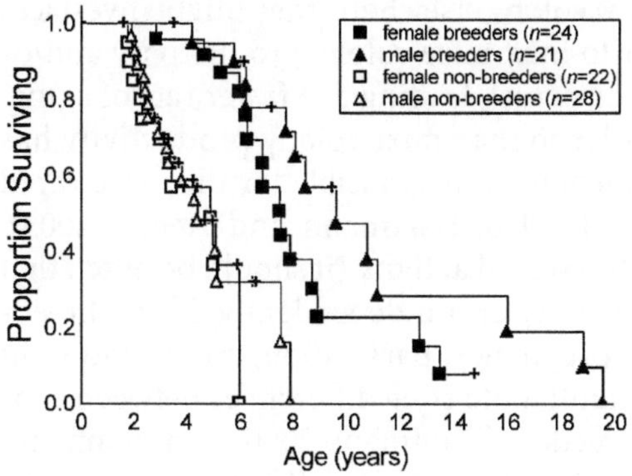

Fig. 18.3. Survival curves of breeding and non-breeding *Fukomys anselli*. *Crosses* indicate censored observations. Mean survival ± SE, breeders: ♀♀: 8.3±0.8 years, ♂♂: 11.1±1.4 years; non-breeders: ♀♀: 4.3 ± 0.4 years, ♂♂: 4.8 ± 0.5 years. The differences were significant between breeders and non-breeders in both genders (log rank test pairwise over strata, ♀♀: $\chi^2 = 16.76$, $p < 0.0001$; ♂♂: $\chi^2 = 17.84$, $p < 0.0001$), but not between the genders in either reproductive caste (♀♀ vs. ♂♂ breeders: $\chi^2 = 2.92$, $p > 0.10$; ♀♀ vs. ♂♂ non-breeders: $\chi^2 = 0.08$, $p > 0.78$). Reprinted in changed form from Current Biology, Vol 16, Dammann and Burda, 'Sexual activity and reproduction delay aging in a mammal', pages R117–R118, Copyright (2006), with permission from Elsevier

ity and/or reproduction per se prolongs life in this species. *Fukomys anselli* thus is the first demonstration of a survival benefit of reproduction in a vertebrate. A very similar aging pattern has been found also in captive giant mole-rats (*F. mechowii*). Again, breeders lived on average approximately twice as long as non-breeders, whereas no difference was found between males and females within either reproductive "caste" (Dammann 2006). Hence, a reversal of the common trade-off between longevity and fertility seems to be characteristic of the genus *Fukomys*.

It still remains to be revisited what exactly goes on in *Heterocephalus glaber*. In this species, maximum longevity of breeding animals appears to equal that of non-breeders (Sherman and Jarvis 2002; Buffenstein 2005). Whether reproductive animals live *on average* longer than non-breeders, as is the case for *Fukomys*, is not known to date, because until now only data on maximum (and not mean) life span have been published.

18.5 "Caste"-Specific Aging: Possible Explanations

Do non-breeders have shorter lives than breeders, or do breeders live for longer than non-breeders? The shorter lives of non-breeders could be adaptive if longevity was traded against some other fitness trait. Since non-breeding mole-rats can maximize their inclusive fitness by either supporting their kin or founding a new colony elsewhere, they might invest less in somatic maintenance in order to maximize colony productivity and/or their own chances for successful dispersal, leading to a faster rate of aging.

In ants, higher workload in order to maximize colony productivity has been proposed as a proximate mechanism responsible for the more rapid senescence in workers (Tsuji et al. 1996; Hartmann and Heinze 2003). However, apart from engagement in sexual activity (it should be noted that breeding mole-rats copulate frequently and independently of any breeding cycle, Burda 1989, 1995) and pregnancy, parturition, and lactation in females there seem to be no substantial life style differences between captive *Fukomys* breeders and non-breeders (Dammann 2006; Dammann and Burda 2006). It seems therefore unlikely that a trade-off between workload and longevity is the cause for the differential aging rates we found, at least not on a simple wear and tear basis. It should however be pointed out that although the ultimate explanations refer to wild animals living under natural conditions, the longevity records and analyses of aging were done on animals living in captivity. Although the breeding and non-breeding mole-rats did not differ in their activities in our lab, there may be such differences in the field. Should the non-breeders under natural circumstances be the workers in the colony, they might spend more energy than the breeders. Behaviourally one might see a difference only under natural conditions, not in captivity where food is provided ad libitum. It is not excluded however that the non-breeders maintain a larger or more energetical demanding metabolic apparatus (heart, liver, kidneys) which would subserve increased work rates in nature, but in captivity still contribute to increased metabolism. This preparedness for higher activity of non-breeders, which would be suppressed in breeders, might then lead to the significant survival differential observed in our study (Serge Daan, personal communication). However, times allocated to feeding do not suggest an increased energetic turnover in non-breeders. Also, our observations suggest that locomotor activity exhibited in captivity is an individually constant quantity which is not affected by availability or non-availability of food. It looks like as if there was a certain daily stint of work (locomotor activity, energy turnover) an animal does independently of availability of food. Surely, differential metabolism in animals of different sexes, ages, sizes, and breeding status has to be measured to recognize the role of energy turnover in determining the life span.

Besides colony productivity, non-breeding mole-rats might trade longevity against one or both of the other fitness traits, i.e. competitiveness in order to defend colonies against intruders or to enhance own chances for successful dispersal. In this context, it is interesting that when complete mole-rat colonies are trapped in the field, breeders are usually among the last individuals to be captured. This suggests that they are least likely to visit areas of potential danger in the burrow system, and that non-breeders are more prone to fulfil risky tasks such as burrow maintenance or colony defence (Bennett and Faulkes 2000). This tendency, besides protecting the colony, has also a selfish component because it enhances the chance to encounter potential mates from adjacent burrow systems. In all *Fukomys* species studied thus far, both sexes disperse (Bennett and Faulkes 2000), and mates are found either through random encounters of individuals from foreign burrow systems or active intrusion of other colonies in order to attain breeding status there (Spinks et al. 2000; Burland et al. 2004). The latter requires aggressiveness and competitiveness, which are also traits that improve defence performance against intruders. Investment in competitiveness might therefore be stronger in non-breeders than in breeders. Thus, if competitiveness somehow impairs somatic maintenance or repair, it would be a good candidate for an alternative trade-off faced more heavily by non-breeders. It is worth mentioning that in large, established laboratory colonies, most aggressive behaviour is initiated mostly by old non-breeders (often, but not always males; own unpublished data).

The second possible explanation, closely combining ultimate and proximate aspects, is that due to their peculiar mating system, mechanisms might have evolved in social mole-rats which directly enhance life span in reproductive animals via mating and/or breeding. Breeders would then indeed live longer than non-breeders, not vice versa. Although very speculative at the moment, such a scenario would be in line with a prediction made by Schrempf et al. (2005) who recently showed that in the ant *Cardiocondula obscurior*, life span of queens is prolonged by mating and/or substances transferred via it. The authors introduced the term "sexual cooperation" for this phenomenon and predicted that it might not be restricted to social insects but could be found also in other animal taxa wherever partners show a lifetime commitment to each other. Schrempf et al. (2005) identified two preconditions favouring the evolution of sexual cooperation in ants: monogamy plus lifetime pair bond, and production of sexual offspring only after the colony has reached a certain size, i.e. a sterile work force has been established. Mole-rats belong to the few mammal species that meet the first criterion; furthermore, although all offspring are physiologically fertile, colony productivity and chances for the colony to survive increase with colony size (Jarvis et al. 1994). Thus, both partners profit from long-lived mates, as they ensure a high lifetime reproductive output. Any trait that helps enhancing the partner's life span, e.g. transfer of antioxidants or other beneficial compounds through copulation, would therefore be under positive selection. It remains to be tested whether such traits have evolved in mole-rats and, if so, which kind they are.

Another important aspect is endocrinology. Sexual activity and pair bonding enjoyed by the breeding pair, but not by the non-breeders, is expected to result in surge of diverse hormones and neurotransmitters, particularly oxytocin, arginine vasopressin, sexual hormones, prolactin, dehydroepiandrostone, and dopamine (cf. Young and Wang 2004). These substances are known to have many positive effects and have been linked with enhancing immunity, reducing risk of cardiovascular diseases, reducing stress effects etc. (cf. Davey Smith et al. 1997). Referring to immunity, however, it should be noted that apparent infections or tumors were not causes of death recorded in our laboratory colonies. Therefore, it is at least questionable in how far the immunosuppressive or immuno-stimulating effects of certain hormones provide a proximate explanation for differential survival in the lab. Besides, it is possible that maturation and breeding in social mole-rats do not coincide with high secretions of the immunosuppressive steroids such as testosterone, corticosterone and progesterone. For example, serum testosterone and testes mass in males of mole-voles (*Ellobius talpinus*), a monogamous social species, are extremely low in comparison with other rodents (Novikov et al. 2004). Furthermore, no differences were found between breeding and non-breeding mole-voles in basal levels of serum and faeces corticosterone and in humoral immune response (Moshkin et al., this volume; Moshkin personal communication). The authors suggested that in the (subterranean) species with low reproductive investment (K-strategists) there is no immune-reproductive trade-off. Analogous studies on some *Fukomys*-species are in progress.

18.6 Conclusions and Perspectives

African mole-rats (Bathyergidae) are suited to address questions about senescence both on the ultimate and the proximate level. Whereas more data on some understudied (solitary) members of the family are required in order to decide whether and how strong social and mating systems ultimately shaped intrinsic aging rates, the good knowledge and the relative ease of maintenance of the social species provides great opportunities already today. Due to their caste-specific aging pattern and their enormous potential for long life spans, eusocial insects have become model organisms for studying the proximate mechanisms underlying the aging process (e.g. Parker et al. 2004; Rueppell et al. 2005). However, extrapolation of results from insects to man is limited, especially on a physiological level. Therefore, for aging studies relevant to humans, the long-lived social bathyergids could become a valuable mammalian substitute. This is particularly true for the genus *Fukomys*, as these long-lived rodents offer the possibility to study differential aging rates within the same genetic configuration. In addition, in contrast to most Hymenoptera (and to some extent also in *Heterocephalus*), there are no morphological, behavioural or nutritional differences between reproductive and non-reproductive animals, which all might influence aging and complicate the identification of factors delaying or accelerating it. The best way to understand the aging process is to manipulate it (Davies 1983). In *Fukomys*, this can easily be done by giving the animals the chance to mate or not. Future studies should aim to elucidate potential changes in gene expression patterns, endocrine profiles, immune response and energy metabolism that might occur when an animal changes from non-breeding to breeding condition.

Acknowledgements. We thank Sabine Begall, Gil Dryden and Cristian Schleich for constructive comments on the manuscript. We also thank Marie-Therese Bappert for her valuable assistance in improving the technical quality of the graphs.

References

Alexander RD (1974) The evolution of social behaviour. Ann Rev Ecol Syst 5:325–383

Andziak B, O'Connor T, Buffenstein R (2004) High levels of oxidative damage in the longest-lived rodent, *Heterocephalus glaber*; age-related changes in lipid peroxidation and antioxidant activity in mice and naked mole-rats. Faseb J 18, Experimental Biology Meeting Abstracts, Abstract No. 679.15 (accessed at: http://select.biosis.org/faseb)

Andziak B, O'Connor TP, Buffenstein R (2005) Antioxidants do not explain the disparate longevity between mice and the longest-living rodent, the naked mole-rat. Mech Ageing Dev 126:1206–1212

Bell G (1980) The costs of reproduction and their consequences. Am Nat 116:45–76

Bennett NC, Faulkes CG (2000) African mole-rats: ecology and eusociality. Cambridge Univ Press, Cambridge

Buffenstein R (2005) The naked mole-rat. A new long-living model for human aging research? J Gerontol A 60:1369–1377

Buffenstein R, Jarvis JUM (2002) The naked mole mat – a new necord for the oldest living rodent. Sci Aging Knowledge Environ 21:7

Burda H (1989) Reproductive biology (behaviour, breeding, and postnatal development) in subterranean mole-rats, *Cryptomys hottentotus* (Bathyergidae). Z Säugetierk 54:360–376

Burda H (1995) Individual recognition and incest avoidance in eusocial common mole-rats rather than reproductive suppression by parents. Experientia 51:411–413

Burda H, Honeycutt RL, Begall S, Locker-Grütjen O, Scharff A (2000): Are naked and common mole-rats eusocial and if so, why? Behav Ecol Sociobiol 47:293–303

Burland TM, Bennett NC, Jarvis JUM, Faulkes CG (2004) Colony structure and parentage in wild colonies of co-operatively breeding Damaraland mole-rats suggest incest avoidance alone may not maintain reproductive skew. Mol Ecol 13:2371–2379

Calder WA III (1983) Body size, mortality, and longevity. J Theor Biol 102:135–144

Carey JR, Judge DS (2000) Longevity records: life spans of mammals, birds, amphibians, reptiles and fish. Odense Univ Press, Odense

Corbet GB, Hill JE (1980) A world list of mammalian species. British Museum and Cornell Univ Press, London Ithaca

Dammann P (2006) Seneszenz bei Afrikanischen Sandgräbern (Bathyergidae, Rodentia) unter besonderer Berücksichtigung der Gattung *Fukomys*. PhD Thesis, Univ Duisburg-Essen, Essen

Dammann P, Burda H (2006) Sexual activity and reproduction delay ageing in a mammal. Curr Biol 16:R117–R118

Davey Smith G, Frankel S, Yarnell J (1997) Sex and death: Are they related? Findings from the Caerphilly cohort study. Brit Med J 315:1641–1644

Davies I (1983) Ageing. In: Studies in biology 151. Edward Arnold, London

de Magalhaes JP, Costa J, Toussaint O (2005) HAGR: the Human Ageing Genomic Resources. Nucleic Acids Res 33(Database Issue):D537–D543

Ebensperger LA (2001) A review of the evolutionary causes of rodent group-living. Acta Theriol 46:115–144

Finkel T, Holbrook NJ (2000) Oxidants, oxidative stress and the biology of ageing. Nature 408:239–247

Hartmann A, Heinze J (2003) Lay eggs, live longer: division of labor and life span in a clonal ant species. Evolution 57:2424–2429

Hölldobler B, Wilson EO (1990) The ants. Springer, Berlin Heidelberg New York

Ingram CM, Burda H, Honeycutt RL (2004) Molecular phylogenetics and taxonomy of the African mole-rats, genus *Cryptomys* and the new genus *Coetomys* Gray, 1864. Mol Phyl Evol 31:997–1014

Jarvis JUM, O'Riain MJ, Bennett NC, Sherman PW (1994) Mammalian eusociality: a family affair. Trends Ecol Evol 9:47–51

Kirkwood TBL (1977) Evolution of aging. Nature 270:301–304

Kock D, Ingram CM, Frabotta LJ, Burda H, Honeycutt RL (2006) On the nomenclature of Bathyergidae and *Fukomys* n. gen. (Mammalia: Rodentia). Zootaxa 1142:51–55

Medawar P (1952) An unsolved problem of biology. HK Lewis, London

Nevo E (1999) Mosaic evolution of subterranean mammals: regression, progression, and global convergence. Oxford Univ Press, Oxford

Novikov E, Petrovski D, Kolosova I, Steinlechner S, Moshkin M (2004) Interactions between intruders and residents in the mole vole *Ellobius talpinus*. Acta Zool Sin 50:19–26

Parker JD, Parker KM, Sohal BH, Sohal RS, Keller L (2004) Decreased expression of Cu-Zn superoxide dismutase 1 in ants with extreme lifespan. Proc Natl Acad Sci USA 101:3486–3489

Partridge L (2001) Evolutionary theories of ageing applied to long-lived organisms. Exper Gerontol 36:641–650

Promislow DEL, Harvey PH (1990) Living fast and dying young: a comparative analysis of life-history variation among mammals. J Zool Lond 220:417–437

Rueppell O, Fondrk MK, Page RE (2005) Biodemographic analysis of male honey bee mortality. Aging Cell 4:13–19

Schrempf A, Heinze J, Cremer S (2005) Sexual cooperation: mating increases longevity in ant queens. Curr Biol 15:267–270

Sherman PW, Jarvis JUM (2002) Extraordinary life spans of naked mole-rats (*Heterocephalus glaber*). J Zool Lond 258:307–311

Spinks AC, Bennett NC, Jarvis JUM (2000) A comparison of ecology of two populations of common mole rat, *Cryptomys hottentotus hottentotus*: the effect of aridity on food, foraging and body mass. Oecologia 125:341–349

Stearns SC (1992) The evolution of life histories: Oxford Univ Press, Oxford, UK

Tsuji K, Nakata K, Heinze J (1996) Lifespan and reproduction in a queenless ant. Naturwiss 83:577–578

Weigl R (2005) Longevity of mammals in captivity; from the living collections of the world. Kleine Senckenberg-Reihe 48, Frankfurt

Williams CE (1957) Pleiotropy, natural selection, and the evolution of senescence. Evolution 11:389–411

Young LJ, Wang Z (2004) The neurobiology of pair bonding. Nature Neurosci 7:1048–1054

Part IV
Environmental and Economic Aspects

19 The Influence of Subterranean Rodents on the Environment

O. J. Reichman

The impact of subterranean rodents can be arrayed along several axes, including their direct influence on plants and plant communities, their role in altering the physical environment and their effect on human dominated systems. In all of these situations their presence is obscured by the dense, opaque medium of the soil. Thus, the nature and scope of their activities may be less appreciated than other, more obvious biotic and abiotic factors.

The chapters in this section present ample evidence, gathered by scientists and other observers, of the importance of subterranean, herbivorous rodents in ecosystems. Perhaps the most widely studied taxa of subterranean rodents in this respect are pocket gophers (Geomyidae), residents across broad parts of North America. Reichman documents in Chap. 20 the wide range of influences engendered by pocket gophers, some of which extensive and dramatic. Focusing on the same group, Witmer and Engeman (Chap. 21) provide an in-depth review of the impact of geomyids on a variety of human endeavors, from agriculture to dangerous buried wastes, underground cables, and even archeological sites. The results presented on geomyids are interesting and provide a review of a well-know taxon, but the results in the remaining chapters provide important new insights on less-known, but equally important, taxa. In Chap. 22, Zhang characterizes how the consumption of plants and the excavation of soil by zokors, (*Eospalax fontanierii*) myospalacine rodents that range across Siberia and northern China. And finally in Chap. 23, Arjo describes some of the intriguing aspects of the burrowing ecology and behavior of a phylogenetically basal subterranean rodent, the mountain beaver (*Aplodontia rufa*).

A brief look at the characteristics of other, cognate, taxa, including the sub-Saharan mole-rats (Bathyergidae), South American tuco-tucos (Ctenomyidae) and coruros (Octodontidae), Eurasian blind mole-rats (Spalacidae), and members of holarctic voles (Arvicolinae), suggest general similarities among subterranean forms (Lacey et al. 2000). For example, *Spalax*, one of the best known genera of rodents (thanks to the extraordinary work by Nevo, his collaborators, and many others (see, e.g. Nevo 1999) appears to have significant impact on their habitats (Heth 1991).

O. J. Reichman: National Center for Ecological Analysis and Synthesis, 735 State Street, Suite 300, Santa Barbara, CA 93103, USA, E-mail: Reichman@nceas.ucsb.edu

Subterranean Rodents: News from Underground
S. Begall, H. Burda, C.E. Schleich (Eds.)
© Springer-Verlag Berlin Heidelberg 2007

Another group of subterranean rodents, the Ctenomyidae, inhabit a wide range of habitats in sub-Amazonian South America and exhibit similar patterns of impact, although they appear to consume a significantly greater proportion of aboveground plant parts than the other subterranean genera (Comparatore et al. 1995).

The bathyergids are among the best known subterranean rodents. In some locations in South Africa three genera of bathyergids are sympatric, segregated by depth, and cumulatively have a large impact on plants and the soils from depths exceeding 60 cm up to the surface (Reichman and Jarvis 1989). Numerous geophytes (bulb-forming plants) occur in the regions of southern Africa inhabited by the mole rats. These serve as important food items for the rodents, with the smallest bulbs eaten first and the larger ones cached for future use. In some cases, the largest bulbs are only partially eaten and continue to grow, providing a renewable resource for the consumer (Spinks et al. 1999). Little is known about the impact of the eusocial *Heterocephalus glaber* on the environment, but since it occurs in large colonies with tunnels extending several kilometers (Sherman et al. 1991; Bennett and Faulkes 2000) there is the potential for significant impact even thought the habitats they occur in are dry and exhibit low vegetational diversity.

Although soils vary in composition, moisture, hardness, and their capacity to host plant populations, burrows are buffered from many ambient environmental extremes. Thus, the environment experienced by subterranean mammals tends to be relatively uniform compared to above ground environments (Nevo 1999; Burda et al. this volume). The rodents exhibit most digging activity near the surface of the soil where root density is the greatest (Jackson et al. 1996), further homogenizing the environmental conditions they experience. Given this pattern of root distribution by depth, it seems likely that the rodents' burrows tend to be as shallow as possible while maintaining structural integrity (which should vary with soil type). This has led to the observation that the subterranean niche of these types of rodents is indivisible (Lessa 2000), although there are places where as many as three species are segregated by depth (see above).

The conditions in burrows might be relatively uniform, but some of them can be quite extreme (cf. Burda et al., this volume; Roper et al. 2001). Soil hardness can affect seasonal burrowing activity patterns (e.g. Romañach et al. 2005, and literature cited therein), and burrow geometry and microclimate (e.g. Šumbera et al. 2003, and literature cited therein). These circumstances – uniform and often extreme selection pressures – generate evolutionary convergence (Nevo 1999). For example, the general pattern of burrow excavation and mound deposition produce similar impacts on the environment (Witmer and Engeman, this volume). At the level of the individuals, mounds cast on the surface affect many biotic and abiotic features of the environment as the mounds come and go over the landscape (Inouye et al. 1997; Reichman and Seabloom 2002). At larger landscape scales, and over longer periods of time, mima mounds appear to be long-term features of certain areas where subterranean rodents live (Cox 1990; Lovegrove 1991).

While subterranean rodents are important components of many ecosystems, they are extremely difficult to study. They are hard to capture, to observe, and maintaining them in captivity can be complicated. However, they can serve as important research subjects for such general ecological and evolutionary features as adaptation to environmental extremes, the evolution of social behavior, and role of keystone consumers and ecosystem engineers in natural and human dominated systems. In addition, because of their significant impact in agricultural settings and other human endeavors, means to safely ameliorate their effects. These and other distinctive characteristics of herbivorous rodents in the subterranean niche will require the application of creative research approaches including field studies and modeling approaches (Seabloom and Reichman 2001).

References

Bennett NC, Faulkes CG (2000) African mole-rats: ecology and eusociality. Cambridge Univ Press, Cambridge

Comparatore VM, Cid MS, Busch C (1995) Dietary preferences of two sympatric subterranean rodent populations in Argentina. Rev Chil Hist Nat 68:197–206

Cox GW (1990) Soil mining by pocket gophers along topographic gradients in a mima moundfield. Ecology 71:837–843

Heth G (1991) The environmental impact of subterranean mole rats (*Spalax ehrenbergi*) and their burrows. Symp Zool Soc Lond 63:265–280

Inouye RS, Huntly N, Wasley GA (1997) Effects of pocket gophers (*Geomys bursarius*) on microtopographic variation. J Mammal 78:1144–1148

Jackson RB, Canadell J, Ehleringer JR, Mooney HA, Sala OE, Schulze ED (1996) A global analysis of root distributions for terrestrial biomes. Oecologia 108:408–411

Lacey EA, Patton JL, Cameron GN (2000) Life underground: the biology of subterranean rodents. Univ Chicago Press, Chicago

Lessa E (2000) The evolution of subterranean rodents: a synthesis. In: Lacey EA, Patton JL, Cameron GN (eds) Life underground: the biology of subterranean rodents. Univ Chicago Press, Chicago, pp 389–420

Lovegrove BG (1991) Mima-like mounds heuweltjies of South Africa: the topographical, ecological, and economic impact of burrowing animals. In: Meadows PD (ed) The environmental impact of burrowing animals and animal burrows. Symp Zool Soc London, No. 63, Oxford Univ Press, Oxford, pp 183–189

Nevo E (1979) Adaptive convergence and divergence of subterranean mammals. Ann Rev Ecol Syst 10:268–308

Nevo E (1999) Mosaic evolution of subterranean mammals: regression, progression and global convergence. Oxford Univ Press, Oxford

Reichman OJ, Jarvis JUM (1989) The influence of three sympatric species of fossorial mole-rats (Bathyergidae) on vegetation. J Mammal 70:763–771

Reichman OJ, Seabloom EW (2002) The role of pocket gophers as subterranean ecosystem engineers. Trends Ecol Evol 17:44–49

Romañach S, Reichman OJ, Seabloom EW (2005) Seasonal influences on burrowing activity of a subterranean rodent, *Thomomys bottae.* J Zool Lond 319–325

Roper TJ, Bennett NC, Conradt L, Molteno AJ (2001) Environmental conditions in burrows of two species of African mole-rat, *Georhychus capensis* and *Cryptomys damarensis.* J Zool Lond 254:101–107

Seabloom EW, Reichman OJ (2001) Simulation models of the interactions between herbivore foraging strategies, social behavior, and plant community dynamics. Am Nat 157:76–96

Sherman PW, Jarvis JUM, Alexander RD (1991) The biology of the naked mole-rat. Princeton Univ Press, Princeton, NJ

Spinks AC, Branch TA, Croeser S, Bennett NC, Jarvis JUM (1999) Foraging in wild and captive colonies of the common mole-rat *Cryptomys hottentotus hottentotus* (Rodentia: Bathyergidae). J Zool Lond 249:143–152

Šumbera R, Burda H, Chitaukali WN, Kubová J (2003) Silvery mole-rats (*Heliophobius argenteocinereus*, Bathyergidae) change their burrow architecture seasonally. Naturwiss 90:370–373

20 The Influence of Pocket Gophers on the Biotic and Abiotic Environment

O. J. Reichman

20.1 Introduction

Subterranean, herbivorous rodents are widespread in the temperate regions of all but two continents (Australia and Antarctica; Andersen 1987). The subterranean habit has evolved independently in at least seven families, some of which are composed entirely of subterranean forms. Their impacts on the environment are profound (Reichman and Seabloom 2002) but because they occur in the dense, opaque medium of the soil the rodents' influence is often underappreciated.

The rodents affect both the biotic and abiotic elements of the environment. They consume large quantities of vegetation, directly affecting individual plants and plant communities. They also have a significant effect on the physical environment, substantially altering many aspects of the soil. The cumulative effect of subterranean rodents is to profoundly alter the structure and function of the ecosystems they inhabit (Reichman and Seabloom 2002).

This chapter will focus on pocket gophers (family Geomyidae), a North American taxon composed entirely of subterranean forms. These are the most thoroughly studied subterranean herbivorous rodents and exhibit many of the traits that convergent forms on other continents possess.

20.2 Burrowing Activity

The overarching feature of pocket gophers that affects their impact on the environment is the extensive underground burrows they excavate. The burrows can be from a few meters to scores of meters long (Reichman and Smith 1990; Romañach et al. 2005b) and, across several species, a population of gophers can excavate between 3.4 and 57 $m^3\,ha^{-1}\,year^{-1}$

O. J. Reichman: National Center for Ecological Analysis and Synthesis, 735 State Street, Suite 300, Santa Barbara, CA 93103, USA, E-mail: Reichman@nceas.ucsb.edu

Subterranean Rodents: News from Underground
S. Begall, H. Burda, C.E. Schleich (Eds.)
© Springer-Verlag Berlin Heidelberg 2007

(mean = 17.8 m^3 ha^{-1} $year^{-1}$; Smallwood and Morrison 1999) of soil. Pocket gopher burrows may underlie up to 7.5% of an area at one time (Reichman et al. 1982). This snapshot of their impact is an underestimate because they are constantly digging new burrows.

The majority of the tailings from their excavations are deposited on the surface. Mounds generally range from 20 to 50 cm at the base and from 5 to 25 cm in height (Inouye et al. 1997; Smallwood and Morrison 1999). When snow cover is present gophers will tunnel into the snow on the soil surface and densely pack the loosened soil into the openings. Tubes of soil, similar to the geometry of underground burrows, are left on the surface when the snow melts in the spring.

20.3 Geometry of Disturbances

20.3.1 Burrow Spacing

The excavation of burrows is extremely costly (360–3400 times more costly than walking the same distance; Vleck 1979). Consequently, there should be strong selection of the precise placement of burrows, both within an individual system and between neighbors. Reichman et al. (1982) showed that the average spacing between burrows is very uniform within a population. Furthermore, at two sites that differed by twofold in the standing biomass of plants, the spacing between burrows was statistically indistinguishable (although mean burrow length was half as long where the standing biomass was twice as high), suggesting that the spacing might be the outcome of social interactions rather than scaling to resource availability. The geometry of burrows is complex, and recent studies have attempted to develop more sophisticated metrics based on fractal geometry (Romañach and Le Comber 2004).

20.3.2 Mound Spacing

Mounds tend to be very clustered, reflecting a pattern of area restricted searching used by gophers (increasing turn radii and number of branches in rich resource patches). Benedix (1993) showed that *Geomys bursarius* spent 50% of their foraging effort in areas of high plant density even though these areas made up only 33% of the plant community. Area-restricted foraging led to a clustered distribution of mounds over both the sort term (two weeks) and long term (five months). Klass et al. (2000) found a similar clustered pattern at scales < 8 m square over 20-week periods. These clusters tended to move around over two week intervals, but still remained clustered in areas up to 20 m on a side.

20.3.3 Comparisons Between Species, Age, Sex, and Size

Romañach et al. (2005a) analyzed the effects of species, age, sex and habitat on burrow length, area, and shape from six sites. Burrow length and area of the burrow system were associated with age (adult and juvenile) although not size, per se (i.e., the intermediate sized species had the longest burrows). Burrow length and area were negatively correlated with plant biomass and the clay content of soil. All other burrow features were similar among species and locations. Significantly more data from field studies are needed to thoroughly analyze how traits of individuals and species affect patterns of burrowing and mound production.

20.4 Activity Patterns

As would be expected for animals that essentially never see the light of day, pocket gophers remain active through the 24-h cycle (Benedix 1994). They exhibit no particular pattern of circadian activity and no difference between males and females in burrowing activity (Bandoli 1987; Cameron et al. 1988), although Gettinger (1984) noted an increased activity in late afternoon and early evening, and lowered activity at night. In all of the studies, individuals spent substantial amount of time in their nests vs. patrolling their burrows or foraging.

The seasonal patterns of gopher activity are probably more important than daily activity patterns with regard to the impact on plant communities. The general pattern is that seasonal activity increases with soil friability, which is usually associated with soil moisture and thus rainfall (Gettinger 1984; Cox and Hunt 1992). The results of Bandoli (1981) were less clear, with only a slight increase activity associated with rainfall. In the Mediterranean climate of California (dry hot summers, cool moist winters) daily and seasonal activity are curtailed in the summer and Cox and Hunt (1992) present evidence that this may be accompanied by aestivation.

Increased activity with the onset of rains could be a response to increased soil moisture and ease of digging, to benefits associated with extending burrows to locate resources such as food and mates, or as a need to repair and ventilate saturated burrows. Miller (1946) noted that, in California, gophers remained active in the summers in irrigated agriculture fields but not in adjacent dry areas. This could, however, be a response to either soil moisture or increase food production. In an attempt to sort the factors out, Romañach et al. (2005b) watered three treatments (for one, two, and three months) in dry California grasslands and compared the production of surface mounds to unwatered control plots. Immediately after watering, when the soil became moist and well before any plant growth took place or below the soil was saturated, numerous mounds began to appear while none were recorded on control plots. However, the rate of mound production stabilized at a rate lower than the initial pulse and plant production increased during the watering treatments. When the treatments ended, the soil became much drier but plant production continued. The net effect over several months was that mound production was not correlated with soil moisture and food availability had a greater overall effect on burrowing patterns than soil conditions. These results point toward an important impact of gophers on naturally occurring plants and agricultural crops. Specifically, water (from natural rain or irrigation) will prompt increased mound production, which will reach its peak as plant production is increasing. While soil friability is an advantage for subterranean mammals, the opportunity for increased food resources will maintain foraging activity until the plants die or are harvested.

20.5 Diets

Most studies suggest that the diets of gophers match the available resources fairly closely. Andersen and MacMahon (1981) found that 22 of 26 plant species were accepted by *Thomomys talpoides* in cafeteria-style feeding trials. Williams and Cameron (1986a) found that 36 of 51 plants in an area inhabited by *Geomys attwateri* were common in the rodent's diet (with five additional species occurring in the diet of males but not females). The pocket gophers also exhibited seasonal preferences, and the authors suggest this is in response to changes in plant nutritional quality over the year, and differences between males and females with regard to movement patterns (males move father and perhaps encounter a greater variety of items) and dietary needs of pregnant or lactating females.

Williams and Cameron (1986a) found that 60% of the diet of *Geomys attwateri* was composed of roots. Cox (1989) found that *Thomomys talpoides* feed primarily on the shoots of forbs, especially in the summer. Hunt (1992) reports that averaged over the year 70% of the diet of *T. bottae* was shoots, with a peak of 99% in early spring, with the proportion as forbs and grasses shifting seasonally. The authors note that diets for *Thomomys* throughout the west are quite similar when seasons are designated by plant growth stage rather than season. Specifically, forb shoots are preferred, particularly during reproduction with grass shoots, corms, and roots increase in importance during plant dormancy.

20.6 Impact on Plants

20.6.1 Individual Plants

Pockets gophers consume substantial amounts of biomass (Andersen and MacMahon 1981 suggest up to 30% of belowground productivity), directly affecting individual plants. While gophers occasionally consume an entire plant the consumption of even a fraction of the root tissue can significantly affect the mortality and productivity of plants. Reichman and Smith (1991) show that the loss of 25% of root tissue has a greater impact on subsequent plant biomass than the loss of 75% of shoot tissue and the mortality of adult plants directly over active burrows and under new mounds is very high (Reichman et al. 1982; Reichman and Smith 1985). The effect differs, however, between forbs and grasses. Tap-rooted forbs are severely affected when they lose their roots, and when covered by mounds. Grasses, with fibrous roots, are less affected by root herbivory.

Pocket gophers generate a variety of disturbance types that affects the success of individual plants. Burrows can be active or vacant, or refilled with excavated soil, and mounds can be of different ages. Studies of the relationship between these various burrow conditions, plant density (as a measure of competition between plants) and the success of an annual (*Berteroa incana*; Reichman 1988) and a biennial (*Tragopogon dubius*; Reichman 1996) species revealed high densities of competitors had a greater effect on mortality, growth, biomass, and seed production than any of the treatment involving pocket gophers. The mortality of individuals on mounds was high, but those that survived grew larger and produced many more seeds than adjacent control plants. This result was different than found by Wolfe-Bellin and Moloney (2000) where a legume species (*Medicago lupulina*) distribution was positively correlated with mounds.

Pocket gophers can significantly affect large, woody, plants and well as those of smaller stature. Cox and Hunt (1994) document an inverse relationship between the demise of individual ocotillo plants (*Fouqueria splendens*, a large, woody shrub) and pocket gopher abundance. Cantor and Whitham (1989) found that an eightfold increase in pocket gopher activity (as measured by mound production) increased mortality in aspen (*Populus tremuloides*) shoots from < 20% to 90%. Excluding pocket gophers from meadows with rich soil caused an immediate increased in aspen shoots, eventually stabilizing at a level three times higher than out the exclosures. Coggins and Connover (2005) found less of an effect on aspens by pocket gophers, but their study took place where there was a significant, and perhaps overriding, effect of free-ranging ungulates. Borchert et al. (1989), in an investigation of seeding recruitment in blue oaks, discovered that pocket gophers were the major predator on buried acorns and significantly affected mortality of acorns on the soil surface, an impact even more severe than consuming roots.

20.6.2 Plant Defenses

Many studies have documented how plants deter and tolerate shoot herbivory but very little has been done with regard to belowground plant parts, particularly with reference to vertebrate herbivores (Andersen 1987). A recent study by Watts (2005) analyzed deterrence and tolerance of root herbivory between plant species on islands that have never had gophers and the same plant species on the mainland where the plants have been subjected to gopher herbivory for millennia. The results indicate that the roots of individuals from mainland populations are heavily defended chemically whereas those from island populations are much less so. As predicted, gophers strongly preferred the roots of individuals from the island. Furthermore, individuals in mainland populations seem to be more tolerant of root loss than those on the islands, suggesting they have adapted to high probabilities of losing root tissue to herbivores.

20.6.3 Impact on Plant Communities

The impact of pocket gophers is composed of the intersection of two major themes in ecological research – herbivory and disturbance, and the sum of these effects has enormous impacts on plant communities. The most obvious evidence of pocket gopher activity is the mounds they deposit on the surface, and these have been well studied. A list of studies would reveal almost all possible combinations of the influences of mounds on plant communities, from increasing or decreasing germination success, survivorship, growth, and fecundity of plants to patterns of diversity (Foster and Stubbendieck 1980; Hobbs and Mooney 1985, 1991; Davis et al. 1995).

As Hobbs and Mooney (1995) note, the relationship between pocket gophers and plant communities is complex in time and space.

Several general patterns emerge, however, from this complexity. For example, the effects of pocket gopher excavations often show classical patterns associated with disturbance. Pocket activity tends to maintain diversity especially at intermediate levels of disturbance (Tilman 1983; Martinsen et al. 1990; Sherrod et al. 2005, although see Rogers et al. 2001), a pattern driven by the microsite characteristics of mounds (Davis et al. 1995). Persistent pocket gopher foraging and disturbance tends to maintain plant communities at earlier stages of succession (Tilman 1983; Stromberg and Griffin 1996).

An important component of the effects of disturbances is that they are often spatially explicit. Reichman et al. (1982) showed that burrows tend to be very uniformly spaced while mounds are clustered (Benedix 1993). An interesting manifestation of the spatial patterns of burrows occurs in California grasslands where pocket gophers (*T. bottae*) may avoid the fibrous, woody roots of native bunchgrasses and forage instead in the interstices between bunchgrasses, thereby sweeping these areas clear of competing plants (Watts 2005). In another example of broad, community-level influences, the significant reduction of plant biomass directly over burrows and mounds generates a competition-induced wave of biomass centered on the disturbances and extending almost 0.5 m in all directions (Fig. 20.1, Reichman et al. 1993).

Another feature of the complex effects of pocket gopher disturbance is that it interacts with exogenous factors such as plant growth form, fire and

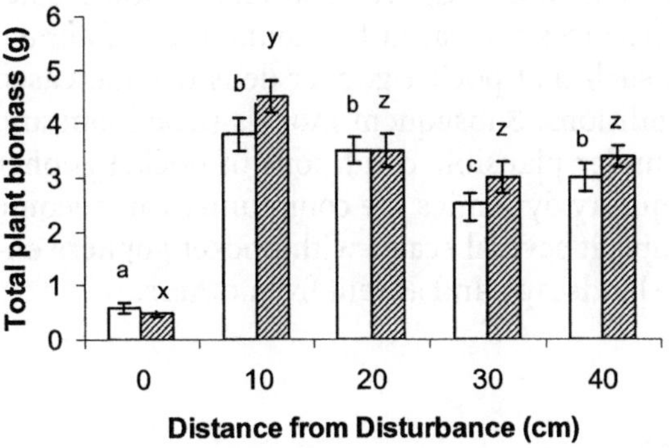

Fig. 20.1. A competition-induced wave of biomass generated by active burrows and recent mounds of *Geomys bursarius*. The pattern extends in 10-cm quadrats from directly over (0 distance) an active burrow or recent mound. Shared letters between the burrow (*solid*) or mound (*diagonals*) sets indicate statistically indistinguishable biomass values (biomass values associated with burrows and mounds did not differ when compared within each distance; Reichman et al. 1993)

climate. Annual and perennial grasses and forbs each have adaptive life history traits that relate to disturbance patterns. For example, Rogers and Hartnett (2001a) showed that the success and influence of difference growth forms is mediated through their means of reproductions (seeds, clonal regeneration, etc.) and that this interacts with mound size. In a related study the authors Rogers and Hartnett (2001b) showed that burrow and mound density and spatial patterns differentially affected grasses and forbs, and that these, in turn, are related to fire regimes. Major climate factors, such as how rainfall and temperature impinge of the growing season, also affect plant community dynamics in the face of disturbance. For example, Hobbs and Mooney (1985) found that mound production at the time of seed set is more important that at times when seeds are not being produced.

Not only are the interactions complex, but outcomes take a long time to become measurable. Williams and Cameron (1986b) and Williams et al. (1986) found very few effects of mounds or pocket gopher removal in a Texas prairie over three years. However, Grant et al. (1980) recorded a 5.5% increase in plant production in an area where pocket gophers had been excluded for decades. Similarly, Forbis et al. (2004) found that that seeds and seedlings on pocket gopher mounds were initially quite rare, but over time (decades) the net effect of disturbances was to facilitate seedling establishment in alpine meadows.

Modeling approaches provide one means to distill some of the complexities on the role of pocket gophers in plant community structure. Using a simulation model, Seabloom and Reichman (2001) analyzed the patch dynamics of annual and perennial species in relation to the spacing of burrow systems, with results showing that the buffer zones between burrows served as refugia for perennials. In addition, over a 50 year period the net effect of pocket gopher disturbances increased the abundance of annual forbs, a preferred food type, such that pocket gopher densities increased compared to the starting conditions. Subsequent models (Seabloom and Richards 2003) suggest that under plausible conditions of pocket gopher demography and plant community dynamics the community can become structured across the landscape at several scales with pocket gophers entrenched in some parts of the landscape and absent from others.

20.7 Complex Biotic Interactions

Pocket gopher activity can have indirect effects on other animals. For example, grasshoppers use pocket gopher mounds as egg-laying sites (Huntly and Inouye 1988) and voles tend to avoid mounds, foraging elsewhere and generating patterns of herbivory related to mound distributions

(Klaas et al. 1998). Ostrow et al. (2002) review a wide range of consequences for plants that have lost root tissue, including death, changes in constituents such as carbohydrates, amino acids, and secondary compounds, and processes such as water stress. Sedentary, herbivorous, sucking insects were more abundant on plants protected from root grazing, suggesting the insects were responding to a decrease in the quality of plant phloem. Vagile, leaf-chewing insects preferred plants exposed to pocket gopher grazing, perhaps because damaged plants produced lower amounts of defensive chemicals.

Experiments by Eviner and Chapin (2003) revealed a complicated relationship between pocket gophers, plants, and fungi. The study, in coastal California grasslands, established that a fungus (*Ulocladium atrum*) enhances the invasive capabilities of an aggressive invader (barbed goatgrass, *Aegilops triuncialis*) by weakening the plants hard seed head and thereby promoting germination. Pocket gophers were most active in concentrations of goatgrass, differentially burying seedlings of this species. Goatgrass seedlings that survived burial were not infected by the fungus, with the net result that pocket gophers reduce the invasive abilities of goatgrass.

20.8 Soil Dynamics

Pocket gophers affect soil dynamics on many scales, from particles and their movement to the formation of gullies and washes. The influences begin with the excavation of burrows and the placement of the tailings on the soil surface or in abandoned burrows. As an example of this process, Thorne and Andersen (1990) followed the development of a single burrow system for 158 days after its initiation. During that time the pocket gopher excavated at least 112 m of burrow (at rates from 2.1 m/week to none per week during the summer). A total of 134 mounds (537 l of soil) were deposited on the surface; there were 68 plugged openings and 77 m of backfilled burrows. The system was 32 m long when it was abandoned. This detailed description, coupled with the summary statistics in Smallwood and Morrison (1999) illustrate both the extent and dynamic nature of the impact pocket gopher foraging has on the soil.

Soil from excavations is placed on the soil surface or into abandoned tunnels. The soil in both is less dense than the adjacent consolidated matrix. Mounds have a bulk density up to 40% lower than undisturbed soil (Sparks and Andersen 1988) and refilled burrows exhibited a soil density 15% lower than adjacent soils (Reichman 1988). The lower density of mounds means that that they both absorb and lose moisture through evaporation more rapidly than nearby soil (Sherrod and Seastedt 2001).

The biogeochemistry (C, N, K, Na, Mg, Zn, Mn, B, Fe, Al as well as pH and cation exchange capacity) of mounds is quite complex and differs between locations based on moisture regimes, underlying soil characteristics, and the embedded plant and microbe community (Reichman and Seabloom 2002; Simkin et al. 2004). For example, depending on the soil nutrient gradient from the surface down, mounds may have more or less particulate carbon than nearby soil (Koide et al. 1987). Mounds also cover, and often kill, vegetation, producing organic hotspots of decomposition (Cortinas and Seastedt 1996).

Pocket gophers strongly affect the soil surface topography where they are abundant (Fig. 20.2). On a small scale, the number of mounds and surface heterogeneity increased over 11 years in relation to pocket gopher abundance (Inouye et al. 1997). Thorne (1978) describes the formation of "terracettes" on slopes where a combination of mounds, collapsed burrows, and snowmelt scours yield a complex, heterogeneous surface. At larger scales, pocket gophers are thought to create mima mounds, large (20 m diameter, 2 m high) mounds of consolidated soil that tend to form in shallow soils (Cox and Hunt 1990a). Cox and Hunt (1990b) report that stone stripes, piles of stones 60–120 m long and 0.5–3.7 m wide found in the Colombia Plateau in the Northwest U.S. are also the result of interactions between pocket gopher excavations, weathering and colluvial dynamics.

The net effect of the soil movement processes generated by pocket gopher excavation and physical processes is vertical mixing (mounds formed from bringing material up from below and burrows collapsing) and horizontal patchiness (spatially explicit patterns of burrows and mounds Reichman et al. 1982). One study that was able to quantify this pattern of movement was undertaken as part of an archeological excavation (Bocek 1992). In 1981 a 1 m by 2 m pit was excavated in 10 cm deep intervals down to 1 m.

Fig. 20.2. Aerial photograph of surface mounds on Konza Prairie, Kansas (USA). The *white corner lines* are 5 m on each side

All of the archeologically pertinent material was sifted from the samples and the soil was placed back in the pit. Seven years later the pit was re-excavated and the soil sifted, providing a measure of the movement into the items into the pit as a result of pocket gophers excavations. Across all depths the total number of new items found was 8% of what was extracted seven years earlier. In the upper three 10-cm levels (the "rodent zone"), starting at the surface, the new material was 41%, 16% and 12% of what had been extracted initially. Bocek (1992) indicates that the entire pit could be replenished in 88 years, a testament to the effects of pocket gophers on soil turnover rates.

Bioturbation by pocket gophers significantly affects geomorphic and hydrologic processes (Gabet et al. 2003). The downslope movement of soil (rate or amount) has generally been thought to be a linear function of slope. Gabet (2000) showed that downslope movement of soil increases significantly at lower gradients in the presence of pocket gophers. At mid-gradients downslope movement declines as mounds begin to function as small terraces. Because particulate carbon is part of the soil matrix, pocket gopher activity increases the movement of carbon downslope and into drainages (Gabet et al. 2005).

Mounds and burrows accelerate other physical processes, including erosion (Sherrod and Seastedt 2001). In particular, pocket gopher disturbances can concentrate water flows (Thorne 1978) in some cases to the extent that gullies form as an extension of water channeling through burrows (Swanson et al. 1989).

20.9 Economic Impacts

When pocket gopher impacts extend into human-dominated systems the economic effects can be extensive. Scores of studies have been conducted on the role pocket gophers play in reducing the survival and yield of plant products. Many of these have focused on reforestation efforts, particularly in the western U.S. where pocket gophers are abundant. Engeman and Witmer (2000) suggest that pocket gophers probably account for more damage to regenerating conifers (natural or planted) than all other animals combined. The authors summarize risk factors from pocket gophers in relation to site characteristics and summarize abatement approaches. Smallwood (1999) effectively outlines traditional abatement methods (primarily poisoning) and the problems associated with them. For example, populations reduced by poisoning may reach higher densities when they return than before they were reduced, suggesting a net detriment to reforestation from poisoning efforts. The author notes that pocket gophers usually reinvade from adjacent areas such that small scale eradication is ineffective (as well as expensive), a pattern found in intensively trapped orchards as well (Sullivan et al. 2001). This points to an obvious but underappreciated circumstance that as long as the burrows remain, even if the current resident is exterminated, the burrows are open for rapid recolonization (Engeman and Campbell 1999). As Smallwood and Geng (1997) note, the relationship between pocket gophers and economics impacts is complicated by the observation that soil tilling by the rodents may actually improve productivity in the long run.

Other, less obvious, forms of pocket gopher damage occur. Connolly and Landstrom (1969) document attempts to armor buried cable and wire to prevent pocket gopher damage. The authors note that damage can occur many years after installation and describe what type of physical and chemical protection is effective. Bioturbation into hazardous waste sites can be an especially dangerous phenomenon. While the net environmental effects of bioturbation into waste sites remains unknown (Smallwood et al. 1998), Bowerman and Redente (1998) indicate that a wide range of animals (and plant roots), including pocket gophers are capable of burrowing through all commonly used barriers, including gravel, cobble, tuff, crushed rock and asphalt.

A less dangerous, but significant, influence of pocket gopher burrowing is its effect on the distribution of archeological material. Pocket gophers redistribute items in disproportionate ways, increasing archeological items near the surface, and concurrently causing larger items to settle into deeper strata (Bocek 1986).

20.10 Conclusion

If there are taxa whose members can be considered keystone consumers or ecosystem engineers, it is pocket gophers (Geomyidae) and their ecological cognates around the world. Subterranean mammalian herbivores consume large quantities of vegetation, often damaging roots, tissue crucial to the immediate survival of plants. While foraging they excavate and relocate large quantities of soil, altering its structure and composition. Furthermore, these impacts occur in spatially explicit patterns that reverberate throughout the community. While much remains to be understood about the animals, it is clear that they are of preeminent importance in the environments where they occur.

References

Andersen DC (1987) Below-ground herbivory in natural communities: a review emphasizing fossorial animals. Quart Rev Biol 62:261–286

Andersen DC, MacMahon JA (1981) Population dynamics and bioenergetics of a fossorial herbivore, *Thomomys talpoides* (Rodentia: Geomyidae), in a Spruce-Fir Sere. Ecol Monogr 51:179–202

Bandoli JH (1981) Factors influencing seasonal burrowing activity in the pocket gopher, *Thomomys bottae*. J Mammal 62:293–303

Bandoli JH (1987) Activity and plural occupancy of burrows in Botta's pocket gopher *Thomomys bottae*. Am Mid Nat 118:10–14

Benedix JH Jr (1993) Area-restricted search by the plains pocket gopher (*Geomys bursarius*) in tallgrass prairie habitat. Behav Ecol 4:318–324

Benedix JH Jr (1994) A predictable pattern of daily activity by the pocket gopher *Geomys bursarius*. Anim Behav 48:501–509

Bocek B (1986) Rodent ecology and burrowing behavior: predicted effects on archaeological site formation. Am Antiq 51:589–603

Bocek B (1992) The Jasper Ridge reexcavation experiment: rates of artifact mixing by rodents. Am Antiq 57:261–269

Borchert MI, Davis FW, Michaelsen J, Oyler LD (1989) Interactions of factors affecting seedling recruitment of blue oak (*Quercus douglasii*) in California. Ecology 70:389–404

Bowerman AG, Redente EF (1998) Biointrusion of protective barriers at hazardous waste sites. J Environ Qual 27:625–632

Cameron GN, Spencer SR, Eshelman BD, Williams LR, Gregory MJ (1988) Activity and burrow structure of Attwater's pocket gopher (*Geomys attwateri*). J Mammal 69: 667–677

Cantor LF, Whitham TG (1989) Importance of belowground herbivory: pocket gophers may limit aspen to rock outcrop refugia. Ecology 70:962–970

Coggins ST, Conover MR (2005) Effect of pocket gophers on aspen regeneration. J Wildl Manag 69:752–759

Connolly RA, Landstrom RE (1969) Gopher damage to buried cable materials. Mater Res Stand 9:13–16

Cortinas MR, Seastedt TR (1996) Short- and long-term effects of gophers (*Thomomys talpoides*) on soil organic matter dynamics in alpine tundra. Pedobiologia 40: 162–170

Cox GW (1989) Early summer diet and food preferences of northern pocket gophers in north central Oregon. Northwest Sci 63:77–82

Cox GW, Hunt J (1990a) Form of mima mounds in relation to occupancy by pocket gophers. J Mammal 71:90–94

Cox GW, Hunt J (1990b) Nature and origin of stone stripes on the Columbia Plateau. Landsc Ecol 5:53–64

Cox GW, Hunt J (1992) Relation of seasonal activity patterns of valley pocket gophers to temperature, rainfall, and food availability. J Mammal 73:123–134

Cox GW, Hunt J (1994) Pocket gopher herbivory and mortality of ocotillo on stream terrace, bajada, and hillside sites in the Colorado desert, southern California. Southwest Nat 39:364–370

Davis MA, Ritchie B, Graf N, Gregg K (1995) An experimental study of the effects of shade, conspecific crowding, pocket gophers and surrounding vegetation on survivorship, growth and reproduction in *Penstemon grandiflorus*. Am Mid Nat 134:237–243

Engeman RM, Campbell DL (1999) Pocket gopher reoccupation of burrow systems following population reduction. Crop Prot 18:523–525

Engeman RM, Witmer GW (2000) Integrated management tactics for predicting and alleviating pocket gopher (*Thomomys* spp.) damage to conifer reforestation plantings. Integr Pest Manag Rev 5:41–55

Eviner VT, Chapin FS III (2003) Gopher-plant-fungal interactions affect establishment of an invasive grass. Ecology 84:120–128

Forbis TA, Larmore J, Addis E (2004) Temporal patterns in seedling establishment on pocket gopher disturbances. Oecologia 138:112–121

Foster MA, Stubbendieck J (1980) Effects of the Plains pocket gopher (*Geomys bursarius*) on rangeland. J Range Manag 33:74–78

Gabet EL (2000) Gopher bioturbation: field evidence for non-linear hillslope diffusion. Earth Surf Processes Landforms 25:1419–1428

Gabet EJ, Reichman OJ, Seabloom EW (2003) The effects of bioturbation on soil processes and sediment transport. Ann Rev Earth Planet Sci 31:249–273

Gabet EJ, Fierer N, Chadwick OA (2005) Prediction of sediment-bound nutrient delivery from semi-arid California watersheds. J Geophys Res 110:G02001

Gettinger RD (1984) A field study of activity patterns of *Thomomys bottae*. J Mammal 65:76–84

Grant WE, French NR, Folse LJ (1980) Effects of pocket gopher mounds on plant production in shortgrass prairie ecosystems. Southwest Nat 25:215–224

Hobbs RJ, Mooney HA (1985) Community and population dynamics of serpentine grassland annuals in relation to gopher disturbance. Oecologia 67:342–351

Hobbs RJ, Mooney HA (1991) Effects of rainfall variability and gopher disturbance on serpentine annual grassland dynamics. Ecology 72:59–68

Hobbs RJ, Mooney HA (1995) Spatial and temporal variability in California annual grassland: results from a long-term study. J Veg Sci 6:43–56

Hunt J (1992) Feeding ecology of valley pocket gophers (*Thomomys bottae sanctidiegi*) on a California coastal grassland. Am Mid Nat 127:41–51

Huntly N, Inouye R (1988) Pocket gophers in ecosystems: patterns and mechanisms. BioScience 38:786–793

Inouye RS, Huntly N, Wasley GA (1997) Effects of pocket gophers (*Geomys bursarius*) on microtopographic variation. J Mammal 78:1144–1148

Klaas BA, Danielson BJ, Moloney KA (1998) Influence of pocket gophers on meadow voles in a tallgrass prairie. J Mammal 79:942–952

Klaas BA, Moloney KA, Danielson BJ (2000) The tempo and mode of gopher mound production in a tallgrass prairie. Ecography 23:246–256

Koide RT, Huenneke LF, Mooney HA (1987) Gopher mound soil reduces growth and affects ion uptake of two annual grassland species. Oecologia 72:284–290

Martinsen GD, Cushman JH, Whitham TG (1990) Impact of pocket gopher disturbance on plant species diversity in a shortgrass prairie community. Oecologia 8:132–138

Miller MA (1946) Reproductive rates and cycles in the pocket gopher. J Mammal 27:335–338

Ostrow DG, Huntly N, Inouye RS (2002) Plant-mediated interactions between the northern pocket gopher, *Thomomys talpoides*, and aboveground herbivorous insects. J Mammal 83:991–998

Reichman OJ (1988) A comparison of the effects of crowding and pocket gopher disturbance on mortality, growth, and seed production of *Berteroa incana*. Am Mid Nat 120: 58–69

Reichman OJ (1996) The influence of crowding and pocket gopher disturbance on growth and reproduction of a biennial, *Tragopogon dubius*. Proc 14th North Amer Prairie Conf, pp 123–127

Reichman OJ, Seabloom EW (2002) The role of pocket gophers as subterranean ecosystem engineers. Trends Ecol Evol 17:44–49

Reichman OJ, Smith S (1985) Impact of pocket gopher burrows on overlying vegetation. J Mammal 66:720–725

Reichman OJ, Smith S (1990) Burrows and burrowing behavior by mammals. Curr Mammal 2:197–244

Reichman OJ, Smith S (1991) Effects of simulated leaf and root herbivory on mortality, biomass, reproduction, and nutrient allocation in a biennial, *Tragopogon dubius*. Ecology 72:116–124

Reichman OJ, Whitham TG, Ruffner G (1982) Adaptive geometry of burrow spacing in two pocket gopher populations. Ecology 63:687–695

Reichman OJ, Benedix JH Jr, Seastedt TR (1993) Animal generated disturbances and size hierarchies in a tallgrass plant community. Ecology 74:1281–1285

Rogers WE, Hartnett DC (2001a) Temporal vegetation dynamics and recolonization mechanisms on different-sized soil disturbances in tallgrass prairie. Am J Bot 88:1634–1642

Rogers WE, Hartnett DC (2001b) Vegetation responses to different spatial patterns of soil disturbance in burned and unburned tallgrass prairie. Plant Ecol 155:99–109

Rogers WE, Hartnett DC, Elder B (2001) Effects of plains pocket gopher (*Geomys bursarius*) disturbances in tallgrass-prairie plant community structure. Am Mid Nat 145: 344–357

Romañach SS, Le Comber SC (2004) Measures of pocket gopher (*Thomomys bottae*) burrow geometry: correlates of fractal dimension. J Zool Lond 262:399–403

Romañach S, Reichman OJ, Seabloom EW (2005a) Seasonal influences on burrowing activity of a subterranean rodent, *Thomomys bottae*. J Zool Lond 266:319–325

Romañach S, Seabloom EW, Reichman OJ, Rogers W, Cameron G (2005b) Effects of species, gender, age, and habitat on pocket gopher foraging tunnel geometry. J Mammal 86:750–756

Seabloom EW, Reichman OJ (2001) Simulation models of the interactions between herbivore foraging strategies, social behavior, and plant community dynamics. Am Nat 157: 76–96

Seabloom EW, Richards SA (2003) Multiple stable equilibria within plant communities mediated by herbivore population dynamics and foraging behavior. Ecology 84: 2891–2904

Sherrod SK, Seastedt TR (2001) Effects of the northern pocket gopher (*Thomomys talpoides*) on alpine soil characteristics, Niwot Ridge, CO. Biogeochem55:195–218

Sherrod SK, Seastedt TR, Walker MD (2005) Northern pocket gopher (*Thomomys talpoides*) control of alpine plant community structure. Arct Antarct Alpine Res 37:585–590

Simkin SM, Michener WK, Wyatt R (2004) Mound microclimate, nutrients and seedling survival. Am Mid Nat 52:12–24

Smallwood KS (1999) Abating pocket gophers (*Thomomys* spp.) in regenerate forest clearcuts. Environ Conserv 26:59–65

Smallwood KS, Geng S (1997) Multiscale influences of gophers on alfalfa yield and quality. Field Crops Res 49:159–168

Smallwood KS, Morrison ML (1999) Estimating burrow volume and excavation rate of pocket gophers (*Geomyidae*). Southwest Nat 44:173–183

Smallwood KS, Morrison ML, Beyea J (1998) Animal burrowing attributes affecting hazardous waste management. Environ Manag 22:831–847

Sparks DW, Andersen DC (1988) The relationship between habitat quality and mound building by a fossorial rodent, *Geomys bursarius*. J Mammal 69:583–587

Stromberg MR, Griffin JR (1996) Long-term patterns in coastal California grasslands in relation to cultivation, gophers, and grazing. Ecol Appl 6:1189–1211

Sullivan TP, Sullivan DS, Hogue EJ (2001) Reinvasion dynamics of northern pocket gopher (*Thomomys talpoides*) populations in removal areas. Crop Prot 20:189–198

Swanson ML, Kondolf GM, Boison PJ (1989) An example of rapid gully initiation and extension by subsurface erosion: coastal San Mateo County, California. Geomorphol 2:393–403

Thorn CE (1978) A preliminary assessment of the geomorphic role of pocket gophers in the alpine zone of the Colorado front range. Geogr Ann 60:181–187

Thorne DH, Andersen DC (1990) Long-term soil-disturbance pattern by a pocket gopher, *Geomys bursarius*. J Mammal 71:84–89

Tilman D (1983) Plant succession and gopher disturbance along an experimental gradient. Oecologia 60:285–292

Vleck D (1979) The energy cost of burrowing by the pocket gopher *Thomomys bottae*. Physiol Zool 52:122–136

Watts SM (2005) The ecological and evolutionary influences of belowground herbivory on plant defense and community structure. PhD. Dissertation, Univ California, Santa Barbara, USA

Williams LR, Cameron GN (1986a) Food habits and dietary preferences of Attwater's pocket gopher, *Geomys attwateri*. J Mammal 67:489–496

Williams LR, Cameron GN (1986b) Effects of removal of pocket gophers on a Texas coastal prairie. Am Mid Nat 115:216–224

Williams LR, Cameron GN, Spencer SR, Eshelman BD, Gregory MJ (1986) Experimental analysis of the effects of pocket gopher mounds on Texas coastal prairie. J Mammal 67:672–679

Wolfe-Bellin KS, Moloney KA (2000) The effect of gopher mounds and fire on the spatial distribution and demography of a short-lived legume in tallgrass prairie. Can J Bot 78:1299–1308

21 Subterranean Rodents as Pests: The Case of the Pocket Gopher

Gary W. Witmer, Richard M. Engeman

21.1 Introduction

With over 2000 species, the order Rodentia has more members than any other order of mammals (Nowak 1999). The distribution of rodents is nearly worldwide; their use of habitats is extensive and varied. Most rodent species are relatively small, secretive, prolific, and all have continuously growing incisors. Many rodent species have ecological, scientific, cultural, and/or economic importance.

A variety of economic and health problems result from rodent interactions with humans. These include damage to growing crops, trees, seeds, pastures; damage and contamination of stored foods; damage to structures and property; and disease transmission (Witmer et al. 1995a). Singleton et al. (2003) estimated that in Asia alone, the amount of grain eaten by rodents would provide enough food to feed 200 million Asians for a year.

Notably, few (perhaps 5%) rodent species around the world are serious pests. Examples of genera and species of rodents considered to be serious pests around the world were provided by Prakash (1988) and Witmer et al. (1995a). Hence, when a damage situation occurs, it is very important to determine the species causing the damage, the extent of the damage, and the abiotic-biotic-cultural factors involved before rodent population and damage management strategies are implemented (Singleton et al. 1999).

We use the North American subterranean rodents, pocket gophers, to illustrate the nature of rodent pests, the types and extent of damages, and the management measures used to reduce populations and damage.

Gary W. Witmer: USDA/APHIS/WS National Wildlife Research Center, 4101 Laporte Avenue, Fort Collins, Colorado 80521-2154 USA, E-mail: Gary.W.Witmer@aphis.usda.gov

Richard M. Engeman: USDA/APHIS/WS National Wildlife Research Center, 4101 Laporte Avenue, Fort Collins, Colorado 80521-2154 USA

Subterranean Rodents: News from Underground
S. Begall, H. Burda, C.E. Schleich (Eds.)
© Springer-Verlag Berlin Heidelberg 2007

21.2 Pocket Gophers (Family Geomyidae)

Pocket gophers are endemic to North America from central Canada to Panama. There are seven genera: *Cratogeomys* (sometimes considered a subgenus of *Pappogeomys*), *Geomys*, *Heterogeomys*, *Orthogeomys*, *Pappogeomys*, *Thomomys*, and *Zygogeomys* (Nowak 1999; Baker RJ et al. 2003). There are at least 35 species and about 300 subspecies within this family of rodents.

In general, pocket gophers have thick-set bodies with short necks and a head-body-length of 15–30 cm. They are subterranean to a high degree, living in closed burrow systems. Pocket gophers have small eyes and pinnae, a good sense of touch, and stout forelegs with strong claws. They rarely venture to the surface except to push soil out, to occasionally clip and gather above ground vegetation, and for dispersal purposes. Except for mating and rearing young, most species live solitary lives within their burrow system.

Pocket gophers possess amazing digging and gnawing abilities, and have been extensively studied. Details of their biology, ecology, and environmental effects have been reviewed by Nevo (1999), Nowak (1999), Baker RJ et al. (2003), and Reichman (this volume).

21.3 Damage by Pocket Gophers

Pocket gophers commonly come into conflict with humans. This occurs through crop damage, rangeland damage, forest damage, and physical damage to structures and property (Marsh 1988). It may be a tribute to their ability to cause damage that Marsh (1988), in his chapter on rodent problems on the North American continent, discussed pocket gophers first amongst 21 rodent groups! He lists the damage by pocket gophers in North America as widespread and, on a site specific basis, as generally in the light-to-heavy damage categories. We review the types and extents of damage caused by pocket gophers. Unfortunately, much of the literature is dated and there is insufficient quantification of the damage.

21.3.1 Forest Damage

Pocket gophers are one of the most serious threats to reforestation in North America (Engeman and Witmer 2000). They kill young trees by debarking stems (sometimes called girdling) at the ground surface, pruning roots below ground, clipping seedlings aboveground, and pulling entire seedlings below ground. Damage higher up the boles of trees can occur with snow cover (pocket gophers are active year round). Damage to roots occurs mostly during the winter and early spring when herbaceous plants are less available. A wide array of tree species can be affected. Graham and Kingery (1990) reported that *Thomomys* gophers killed 71% of the pine trees on plantations by year 6 after planting. Pocket gophers readily invade clearcut units, especially once herbaceous ground cover becomes abundant. When foresters replant the unit with seedlings, pocket gopher foraging can remove all or large portions of the seedlings. This requires replanting, often preceded by pocket gopher population control. This process greatly escalates total costs, often prompting foresters to routinely practice pocket gopher control prior to reforestation. Engeman and Witmer (2000) assessed the risk posed by *Thomomys* gophers and provided guidelines for how to avoid serious damage and when to apply pro-active measures. However, pocket gopher populations recover very quickly after control, by reproduction of survivors and by re-invasion from surrounding areas (Engeman and Campbell 1999). Hence, some researchers have recommended that forest management practices be modified to provide less suitable habitat for pocket gophers (Smallwood 1999; Engeman and Witmer 2000).

21.3.2 Rangeland Damage

The substantial impacts that *Geomys* gophers can have to rangelands was reviewed by Foster and Stubbendieck (1980). They documented 18–49% reduction in range plant production on sites occupied by pocket gophers in Nebraska. This included an increase in some species of perennial grasses, but a decrease in coverage of annual forbs and annual grasses. Because of the soil mounding activities of pocket gophers, there was also an increase of 5–25% in bare soil coverage. They also cited studies reporting on range production changes: one reported an herbage reduction of 284 kg/ha in California where *Thomomys* gophers occurred and another reported an herbage increase of 218 kg/ha after *Geomys* gopher control in Colorado. Fitch and Bentley (1949) reported a 25% forage reduction by *Thomomys* gophers in California. In Alberta, Canada, Alsager (1977) reported an 18% decrease in forage production where *Thomomys* gophers were not controlled. He also noted a 16% increase in forage production within 60 days of pocket gopher control.

21.3.3 Fruit Tree Damage

Documentation of *Thomomys* gopher damage to fruit trees has included apple, cherry, and pear trees in the Pacific Northwest (Sullivan et al. 1987; Sullivan and Hogue 1987) and citrus trees in California (Cummings and Marsh 1978). In Guatemala, *Orthogeomys* gophers damage banana trees (Caid 1959) as do *Orthogeomys* and *Pappogeomys* gophers in Mexico (Whisson and Villa Cornejo 1996). Similar to forestry damage, this type of damage mostly involves root gnawing and basal girdling, with young trees ($\leq$ 10 years old) most susceptible. Sullivan et al. (1987) noted that the damage in apple orchards did not seem related to soil type, perhaps because the orchards were irrigated and fertilized, and about 30–40% of surveyed orchards had *Thomomys* gopher damage. Sullivan and Hogue (1987) reported that the incidence of rodent damage (voles and *Thomomys* gophers) dropped from 40.6% to 9.6% once the rodents were controlled. They also achieved better rodent control by greatly reducing low ground vegetation with herbicides rather than using rodenticides.

21.3.4 Alfalfa and Field Crop Damage

Pocket gophers cause substantial damage to alfalfa crops in southern Canada and throughout the midwestern plains and western states of the USA. Case (1989) reported reductions in yield of 17–49% from *Geomys* gophers. They caused declines in yield in Nebraska of 30.2% in hay meadows and 16.7% in alfalfa fields (Hegarty 1984). The main declines were in some late successional perennial grasses and clovers. Losses were directly correlated with pocket gopher density and the percent of their mound cover. He also documented an increase in plant diversity because many annuals (including invasive and "weedy" plant species) germinated on the bare soil of the mounds. Similarly, Luce et al. (1981) documented alfalfa yield reductions of 40% in Nebraska. Jasch et al. (1992) estimated the economic losses in alfalfa production in Nebraska to be about $10 million per year. They noted that fibrous-rooted varieties of alfalfa seem to better withstand pocket gopher damage. Proulx (2002) estimated losses to alfalfa fields in Alberta and Manitoba, Canada, by *Thomomys* gophers at $15–30 million per year. He also compared the costs of various methods of pocket gopher control and noted an 18–28% increase in yield after control. Case and Timm (1984) created a computer model to calculate the dollar loss due to *Geomys* gopher activity so returns on control costs could be weighed against yield increases. In a survey of alfalfa producers in Nevada, Lewis and O'Brien (1990) identified *Thomomys* gophers as the main pest rodent, present on 87% of the farms. The main problems associated with the pocket gophers were, in declining order, 1) loss of yield, 2) equipment damage, 3) an increase in labor and fuel costs, and 4) a decrease in hay quality. Pocket gopher control was practiced by 80% of the farmers, but 53% (the largest proportion of farmers) said it was only "somewhat effective". Next to irrigation costs, they reported vertebrate pest management costs as their greatest expense; even higher than weed and insect management costs and fertilizer costs. Smallwood and Geng (1997) noted the relationship between damage level and *Thomomys* gopher density, but also explained that complex relationships exist: for example, the field may produce a higher yield a year after some gopher damage than it had averaged before gopher infestation, and more productive lands can sustain more damage.

Pocket gophers also damage other field crops. Villa Cornejo (2000) reported over 200,000 ha of sugarcane fields in Mexico had evidence of *Orthogeomys* gopher damage. She examined over 66,000 individual stalks and found about 21% had been damaged by pocket gophers. Whisson and Villa Cornejo (1996) examined over 1400 corn stalks in Mexico and found about 4% had *Orthogeomys* and/or *Pappogeomys* gopher damage. They mentioned other crops damaged, including wheat, potatoes, and cocoa, but gave no specifics.

21.3.5 Wire and Cable Damage

Connolly and Landstrom (1969) provided a good review of pocket gopher damage to buried cables. They considered the various species of *Geomyidae* to be the most significant source of animal damage to buried electrical and communication cables, noting smaller cable diameters sustained greater damage. Gophers were surmised to encounter and damage cables while seeking their main food source, roots. Most damage seems to occur in the rooting zone, 10–30 cm deep. They and other researchers (e.g. Hegdal and Harbour 1991; Ramey and McCann 1997) tested many types of cable coating, but in general, only a hard metal sheath or placement in a large-diameter ($\geq$ 5.5 cm) conduit provided adequate protection from *Geomys* gophers. Isaac (1959) suggested that compacting soils in cable trenches would reduce *Geomys* gopher damage. He also recommended avoiding cable placement in areas of substantial pocket gopher activity. Cables are less likely to be damaged by gophers if they are placed more deeply underground because gopher burrows rarely exceed

2 m in depth (Case and Jasch 1994). Small diameter underground irrigation pipes also sustain pocket gopher damage (Hegdal and Harbour 1991).

21.3.6
Hydraulic Structure Damage, Disturbance of Hazardous Waste and Archaeological Sites

Pocket gophers are truly efficient "digging machines". Smallwood and Morrison (1999) reviewed studies reporting the excavation rate and burrow volumes of various species of pocket gophers. They found much variation which they attributed to the varying species, study locations, and methodologies used. In general, however, a single *Geomys* or *Thomomys* gopher can excavate 18 m^3 per ha per year and its burrow has a volume of at least 0.68 m^3. Hence, it is not surprising that earthen structures (dikes, canals, and levees) can sustain substantial damage from pocket gophers (Hegdal and Harbour 1991). This results in seepage, piping, and eventually, washouts (especially with water surges).

Several researchers have noted that *Thomomys* gophers can disrupt the integrity of waste burial sites and this is especially of concern when hazardous wastes are involved (Winsor and Whicker 1980; Hakonson et al. 1982; Sejkora 1989; Bowerman and Redente 1998; Smallwood et al. 1998). Hakonson et al. (1982) did not believe that much material was being brought to the surface, but a considerable "void space" was being created in the soil cover profile. Winsor and Whicker (1980) found that pocket gophers were bringing buried plutonium to the surface at a waste site in Colorado. The highest radioactivity counts were in the 0–10 cm soil layer and in the pocket gopher mounds. They noted that most burrowing activity was in the top 30 cm of the soil. On the other hand, Sejkora (1989) noted that pocket gopher burrowing activities reduced runoff, soil erosion, and chemical transport of surface materials.

Pocket gopher burrowing can disturb archaeological materials and burial sites. Bocek (1992) reported that *Thomomys* gophers often redistributed items in disproportionate ways; smaller items were moved nearer to the surface while larger items settled to greater depths.

21.3.7
Invasive Plant Establishment and Dispersal

The disturbance of soil and seed dispersal by rodents has been suggested to enhance the establishment of invasive plant species (e.g., Case and Jasch 1994; Hobbs and Mooney 1991). The large increase in bare soil on the surface resulting from pocket gopher burrowing provides abundant germination sites for annual plants. The magnitude of this effect on the establishment and spread of invasive plants has not been well documented and the relationships between abiotic and biotic factors can be quite complex (see Reichman, this volume).

21.4
Pocket Gopher Damage Management

Because of the extensive nature of pocket gopher damage, many tools, methods, and strategies have been developed to reduce populations and damage (Case and Jasch 1994; Engeman and Witmer 2000; Marsh 1992). With rodents, the amount of damage is often related to the population density; hence, one approach to damage reduction is to reduce the pocket gopher population in the area of the damage. A second approach is to reduce the area's carrying capacity for pocket gophers, or to provide protection to specific resources so that they are less likely to be damaged by pocket gophers living in the area. Often a combination of methods is the best way to achieve significant rodent damage reduction. This concept of Integrated Pest Management (IPM) has been elaborately developed for many pest groups (insects, weeds, disease organisms and their vectors), but less so for vertebrate pests. In dealing with rodent pests, Singleton et al. (1999) took this concept a step further and proposed an "ecologically-based rodent pest management" system that goes well beyond the traditional, heavy reliance on toxic rodenticides. We briefly review the various methods employed to reduce damage by pocket gophers.

21.4.1
Pocket Gopher Population Control

Humans around the world have relied on kill traps and rodenticide baits as the main tools to reduce damage by rodents, although many other methods have been used with varying levels of success (Witmer et al. 1995a). The use and effectiveness of traps for pocket gopher control has been reviewed by Marsh (1998), Pipas et al. (2000), Proulx (1997), and Witmer et al. (1999).

One of the main problems with traps is they are very labor intensive to use. Because pocket gophers traps are placed down in the burrow, non-target animal losses are generally relatively low.

Rodenticides currently registered in the USA for use in pocket gopher control include chlorophacinone, diphacinone, strychnine, and zinc phosphide (Case and Jasch 1994). Burrow fumigants (gas cartridges, aluminum phosphide pellets) are also registered for gopher control (Baker RO 2004; Case and Jasch 1994). Marsh (1992) discussed the history of rodenticide use for gopher control in California, including several rodenticides no longer registered for that purpose. Other rodenticides, such as cholecalciferol, have proven effective in gopher control and may be registered for that use in the future (Witmer et al. 1995b). Most rodenticides can be classified as anticoagulants, which are further broken down into first and second generation materials, and the acute toxicants (see Timm 1994). The first generation anticoagulants (e.g., warfarin, chlorophacinone, diphacinone) are relatively lower in toxicity and require multiple feedings over several days before they are lethal to the rodent. The second generation anticoagulants (e.g., bromaliolone, brodifacoum) are more potent and were developed starting in the early 1970s when genetic resistance to first generation anticoagulants began to occur. Anticoagulants are used in relatively low concentrations (0.0025–0.005%) and an antidote (vitamin K) exists in case of accidental intoxication of people, pets or livestock. The acute toxicants (e.g., zinc phosphide, strychnine, cholecalciferol, bromethalin) are toxic to most vertebrates and may kill rodents with a single feeding. Rodenticides are carefully regulated by federal, provincial, territorial, and state agencies to assure proper use and to reduce adverse effects. Concerns with rodenticide use revolve around primary and secondary poisoning hazards, residue bioaccumulation, and environmental persistence. Because pocket gopher rodenticide baits are placed within the burrow which is then sealed over, the hazards to non-target animals are relatively low. In general, the use of rodenticide baits ($30/ha) is more cost effective than the use of traps ($100/ha; Proulx 2002). Use of a tractor-drawn burrow builder that systematically dispenses bait into the burrow as it is created can keep costs even lower ($6/ha; Proulx 2002).

Pocket gopher populations can be quickly reduced with traps or rodenticides, but they typically recover within a year to pre-control levels (Engeman and Campbell 1999; Engeman and Witmer 2000). For this reason, Proulx (2002) recommended the implementation of a "border control" strategy. After an area is cleared of pocket gophers, some trappings (or bait applications) are continued around the perimeter to reduce the re-invasion rate.

21.4.2
Habitat Management and Resource Protection

Some land management activities can help reduce pocket gopher densities and damage. These include the use of less palatable species or varieties of plants, crop rotation, alteration of planting and harvesting dates, flood irrigation, less canopy removal with forest harvest, less ground surface disturbance (to discourage herbaceous plant invasion of the site), use of herbicides or livestock to reduce ground forage, and encouragement of natural predation (Case and Jasch 1994; Engeman and Witmer 2000). For some situations, increasing the seedling stocking rate may be an effective and less costly alternative to other more expensive or legally restricted damage control methods (Engeman et al. 1998). Pocket gophers generally prefer to feed on forbs (dicots) over grasses (monocots) (Keith et al. 1959). Consequently, the selective removal of forbs with the herbicide 2,4-D has reduced *Thomomys* gopher densities (Keith et al. 1959; Tietjen et al. 1967).

Individual trees can also be protected with barriers made of small mesh wire or plastic tubes (Engeman et al. 1999a; Engeman and Witmer 2000). Unfortunately, this is costly because of initial material cost, the cost of maintenance, and overall labor cost. Care must also be taken to assure that the barriers do not affect the growth form of the trees and do not create a lethal thermal microclimate around seedlings.

There has been considerable research on pocket gopher repellents to protect trees, but these have not proven effective (Engeman and Witmer 2000). Some researchers feel predator odors (urine or feces) help keep rodents out of areas, but this has not been well documented (Witmer et al. 1997). Sonic or vibrator devices have not proven effective at driving underground rodents from an area (Timm 2003).

21.4.3
Population Densities and Monitoring

Densities of pocket gophers are highly variable. For *Thomomys*, densities are commonly 40–50 per ha, but can be as high as 153 per ha (Case and Jasch 1994). For *Geomys*, densities are usually no greater than 20 per ha (Case and Jasch 1994). Densities are considered to be influenced by such factors as local climate, soil suitability, body mass, and vegetation types (Baker RJ et al. 2003). Sharp declines in gopher populations have occasionally been noted. These are usually related to a climatic factor such as a rapid snowmelt with a subsequent rise in the ground water level (Case and Jasch 1994). Pocket gopher populations, however, do not exhibit regular cycles in densities like those observed with some microtine species.

Monitoring the level and distribution of activity for pocket gopher populations is essential to the decision-making process on whether and where to apply control technologies. Subterranean animals must be observed indirectly, using signs to reflect abundance, distribution, and level of activity. Forest pocket gopher activity usually is assessed by examining sample plots for new mounds (Anthony and Barnes 1984) or monitoring opened burrows for closure (Richens 1967). Mound building activity fluctuates seasonally, whereas pocket gophers always try to maintain closed burrow systems. *Thomomys* gopher burrow systems have a single occupant during much of the year. Thus, open-hole assessments are most valuable when assessing the effectiveness of direct population control measures (Engeman et al. 1993). Generally, two or three holes are opened in each burrow system and rechecked for closure after 24–48 h. Hole closure indicates an occupied burrow system. Parameters for applying the open-hole method have been optimized to maximize sensitivity of the results relative to in-field labor (number of holes opened and size of activity plots; Engeman et al. 1999b). The open-hole method is more difficult to interpret for species having more than one burrow occupant (Matschke et al. 1994).

21.5 Conclusions and Future Needs

Pocket gophers are subterranean rodents widely distributed across Canada, the United States, and south to Panama. Pocket gophers cause substantial damage to a wide array of resources through their foraging, burrowing, and gnawing habits. Population control measures are frequently implemented to reduce their numbers and subsequent damage. Control requires a diligent, long-term commitment with substantial funds, because pocket gophers can achieve high densities and rapidly reinvade sites having good resources. Thus, control should not be undertaken unless significant damage levels are anticipated.

More research is needed to understand better the nature of pocket gopher damage to crops and other resources, and ways in which to reduce that damage. Evaluations of existing methods are still needed in many places. The ability to develop new, safe, and effective methods of rodent damage management will depend on many factors, besides laboratory, pen, and field trials: funding, actions by regulatory agencies, interest and involvement by commercial developers of methods and products, and pressures by agricultural producers, special interest groups, and the general public.

References

Alsager DE (1977) Impact of pocket gophers on the quantitative productivity of rangeland vegetation in southern Alberta: a damage assessment tool. In: Jackson W (ed) Test methods for vertebrate pest control and management materials, ASTM STP 625, Am Soc Test Mat, Philadelphia, pp 47–57

Anthony RM, Barnes VG Jr (1984) Plot occupancy for indicating pocket gopher abundance and conifer damage. In: Kaukeinen DE (ed) Vertebrate Pest Control and Management Materials: Fourth Symposium, ASTM STP 817, Am Soc Test Mat, Philadelphia, pp 247–255

Baker RJ, Bradley RD, McAliley LR (2003) Pocket gophers. In: Feldhamer GA, Thompson BC, Chapman JA (eds) Wild mammals of North America. Johns Hopkins Univ Press, Baltimore, pp 276–287

Baker RO (2004) Field efficacy of Fumitoxin (55% aluminum phosphide) tablets for controlling valley pocket gophers. Proc Vertebr Pest Conf 21:253–257

Bocek B (1992) The Jasper Ridge reexcavation experiment: rates of artifact mixing by rodents. Am Antiq 57:261–269

Bowerman AG, Redente EF (1998) Biointrusion of protective barriers at hazardous waste sites. J Environ Qual 27:625–632

Caid RD (1959) Taltuza investigations in Guatemala. Res Newsl (Department of Fruit, United Fruit Company, Boston) 6:15–20

Case RM (1989) Managing damage to alfalfa caused by plains pocket gophers. Proc Great Plains Wildlife Damage Control Workshop 9:160–161

Case RM, Jasch BA (1994) Pocket gophers. In: Hygnstrom SE, Timm RM, Larson GE (eds) Prevention and control of wildlife damage. Nebraska Cooperative Extension, Lincoln, pp B-17–B-29

Case RM, Timm RM (1984) Economic model of pocket gopher control. Proc Great Plains Wildlife Damage Control Workshop 6:53–56

Connolly RA, Landstrom RE (1969) Gopher damage to buried cable materials. Mater Res Stand 9:13–16

Cummings MW, Marsh RE (1978) Vertebrate pests of citrus. The citrus industry, vol IV, Publication 4088, Univ California, Div Agr Sci, Davis, pp 237–273

Engeman RM, Campbell DL (1999) Pocket gopher reoccupation of burrow systems following population reduction. Crop Prot 18:523–525

Engeman RM, Witmer GW (2000) Integrated management tactics for predicting and alleviating pocket gopher damage to conifer reforestation plantings. Integr Pest Manag Rev 5:41–55

Engeman RM, Campbell DL, Evans J (1993) A comparison of 2 activity measures for northern pocket gophers. Wildl Soc Bull 21:70–73

Engeman RM, Anthony RM, Barnes V, Krupa HW, Evans J (1998) Double-stocking for overcoming damage to conifer seedlings by pocket gophers. Crop Prot 17:687–689

Engeman RM, Anthony RM, Barnes V, Krupa HW, Evans J (1999a) Evaluations of plastic mesh tubes for protecting conifer seedlings from pocket gophers in three western states. Western J Appl For 14:86–90

Engeman RM, Nolte DL, Bulkin SP (1999b) Optimization of the open-hole method for assessing pocket gopher activity. Can Field Nat 113:241–244

Fitch HS, Bentley JR (1949) Use of California annual-plant forage by range rodents. Ecology 30:306–321

Foster MA, Stubbendieck J (1980) Effects of the plains pocket gopher on rangeland. J Range Manag 33:74–78

Graham RT, Kingery JL (1990) Seedling damage and mortality of conifer plantations on transitory ranges in northern and central Idaho. Proc Vertebr Pest Conf 14:209–213

Hakonson TE, Marinez JL, White GC (1982) Disturbance of low-level waste burial site cover by pocket gophers. Health Phys 42:868–871

Hegarty PJ (1984) Effects of the plains pocket gopher on hay meadows and irrigated alfalfa in the Nebraska sandhills. MSc Thesis, Univ Nebraska, Lincoln

Hegdal PL, Harbour AJ (1991) Prevention and control of animal damage to hydraulic structures. US Department of Interior, Bureau of Reclamation, Washington, D.C.

Hobbs RJ, Mooney HA (1991) Effects of rainfall variability and gopher disturbance of serpentine annual grassland dynamics. Ecology 72:59–68

Isaac RL (1959) Pocket gopher damage to underground telephone wire in Caddo County, Oklahoma. USDA Division of Wildlife Services, Oklahoma City

Jasch B, Case RM, Luchsinger JC (1992) Pocket gopher damage and alfalfa root structure. Proc Nebraska Acad Sci 102:28

Keith JO, Hansen RM, Ward AL (1959) Effect of 2,4-D on abundance and foods of pocket gohers. J Wildl Manag 23:137–145

Lewis SR, O'Brien JM (1990) Survey of rodent and rabbit damage to alfalfa hay in Nevada. Proc Vertebr Pest Conf 14:116–119

Luce DG, Case RM, Stubbendieck JL (1981) Damage to alfalfa fields by plains pocket gopher. J Wildl Manag 45:258–260

Marsh RE (1988) Rodent problems on the North American continent. In: Prakash I (ed) Rodent pest management. CRC Press, Boca Raton, pp 1–11

Marsh RE (1992) Reflections on current (1992) pocket gopher control in California. Proc Vertebr Pest Conf 15:289–292

Marsh RE (1998) One hundred years of pocket gopher traps and trapping. Proc Vertebr Pest Conf 18:221–226

Matschke GH, Sterner RT, Engeman RM, O'Brien JM (1994) Limitations of open-hole and plot occupancy indices in field efficacy studies with Townsend's pocket gophers. Proc 15th Annual SETAC meeting, Denver, 15:245

Nevo E (1999) Mosaic evolution of subterranean mammals: regression, progression and global convergence. Oxford Univ Press, Oxford

Nowak RM (1999) Mammals of the world. Johns Hopkins Univ Press, Baltimore

Pipas MJ, Matschke GH, McCann GR (2000) Evaluation of the efficacy of three types of traps for capturing pocket gophers. Proc Vertebr Pest Conf 19:385–388

Prakash I (1988) Rodent pest management. CRC Press, Inc., Boca Raton

Proulx G (1997) A preliminary evaluation of four types of traps to capture northern pocket gophers. Can Field Nat 111:640–643

Proulx G (2002) Effectiveness of trapping to control northern pocket gophers in agricultural lands in Canada. Proc Vertebr Pest Conf 20:26–31

Ramey A, McCann GR (1997) Evaluating cable resistance to pocket gopher damage – a review. Proc Great Plains Wildlife Damage Control Conf 13:107–113

Richens VB (1967) The status and use of gophacide. Proc Vertebr Pest Conf 3:118–125

Sejkora KJ (1989) Influence of pocket gophers on water erosion and surface hydrology. Dissertation. Colorado State Univ, Fort Collins

Singleton GR, Leirs H, Hinds LA, Zhang Z (1999) Ecologically-based management of rodent pests – re-evaluating our approach to an old problem. In: Singleton G (ed) Ecologically-based management of rodent pests. Austr Centre Intl Agricult Res, Canberra, pp 17–29

Singleton GR, Hind LA, Krebs CJ, Spratt DM (2003) Rats, mice, and people: rodent biology and management. Austr Centre Intl Agricult Res, Canberra

Smallwood KS (1999) Abating pocket gophers to regenerate forest clearcuts. Environ Conserv 26:59–65
Smallwood S, Geng S (1997) Multiscale influence of gophers on alfalfa yield and quality. Field Crops Res 49:159–168
Smallwood KS, Morrison ML (1999) Estimating burrow volume and excavation rate of pocket gophers. Southwest Nat 44:173–183
Smallwood KS, Morrison ML, Beyea J (1998) Animal burrowing attributes affecting hazardous waste management. Environ Manag 22:831–847
Sullivan TP, Hogue EJ (1987) Influence of orchard floor management on vole and pocket gopher populations and damage in apple orchards. J Am Soc Hort Sci 112:972–977
Sullivan TP, Krebs JA, Kluge HA (1987) Survey of mammal damage to tree fruit orchards in the Okanogan Valley of British Columbia. Northwest Sci 61:23–31
Tietjen HP, Halvorson CH, Hegdal PL, Johnson AM (1967) 2,4-D herbicide, vegetation, and pocket gopher relationships, Black Mesa, Colorado. Ecology 48:634–643
Timm RM (1994) Active ingredients. In: Hyngstrom S, Timm RM, Larson GE (eds) Prevention and control of wildlife damage. Nebraska Cooperative Extension Service, Lincoln, pp G-23–G-61
Timm RM (2003) Devices for vertebrate pest control: are they of value? Proc Wildl Damage Manage Conf 10:152–161
Villa Cornejo B (2000) Pocket gopher damage in sugarcane fields in the state of Veracruz, Mexico. Proc Vertebr Pest Conf 19:358–361
Whisson D, Villa Cornejo B (1996) The pocket gopher as a pest in Mexico. Proc Vertebr Pest Conf 17:151–153
Winsor TF, Whicker FW (1980) Pocket gophers and redistribution of plutonium in soil. Health Phys 39:257–262
Witmer GW, Fall MW, Fiedler LA (1995a) Rodent control, research needs, and technology transfer. In: Bissonette J (ed) Integrating people and wildlife for a sustainable future. The Wildlife Society, Bethesda, pp 693–697
Witmer GW, Matschke GH, Campbell DL (1995b) Field trials of pocket gopher control with cholecalciferol. Crop Prot 14:307–309
Witmer GW, Sayler RD, Pipas MJ (1997) Repellent trials to reduce reforestation damage by pocket gophers, deer, and elk. In: Mason JR (ed) Repellents in wildlife management, USDA National Wildlife Research Center, Fort Collins, pp 321–332
Witmer GW, Marsh RE, Matschke GH (1999) Trapping considerations for the fossorial pocket gopher. In: Proulx G (ed) Mammal trapping. Alpha Wildlife Research and Management, Ltd., Sherwood Park, pp 131–139

22 Influence of Plateau Zokors (*Eospalax fontanierii*) on Alpine Meadows

Yanming Zhang

22.1 Introduction

Studies of the influence of animals on ecosystems have until now almost always focused on their roles as consumer organisms and on their effects on energy flow rather than on the part they play in modifying community structure, regulating the competition of plant growth, and transporting soil nutrients (Grant et al. 1980; Reichman and Smith 1990; Huntly and Reichman 1994; Reichman and Seabloom 2002).

The plateau zokors (*Eospalax fontanierii*), the only subterranean small herbivorous mammal on the Tibetan plateau of China, may have effects on their alpine ecosystem in a variety of ways due to their specific life-style, behavioral characteristics, and population structure. Such effects could range from herbivory, possibly altering plant species composition and reproduction strategies, to soil movement and burying plants during mound formation. Similar to other entirely subterranean herbivores (Nevo 1979; Andersen and MacMahon 1981), plateau zokors excavate and inhabit burrow systems and deposit the tailings in abandoned tunnels or as mounds on the soil surface. These activities may reduce standing crop biomass or continually provide colonization sites for early successional species (Wang and Du 1990). Above- and below ground biomass decreases significantly in areas which zokors have occupied for more than 10 years compared to areas where zokors have been removed or are absent. Nutrient cycling may also be affected by long-term occupation of plateau zokors (Zhang and Liu 2003). The soils in such areas often show significantly lower nutrient concentrations than in areas where plateau zokors had been removed for five years (Zhang and Liu 2003).

An overall goal of this chapter is to evaluate the impact of plateau zokors on the alpine meadow in Tibet. Specific objectives are to introduce the effects of burrow activity by zokors on soil characteristics, plant diversity, plant abundance, and defense characteristics of dicotyledonous plants in alpine meadows.

Yanming Zhang: Northwest Plateau Institute of Biology, Chinese Academy of Sciences, Xining, Qinghai, 810001, People's Republic of China, E-mail: zhangym@nwipb.ac.cn

Subterranean Rodents: News from Underground
S. Begall, H. Burda, C.E. Schleich (Eds.)
© Springer-Verlag Berlin Heidelberg 2007

22.2 Effects on Soil and Nutrients

Huntly and Inouye (1988) suggest that the changes subterranean herbivores induce in the vertical and horizontal distribution of both the physical and chemical soil characteristics can affect the resources available to plants in several ways: 1) the average level of nutrients is changed; 2) the spatial pattern of soil nutrients is changed – soil nutrients become more heterogeneous; and 3) the relationship between soil nutrients and light, another resource that limits plant growth, is altered.

The most conspicuous effects of plateau zokors on the soil arise from the tailings that they produce while building and maintaining their tunnel systems. It is estimated that a plateau zokor deposits at least 1,024 kg/year at the soil surface of alpine meadows (Wang and Fan 1987). Studies conducted at the Haibei Research Station (Menyuan County, Qinghai Province) have shown that new mounds were formed at a rate of 22.2/zokor/month in May and 12.6/zokor/month in September, respectively with a mean volume of mounds of 7,089.3 cm^3 on alpine meadows. The soil in zokor mounds frequently differs in hardness, moisture and organic matter from the surrounding undisturbed soil (Zhang et al. 2003). The soil moisture of fresh mounds amounts to 14.4%, being much lower than in random samples in adjoining meadow (30.4%). Similarly, levels of soil nutrients, including nitrogen and phosphorus vary significantly among zokor mounds of differing ages and from undisturbed soil. The percentage of total nitrogen concentration in the 0–5 cm soil layer of fresh mounds averages 0.15 ± 0.02, representing a lower value than that of undisturbed areas (0.28 ± 0.07). However, the soil of fresh mounds carries a higher content in available nitrogen and phosphorus than in randomly collected soil samples (Zhang 2000).

The effect of plateau zokors on soil nutrients results from the manner in which zokors forage and from the vertical distribution of nutrients in the soil. The nitrogen and phosphorus content of alpine meadow soils drops rapidly with increasing depth (Le et al. 1989); zokors thus deposit mounds of nutrient-poor soil on the ground surface when excavating their foraging tunnels (Wang et al. 1993; Zhang 2000). However, once at the surface, the temperature of soil found in fresh mounds increases, leading to a concomitant mineralization promotion of organic phosphorus and nitrogen, and increasing the abundance of soil microorganisms (Li 1989). Over time, however, the soil of old mounds significantly loses available nitrogen and phosphorus content (Zhang 2000).

22.3 Effects on Vegetation and Diversity

The creation of mounds changes soil nutrients as well as surface light, thus affecting plant biomass and species composition. The above-ground biomass of plants surrounding zokor mounds is higher than in control areas (Wang et al. 1993; Zhang et al. 2003), with the flora typically differing from the surrounding vegetation (Wang and Du 1990; Bian et al. 1991; Zhang et al. 1994; Zhang 2000): frequently, forbs are more abundant on mounds (Bian et al. 1991; Zhang et al. 1994). Mounds of gophers have been assumed to affect plant community diversity and species composition simply by providing space for colonization by so-called fugitive plant species that are competitively eliminated over time by more advanced successional species (McDonough 1974; Hobbs and Hobbs 1987; Hobbs et al. 1988). Many studies about zokors showed that the competitive relations of plants can be changed by creating mounds and the consumption of plants (Wang and Du 1990; Zhang et al. 1994).

Plateau zokors can also have long-term effects on plant biomass and species composition. Our study results at Haibei Research Station indicate that above- and below-ground biomass, height and cover of vegetation significantly decrease in a habitat occupied by plateau zokors over the course of a decade (Zhang and Liu 2003; Zhang et al. 2003). However, no significant difference in biomass of dicots between occupied and unoccupied areas can be detected, although the percentages of dicots in the proportional composition of the community vary substantially in these areas (Zhang and Liu 2003). Growth and development of monocotyledons is restrained. Long-term zokor colonization also results in fewer plant species and lower plant species diversity (Zhang and Liu 2003), while the community on the occupied sites becomes dominated by those plants containing secondary chemical compounds such as *Ajania tenuifolia* (containing monoterpenes) and *Elsholtzia calycocarpa* (containing pennyroyals) (Yang 1991). There are many differences in the dominant species among mounds, arising from different patterns of age and habitat colonization. Spatial and temporal patterning in soil nutrient and plant composition may contribute to large-scale vegetational diversity and increase the landscape complexity (Zhang 2000).

22.4 Effects on Monocotyledons

Huntly and Reichman (1994) pointed out that the effects of subterranean herbivores on plant species could arise from diet selection, foraging behavior, and dug up mounds.

Plateau zokors, like other subterranean rodents, are diet generalists in their habitat, the plateau meadow, and consume a wide variety of plants (Zhang 2000). An analysis of stomach contents illuminated that both the below- and aboveground parts of monocots are common in diets of plateau zokors (Zhang 2000). Plateau zokors forage on many plants as long as they have no negative effects on digestion – the special life style with limited energy supplies underground does not allow for any choosiness. Obviously, consumption of monocots by plateau zokors is an important factor in causing the biomass to decrease (Zhang et al. 2004).

Reichman and Seabloom (2002) reported that the loss of roots affects plants more significantly than does the loss of leaves. Grasses with diffuse root systems seem less affected than forbs, which tend to have a taproot, as subterranean herbivorous mammals feed primarily on rich underground storage organs. An earlier study of our working group found that the burrowing activities of plateau zokors negatively affect monocots in alpine meadows since samples taken from overlying burrows yielded less monocot biomass than removal and control samples (Zhang et al. 2004), suggesting that soil excavating and burrow system maintenance by plateau zokors could result in the decrease of monocots biomass.

Deposition of considerable amounts of soil on the surface was always thought to be the main reason for altering species composition, abundance, or plant biomass (Spencer et al. 1985; Williams et al. 1986; Huntly and Inouye 1988; Huntly and Reichman 1994; Zhang et al. 1994; 2003). Plateau zokors pile up most mounds during early spring and late summer to autumn (Wang and Fan 1987). This mound formation obviously also indirectly effects plant abundance and biomass, which agrees with the conventional wisdom that physical factors are more important than biotic factors in vegetation succession. One obvious effect of the mounds is to smother virtually all vegetation they cover, indicated by significant differences in survival rates of plants covered by mounds due to the difference of life history characteristics. Soil burials increase monocot mortality, as many species of alpine meadows (including grasses and sedges) cannot grow through accumulated soil (Bian et al. 1991).

On the other hand, mounding activities provide suitable sites for some annual species that seed and germinate on bare ground. Although many monocots have evolved several survival strategies including cloning and sexual reproduction, the clone strategy can be considered primarily due to the cold climate and short growing season of the alpine meadow leading to much lower germination rates (Deng et al. 1999). Thus, the environment of intensive mound-building activities and foraging caused by plateau zokors is not conducive for the survival and growth of monocots.

22.5 Effects on Dicotyledons

Plateau zokors do not significantly influence biomass changes of dicotyledons, mainly because different species responses cancel each other out (Zhang and Liu 2003).

The impacts of underground foraging by plateau zokors in alpine meadows on dicot species might depend on their palatability. Many nonpalatable species, such as *Ajania tenuifolia*, *Polygonum sibiricum*, and *Elsholtzia calycocarpa* may even benefit from underground foraging and increased biomass, because they are avoided by the plateau zokors. These species finally dominate the plant community, and obviously show chemical defense characteristics in long-term occupied areas. For instance, *Polygonum sibiricum* and *Ajania tenuifolia* have rhizomes, but are avoided by plateau zokors because they contain secondary chemical compounds (Wang et al. 2000; Zhang 2000). However, the stem and roots of these plants are dug out and cached separately in shallow chambers which are located 10–20 cm below the surface. Generally, these storage sites are not available for plateau zokors in winter, and their contents may partly sprout and grow through the soil layers in the next spring. In the long run, the foraging activities alter the disturbance of these plants, and promote their dispersal. This circumstance gives rise to an increase in biomass of these unfavoured plants and their dominance in occupied areas.

Some annual species, such as *Elsholtzia calycocarpa* and *Pedicularis kansuensis* increase in biomass during excavating and mound-building, as the mounds formed by plateau zokors generate bare ground and thus provide a suitable substrate for seeding. It is also possible that these annual dicots species are more quickly able to colonize areas with lower plant densities. Of course, the diffuse root systems of these plants avoided by plateau zokors may be another important factor that reduces the influence of zokor activities in alpine meadows.

However, in the case of *Potentilla anserina,* biomass does not decrease with constant foraging and mound-building, though this species is a favoured food item for zokors (Zhang 2000). Avoided by livestock, this dicot represents even the most preferred food for plateau zokors, although it contains tannins. Zokors tend to cache the belowground storage organs during fall and consume these in winter when cold temperatures preclude any plant growth (Zhang 2000). Several reports suggest that fossorial animals negatively influence the plant species upon which they harvest and cache, because they heavily utilize plant storage organs (Huntly and Reichman 1994; Begall and Gallardo 2000). For *Potentilla anserina*, however, digging and foraging may provide even more space for growing because of the plant's extensive creeping form. This growth strategy seems to respond well to being buried under a mound, as it allows the plant to grow back to the surface and result in increasing biomass. Obviously, this mutualistic relationship has coevolved between plateau zokors and *Potentilla anserina* in alpine meadows.

Foraging underground affects much stronger those plants that are preferred by plateau zokors and have taproots, such as *Taraxacum mongolicum*, *Morina chinensis*, *Oxytropis kansuensis*, *Gentiana straminea*, and *Notopterygium forbesii*. Reichman and Smith (1990) found that the removal of root tissue (simulating foraging by pocket gophers) had a substantially greater effect on *Tragopogon dubius* than did removal of a similar amount of leaf material. Thus, foraging or damaging by plateau zokors could cause the mortality of plants with taproots to increase, and, in turn, the biomass to dramatically decrease. Another reason why these dicot species are unable to survive high zokor activity is their inability to cope with drastic burial in the occupied areas (Zhang 1999).

A previous study of our group showed that both biomass and percentage of proportional composition within the plant community of *Leontopodium nanum* increased significantly and resulted in dominance of this species after plateau zokors had been removed for five years. This small plant may reduce the ability of grazing aboveground living animals leading to lower losses of flowers, seeds and leaves. It is the small stature that appears to be an important trait for persisting under heavy aboveground grazing (Bach 2001; Dorrough et al. 2004). Similarly, the biomass of taproot plants, especially *Morina chinensis* with many thorns on the leave rims, increases significantly in areas where zokors were removed for 5 years compared with areas that zokors had occupied for 10 years. Apparently, the characteristic of chemical defense was replaced by physical defense after plateau zokors had been removed.

22.6 Conclusion

In the alpine grassland of Tibet, plateau zokors (*Eospalax fontanierii*) have a variety of effects upon the alpine ecosystem. Plateau zokors may act to increase local environmental heterogeneity at the landscape level, altering nutrients, aeration and mixing of soil. Underground foraging appears to have led to a decline in biomass and in the monocot share of plant communities. The response of individual dicot species to underground feeding depends on palatability, root type, growth form, and life styles. Areas with long-term occupancy by plateau zokors yield plant communities with lower biomass and proportionally fewer preferred dicot species, but with higher biomass and proportionally higher representation of dicot species containing secondary chemical compounds. Also, the foraging activities by plateau zokors alter the disturbance of unfavoured plants, and might promote their dispersal. A mutualistic relationship can be considered as having evolved between plateau zokors and *Potentilla anserina* in alpine meadows. Other annual dicot species may also benefit from zokor activities, as mounding provides bare substrate for quickly colonizing the areas of lower plant densities. Small species and species with defense mechanisms dominate the plant community, with chemical defense being replaced by physical defense after plateau zokor removal.

Acknowledgements. This work was supported by the National Science Foundation of China (No.30470311, and 30670357), The Chinese Key Technological Research and Development Program (No.2005BA529A05), and The Western Light Projects of China.

References

Andersen DC, MacMahon JA (1981) Population dynamics and bioenergetics of a fossorial herbivore, *Thomomys talpoides* (Rodentia: Geomyidae), in a spruce-fir sere. Ecol Mon 51:179–202

Bach CE (2001) Long-term effects of insect herbivory and sand accretion on plant succession on sand dunes. Ecol 82:1401–1416

Begall, S, Gallardo MH (2000) *Spalacopus cyanus* (Octodontidae, Rodentia): an extremist in tunnel constructing and food storing among subterranean mammals. J Zool Lond 251:53–60

Bian J, Wang Q, Shi Y (1991) Experimental analysis of the relationship between primary plant component on zokor mounds and precede cover vegetation, In: Liu J, Wang Z (eds) Alpine meadow ecosystem fascicle 3. Science Press, Beijing, China, pp 181–188 [In Chinese with English summary]

Deng Z, Wang W, Wang Q, Zhou X, Li Y (1999) The study of reproductive strategies of *Elymus nutans* population in alpine meadow. Acta Biol Plat Sin 14:101–109 [In Chinese with English summary]

Dorrough J, Ash J, McIntyre S (2004) Plant responses to livestock grazing frequency in an Australian temperate grassland. Ecography 27:798–810

Grant WE, French NR, Folse LJ (1980) Effects of pocket gopher mounds on plant production in shortgrass prairie ecosystems. Southwest Nat 25:215–224

Hobbs RJ, Hobbs VJ (1987) Gophers and grasslands: a model of vegetation response to patchy soil disturbance. Vegetatio 69:141–146

Hobbs RJ, Gulmon SL, Hobbs VJ, Mooney HA (1988) Effects of fertilizer addition and subsequent gopher disturbance on a serpentine annual grassland community. Oecologia 75:291–295

Huntly N, Inouye R (1988) Pocket gophers in ecosystems: patterns and mechanisms. BioScience 38:786–793

Huntly N, Reichman OJ (1994) Effects of subterranean mammalian herbivores on vegetation. J Mammal 75:852–859

Le Y, Zhang J, Wang Z, Chen W (1989) Mineralization of soil organic nitrogen in alpine meadow soils. In: Xia W (ed) Proceedings of the International Symposium of Alpine Meadow Ecosystem. Science Press, Beijing, China, pp 155–167 [In Chinese with English summary]

Li D (1989) Economically important vertebrates of Qinghai Province. Qinghai People's Publishing House, Xining, China. [In Chinese with English smmary]

McDonough WT (1974) Revegetation of gopher mounds on aspen range in Utah Great Basin Nat 34:267–274

Nevo A (1979) Adaptive convergence and divergence of subterranean mammals. Ann Rev Ecol Syst 10:269–308

Reichman OJ, Seabloom EW (2002) The role of pocket gophers as subterranean ecosystem engineers. Trends Ecol Evol 17:44–49

Reichman OJ, Smith SC (1990) Burrows and burrowing behavior by mammals. Curr Mammal 2:197–244

Spencer SR, Cameron GN, Eshelman BD, Cooper LC, Williams LR (1985) Influence of pocket gopher mounds on a Texas coastal prairie. Oecologia 66:111–115

Wang G, Du G (1990) Niche analysis of species in vegetative succession on zokor mounds. J Ecol (China) 9:1–6 [In Chinese with English summary]

Wang Q, Fan N (1987) Studies on the digging activities and exploration about the method of number estimation of plateau zokor. Acta Theriol Sin 7:283–290 [In Chinese with English summary]

Wang Q, Bian J, Shi Y (1993) Influence of plateau zokor mounds on the vegetation and soil nutrients in an alpine meadow. Acta Theriol Sin 13:31–37 [In Chinese with English summary]

Wang Q, Zhang Y, Wei W, Bian J (2000) Food habit of the plateau zokor. Acta Theriol Sin 20:198–210 [In Chinese with English summary]

Williams LR, Cameron GN, Spencer SR, Eshelman BD, Gregory MJ (1986) Experimental analysis of the effects of pocket gopher mounds on Texas coastal prairie. J Mammal 67:672–679

Yang Y (1991) Tibetan medicine. Qinghai People's Publishing House, Xining, China [In Chinese with English summary]

Zhang Y (1999) Effect of plateau zokor on characters and succession of plant communities in alpine meadow. Zool Res (China) 20:435–440 [In Chinese with English summary]

Zhang Y (2000) Studies on the pattern of animal-plant interaction: the effects of plateau zokor on the biogeochemical cycling of alpine meadow ecosystem and its response to the chemical defense of plants. PhD Dissertation, The Chinese Academy of Sciences, Beijing, China [In Chinese with English summary]

ZhangY, Liu J (2003) Effects of plateau zokors (*Myospalax fontanierii*) on plant community and soil in an alpine meadow. J Mammal 84:644–651

Zhang Y, Zhou X, Cincotta RP (1994) Study of diversity and succession pattern of plant communities on zokor mound. Acta Ecol Sin 14:42–49 [In Chinese with English summary]

ZhangY, Zhang Z, Liu J (2003) Burrowing rodents as ecosystem engineers: the ecology and management of plateau zokor (*Myospalax fontanierii*) in alpine meadow ecosystems on the Tibetan Plateau. Mammal Rev 33:284–294

Zhang Y, Liu J, Du Y (2004) The impact of plateau zokor (*Myospalax fontanierii*) burrows on alpine meadow vegetation on the Qinghai-Xizang (Tibetan) plateau. Acta Theriol 49:43–51

23 Mountain Beaver: A Primitive Fossorial Rodent

Wendy M. Arjo

23.1 Introduction

As the largest mammalian order, rodents are nearly cosmopolitan in distribution, can exploit a broad spectrum of foods, and can often reach high population densities. One sole representative of the most primitive family of rodents, Aplodontidae, does not share some of these common rodent characteristics. The aplodontoid rodents in the family Aplodontidae and Mylagsulidae radiated during the Miocene from the Allomyinae family (Carraway and Verts 1993). The extinct Mylagaulidae represents the earlier radiation of these rodents who exhibited great specialization (Carraway and Verts 1993). Unlike the other members of the rodent order, mountain beavers are not prolific breeders; nor are they broad-spectrum habitat invaders, retaining in their morphology the primitive condition of the masseter muscle originating entirely on the zygomatic arch. The mostly extinct Aplodontidae family is now made up of the monotypic genus *Aplodontia* which has been able to survive since the early Oligocene and in some areas is even considered a pest.

Mountain beavers (*Aplodontia rufa*) are known from a variety of common names including boomer, whistler, and the Native American names sewellel and showt'l. Although the common name suggests relationship to true beaver (Castoridae), this semi-fossorial rodent shares only the behavior of tree clipping with the stream beaver and is usually more abundant at lower elevations than in mountainous areas. In Oregon, mountain beavers have retained the designation of boomer even though their vocalizations do not include booms or whistles. Very few people actually know about, much less have seen, this compact fossorial rodent. In this chapter I will present information on the ecology and human-wildlife conflicts of this little-known rodent species.

Wendy M. Arjo: USDA/APHIS/WS/National Wildlife Research Center, 9730-B Lathrop Industrial Dr., SW. Olympia, WA 98512, USA, E-mail: Wendy.M.Arjo@aphis.usda.gov

Subterranean Rodents: News from Underground
S. Begall, H. Burda, C.E. Schleich (Eds.)
© Springer-Verlag Berlin Heidelberg 2007

23.2 Distribution of the Aplodontidae Family

The mountain beaver, endemic to western North America, is the only extant member of the family Aplodontidae. Considered the most primitive rodent species, mountain beavers have a restricted geographic distribution to western North America. Currently seven subspecies are recognized (Dalquest and Scheffer 1945; Hall 1981), with the largest distribution of a single subspecies (*A. r. rufa*) ranging from southern British Columbia, Canada through coastal Washington and into central Oregon. Contiguous with this population are three other subspecies including *A. r. rainieri* (Cascade Range of Washington, Oregon, and northern California), *A. r. pacifica* (Oregon and northwestern California coast), and *A. r. humboldtiana* (northern coastal California; Feldhamer et al. 2003). Three isolated mountain beaver populations occur in California. *A. r. californica* extends through much of the Sierra Nevada Range in eastern California into the western extreme portion of Nevada. The remaining two California populations are isolated along the northern coast at Point Arena, *A. r. nigra*, ($62\,km^2$) in Mendocino County (Steele 1986; Zielinski and Mazurek 2006) and Point Reyes, *A. r. phaea*, ($285\,km^2$) in Marin County (Fellers et al. 2004; Wake 2006).

Very little change in geographic distribution of the mountain beaver has occurred since the late Oligocene (Carraway and Verts 1993); however, climatic changes, physical geographic changes to the coast line, and fragmentation of boreal forest communities have likely contributed to the isolation of the southern coastal *A. rufa* subspecies at Point Arena and Point Reyes (Wake 2006). Recent recoveries of *A. rufa* in Duncan's Point Cave documented the historical distribution of the mountain beaver during the mid-Holocene likely extended further north of Point Reyes (Wake 2006). Currently the subspecies at Point Arena (*A. r. nigra*) is listed as federally endangered (United States Federal Registry, 56 FR 64716). In Canada, mountain beavers (*A. r. rufa* and *A. r. rainieri*) are designated as species of concern by the Committee on Status of Endangered Wildlife in Canada (sararegistry.gc.ca/species/).

23.3 Mountain Beaver Ecology

The general body conformation of the mountain beaver is cylindrical and compact (Feldhamer et al. 2003). Average adult length of *A. rufa*, without the stump-like tail, is 32 cm, but can range from 30 to 50 cm (Feldhamer et al. 2003). Maximum adult body weight in the wild was reported to be 1.4 kg (Maser et al. 1981); however, in captivity weights up to 1.6 kg have been documented (Arjo, personal observation). In the field, average adult weight of the *A. rufa* subspecies is 1.0 kg (Arjo, unpublished data) which is also similar to weights for *A. r. nigra* (Zielinski and Mazurek 2006). Although mountain beavers spend a large portion of their time below ground, movement above ground does occur both during the day and at night (Ingles 1959; Arjo, unpublished data). Mountain beavers have low basal metabolic rates (McNab 1979) and their activity is often short in duration, followed by longer periods of inactivity (Ingles 1959; Arjo, personal observation).

Although the extensive burrow system used by the mountain beaver may offer some protection from predators, mountain beavers are prey to a large variety of predators. Striped (*Spilogale gracilis*) and spotted skunks (*Mephitis mephitis*), mink (*Mustela vison*), gray fox (*Urocyon cinereoargenteus*), raccoons (*Procyon lotor*), badgers (*Taxidea taxus*), bobcats (*Lynx rufus*), coyotes (*Canis latrans*), fisher (*Martes pennanti*), long-tailed weasels (*Mustela frenata*), cougars (*Puma concolor*), and northern goshawks (*Accipiter gentiles*) have all been documented as predators on mountain beaver (Taylor 1918; Toweill and Maser 1985; Witmer and deCalesta 1986; Thrailkill et al. 2000; Arjo, unpublished data).

23.3.1 Habitat

Mountain beavers can be found up to 3,000 m elevation in portions of the Sierra Nevada Range; however, they are more commonly found at lower elevations in humid, densely vegetated understory areas (Feldhamer et al. 2003). Physiological constraints (e.g., inefficient kidneys; Pfeiffer et al. 1960) and the need for a constant water source likely limit the mountain beaver's geographical range to humid temperate climates. Open canopy habitats, often produced after forest harvesting, are preferred by Pacific Northwest mountain beavers, and populations tend to be less numerous in closed canopy habitat (Hooven 1977; Neal and Borrecco 1981). Sword fern (*Polystichum munitum*) and salal (*Gaultheria shallon*) are clipped year-round as a food and bedding source (Neal and Borrecco 1981) and are considered to be important resources for Pacific Northwest mountain beavers (Voth 1968; Allen 1969). Similar Douglas-fir forested habitats in addition to redwood (*Sequoia sempervirens*) forests are occupied by *A. r. humboldtiana* in northern California (Steele 1986). Good forage cover (e.g., ferns, forbs, and shrubs) as well as large amounts of small diameter woody debris or uprooted stumps are areas usually selected by mountain beaver (Todd 1992; Hacker and Coblenz 1993; Arjo, unpublished data). Beier (1989) argues that mountain beavers have flexible food requirements and that habitat is chosen based on adequate soil drainage and appropriate thermal regimes. Willow (*Salix* sp.), alder (*Alnus* sp.) and fir (*Abies* sp.) dominate areas preferred by mountain beavers in the higher elevations of the Sierra Nevada Mountains.

Unlike the other four subspecies that mostly occur in forested environments, the two southern mountain beaver subspecies occur in primarily coastal scrub, dune scrub and some forested habitats along the northern California coast. Coyotebrush (*Baccharis pilularis*), salal, blackberry (*Rubus* spp.), and poison oak (*Toxicodendron diversilobum*) are common shrubs found in these drier scrub habitats. Associated coniferous forest habitat used by *A. r. nigra* and *A. r. phaea* include Douglas-fir, grand fir (*Abies grandis*) and bishop pine (*Pinus muricata*). Sword fern and bracken fern (*Pteridium aquilinum*) are understory plants common in most habitat types occupied by mountain beaver. Habitat degradation from livestock grazing, invasion of exotic plants, fire, and human activities in fragile coastal habitats are factors contributing to the endangered population status of *A. r. nigra* and the small population size and distribution of *A. r. pheae* (Fitts et al. 2002; Fellers et al. 2004; Wake 2006).

23.3.2 Burrow and Nest Structure

Extensive burrow systems, containing a highly variable number of openings, with high humidity and good soil drainage are used year-round (Voth 1968; Beier 1989). Microclimates within the burrow system are cool and stable with an annual temperature range between 2 and 14°C (Johnson 1971) (no SD is provided) and weekly temperature variation rarely fluctuates more than 4°C (Johnson 1971). Burrow systems are usually < 120 cm deep (Camp 1918; Feldhamer et al. 2003); however, older tunnel systems have been documented as deep as 2 m below the surface. Shallow tunnel systems lead to burrow openings, 15–18 cm in diameter, which vary in frequency between individual animals. Active burrow systems can be distinguished by fresh pushed dirt, debris, or rocks piled at burrow entrances, and well-established runways firmly packed and smooth from use. In addition, piles of sword fern or other clipped vegetation at burrow entrances are also good indicators of mountain beaver presence. Each mountain beaver system usually contains at least one runway system that leads to an above-ground water source. Mountain beavers also make extensive use of large woody debris piles to supplement above-ground runway systems.

Voth (1968) described five different use chambers within a burrow system: nest, feeding, refuse, fecal pellet, and earth balls. Compacted dirt and rocks are often stored in the earth ball chamber and may be used for plugging of burrow entrances (Voth 1968). Mountain beavers are highly dependent upon the nest (or den) chamber which fulfills both reproductive and non-reproductive functions. Nests are usually located 70 cm below ground and usually under logs, uprooted stumps, logging slash, or thick vegetation (Martin 1971; Arjo, unpublished data). Nests of *A. r. nigra* were found to be primarily composed of grass and herbaceous vegetation that culminated in a nest cup of iris (*Iris douglasiana*) leaves (Zielinski and Mazurek 2006). In the Pacific Northwest, mountain beaver nests are composed primarily of sword fern and salal (Camp 1918; Scheffer 1929; Martin 1971; Hooven 1977; Arjo, unpublished data) with the addition of moss to act as a desiccant. These dry, circular nests average 50–60 cm in diameter and 36–41 cm in height (Martin 1971; Arjo, unpublished data). Only one tunnel leads directly into the nest, and burrow openings are seldom in close proximity to the nest. Adjacent to the nest is usually a feeding and fecal chamber. The feeding chamber contains caches of recently cut or wilted vegetation which is deposited in the refuse chamber when plant matter starts to decay (Feldhamer et al. 2003). Voth (1968) stated that several fecal pellet chambers may be associated with one burrow system and that each may contain pellets at different stages of decomposition.

23.3.3 Reproductive Behavior

Current understanding of mountain beaver social systems documents a solitary existence, with males and females only tolerating one another during breeding season. Male onset of reproductive activity for mountain beaver occurs in November and December (Hubbard 1922; Pfeiffer 1956; Lovejoy et al. 1978; Carraway and Verts 1993). Mountain beavers are monestrous and considered reproductively capable after their second year, and have low reproductive rates (Pfeiffer 1958). In both the *A. r. pacifica* and *A. r. rufa* subspecies, breeding usually occurs from late January to early February. Estimated parturition of 2–4 pups after a 28–30 day gestation is late March to early April, with weaning occurring at the beginning of June (Lovejoy and Black 1974; Arjo, unpublished data). Although little data is available on reproduction of the southern subspecies, the large latitudinal range in distribution of mountain beaver may suggest local variation in reproductive traits (Pfeiffer 1958; Zielinski and Mazurek 2006). Early signs of estrous were observed in *A. r. nigra* which may suggest parturition as early as February (Zielinski and Mazurek 2006). The mating system of the mountain beaver is currently unknown.

23.3.4 Population Dynamics

In older aged forest stands, densities of mountain beaver seldom exceed 4 per ha (Borrecco and Anderson 1980); however, after forest harvest, densities can approach 15–20 per ha (Hooven 1977; Neal and Borrecco 1981). Most of the available data on mountain beaver populations are limited to a few studies conducted on 8–20 year-old reforestation units. Populations in these regeneration sites averaged 3.4–4.6 mountain beaver per ha (Lovejoy and Black 1979b; Neal and Borrecco 1981). Conflict between humans and wildlife occurs mainly in harvested forest units in the Pacific Northwest where management of reforested areas is difficult where mountain beaver populations thrive. Gyug (2000) documented more mountain beaver activity in clear-cuts based on forest harvest method. Units that were harvested with methods that created less ground disturbance often contained more activity. Ground disturbance can cause collapse of the burrow systems and nest chambers that are vital for mountain beaver survival. Recent observations on a unit in western Washington documented a greater population of mountain beaver post-forest harvest than was present prior to harvest in the > 30-year-old Douglas-fir (*Pseudotsuga menziesii*) stand (Arjo, unpublished data). In newly reforested units, populations averaging 0.49–2.13 mountain beaver per ha were documented (Arjo and Nolte 2006). These numbers decreased on one unit the following year after herbicide treatments became effective, but doubled on another unit to 4.38 mountain beavers per ha that was not treated (Arjo and Nolte 2006).

Population estimates for mountain beaver in the southern and eastern extremes of their range, where habitat encompasses non-forested areas, are few (Fellers et al. 2004; Zielinski and Mazurek 2006). Population estimate for *A. r. nigra* in the 62 km^2 range was estimated at 200–500 individuals (USFWS 1998). Limitation to the amount of disturbance these small, isolated populations can sustain precludes more extensive monitoring methods, and limits monitoring of populations to burrow counts which can be highly variable but is currently the best available method.

23.3.5 Movements and Home Ranges

Space use by mountain beavers seems to be influenced by available habitat. Neal and Borrecco (1981) documented smaller home ranges in areas with greater quantities of salal and herbaceous vegetation due to the presence of more stand openings that allowed for growth of the understory. Martin (1971) estimated home ranges in a regenerating forest unit to vary from

0.03 to 0.2 ha and averaging 0.12 ha, and home ranges in late seral stage units were estimated at 0.26 ± 0.04 ha (Lovejoy and Black 1979a). Arjo and Nolte (2006) documented home ranges and core areas in newly reforested areas significantly larger than previously reported. Male home ranges averaged 4.16 ± 0.69 ha and core areas 0.73 ± 0.13 ha in areas where herbicide treatments were used prior to planting seedlings. Female home ranges were similar, although core use areas were slightly larger (1.02 ± 0.44 ha; Arjo, unpublished data). In areas were herbicide treatments were not conducted prior to planting, home ranges were smaller averaging 1.26 ± 0.5 ha for males and 1.7 ± 0.82 ha for females. Home ranges can overlap extensively depending on habitat availability with nest sites being defended (Martin 1971; Arjo, unpublished data). Mountain beaver continually explore neighboring runway systems (Nolte et al. 1993). Although telemetry data shows that only one animal occupies a given core area, upon removal of the resident animal, other animals can quickly reinvade a system (Arjo, unpublished data).

23.3.6 Forage Preference

Mountain beavers are strictly herbivorous and are coprophagic, reingesting certain fecal pellets for maximum nutrition (Feldhamer et al. 2003). Areas occupied by mountain beavers can often be identified by stacks of clipped vegetation, also termed "haystack", outside of burrow entrances, called feeder holes. Voth (1968) suggested that this behavior might improve the nutritional quality or succulence of the vegetation, and forage preference is often determined from these haystacked piles of vegetation. In coastal Oregon, food preference for males and females is similar in the summer and fall when bracken fern is preferred, but varies in the winter and spring (Voth 1968). In the winter months, males usually prefer sword fern and lactating females, conifers. Arjo et al. (2004) have documented that preference for conifers is not dependent upon reproductive status. At higher elevations in the Sierra Nevada Range, conifers, red fir (*Abies magnifica*), dogwood (*Cornus* spp.), willow (*Salix* spp.) and corn lily (*Veratrum californicum*) are preferred, as well as bark and twigs during periods of deep snow (O'Brien 1988; Todd 1992). Vine maple (*Acer circinatum*), red huckleberry (*Vaccinium parvifolium*), red alder (*Alnus ruba*), sword fern, bracken fern, salal, elderberry (*Sambucus racemosa*), salmonberry (*Rubus spectabilis*), and Oregon grape (*Mahonia aquifolium*) are often heavily clipped plant species in the Pacific Northwest (Crouch 1968; Allen 1969; Arjo, personal observation). Angelica (*Angelica hendersonii*) and iris appear to be a preferred plant species clipped by *A. r. nigra* and iris was the most common species observed in haystacks (Zielinski and Mazurek 2006). Due to their inability to concentrate urine, mountain beaver require a large daily intake of water and succulent vegetation (Pfeiffer et al. 1960; Cafferata 1992). Since mountain beaver require almost one-third of their body weight in water daily (Nungesser and Pfeiffer 1965), it is more likely that haystacks represent drying nesting material and not a food resource.

23.4 Management and Economic Impacts

Mountain beavers are a unique species that fill a biological niche as well as possess intrinsic value as a native animal to the Sierras of California and Pacific Northwest (Steele 1986). Maintaining biological diversity to include an archaic and monotypic species such as the mountain beaver can provide ecological stability. Conservation efforts for the isolated subspecies in the extremes of the range offer challenges to land managers in balancing habitat needs of a single species with recreational opportunities, agriculture and timber production, and urban development. Burrow disturbance in both California, due to human recreation, and Canada, from logging activities (Gyug 2000), can adversely affect mountain beaver populations. However, balancing economic and aesthetic benefits of maintaining recreation areas in Point Arena or Point Reyes, with the need to protect fragile burrow areas, can be difficult. Invasion of exotic plant species threatens areas of coastal scrub habitat vital to mountain beaver; however, little standing water is available most of the year in the Point Arena area and succulent plants provide a readily available source of water. Management of exotic plant species, such as ice plant (*Carpobrotus edulis*), at Point Arena must be balanced with the reliance of this plant by mountain beaver and potential further disturbance to the habitat (Fitts 1996).

In the northern portion of the range with the exception of Canada, mountain beavers are economically important because of the damage they cause to seedlings, especially Douglas-fir (Hooven 1977; Borrecco and Anderson 1980; Campbell and Evans 1988). Borrecco and Anderson (1980) documented that the majority of recorded mountain beaver damage occurred from the Olympic peninsula to the Puget Sound Trough and the Coast Range to the Willamette Valley, with very little damage occurring in northeast California. In a 1977 survey, over 111,000 ha in the Pacific Northwest, 75% of that in Douglas-fir stands alone, was damaged by mountain beavers (Borrecco et al. 1979). Depending on tree size, damage by mountain beavers can cause suppression of height growth, reduce stocking in plantations, or delay forest regeneration (Borrecco and Anderson 1980). The most prevalent problem is clipping of seedlings after planting (Hooven 1977;

Borrecco et al. 1979), where stems up to 19 mm can be clipped, resulting in a continual loss of trees of up to four years after planting (Lawrence et al. 1961). In addition, basal girdling of 10- to 20-year-old trees and undermining of roots in sapling stands can also occur in high density mountain beaver areas (Cafferata 1992).

Historically, managers have implemented both lethal and non-lethal methods to control mountain beaver populations. Lethal trapping is the preferred method used to control mountain beaver populations, and occurs from October through February in both Oregon and Washington prior to seedling planting. Most non-lethal methods, such as box traps, individual tree protectors, and fencing have been only marginally effective. Fencing is ineffective unless wire is buried at least 1.5 m underground, and even then, some mountain beaver have been known to have deeper burrow systems. Installation of tree barriers can be labor intensive because tubes are placed on the seedlings prior to planting or with an additional crew after planting, and maintenance of the tubes is required to insure integrity. Borrecco and Anderson (1980) documented a significant decrease in damage of seedlings from 3% to 44% with the application of tree barriers. However, even with barriers, damage to seedlings can occur. Tubes can be penetrated by mountain beavers, especially those tubes with perforations or seams that allow the mountain beaver to hold onto the plastic (Runde, personal communication). Mountain beavers have also been documented to climb larger tree tubes in order to clip individual seedlings inside the tubes, as well as undermining the tubes (Cafferata 1992).

23.5 Conclusion and Future Research Needs

Mountain beavers offer a unique challenge to land managers where southern and extreme northern populations are protected, yet populations in the largest contiguous range in Oregon and Washington are managed as pests. Understanding mountain beavers, whether in dealing with an endangered or pest species, must begin with a better understanding of the ecology and life history traits across the species' range. The limited, yet diverse distribution of the species offers researchers a rare opportunity for comparative research. For example, Pfeiffer (1958) documented potentially earlier onset of breeding in southern subspecies compared to the Pacific Northwest subspecies. Reproductive information is limited in these small isolated populations, due to their protected status, yet information on exact parturition dates would assist mangers in determining when extra precautions should be implemented for the species' protection. Although, mountain beavers are thought to not be reproductively capable until their second year

(Pfeiffer 1958), others have documented reproduction in younger individuals (Hacker 1992) in the Pacific Northwest in preferred habitats. Population growth is affected by age of first reproduction and can have significant impacts on management practices when dealing with a pest species, or in trying to manage an isolated, remnant population.

One of the main tools biologists are lacking in working with mountain beavers is an accurate way to determine population size. Current population estimates have been limited to burrow counts usually of active burrows determined from digging or haystacked vegetation (Gyug 2000; Fellers et al. 2004). One of the problems associated with this method is that burrows within an individual home range are highly variable, as is the number of haystack piles associated with an animal. In addition, dense vegetation or large debris piles may preclude monitoring in some areas and therefore bias population estimates. Sword fern monitoring can be used to show activity at burrows, but ferns may not always be taken by resident animals (Engeman et al. 1991; Zielinski and Mazurek 2006; Arjo, unpublished data). Hair snares to determine individual occupancy of burrows has been explored for the Point Arena mountain beaver (Zielinski and Mazurek 2006) and may be successfully used to fragile populations with minimal disturbance.

In addition to population size, dispersal capabilities and limitation to dispersal for mountain beaver, the effects on population are also poorly understood. Timber harvesting appears to limit suitable habitat for immigrating mountain beaver in Canada (Gyug 2000), yet newly harvested areas are preferred by mountain beaver in Washington and Oregon. Understanding if this limitation is due to limited available ecological requirements (e.g., soils, water, habitat types) or a limited source population for immigrants may assist in understanding the limitation of the northern extent of the mountain beaver range.

The diverse distribution of the mountain beaver to areas that in winter are often covered with meters of snow offers a unique opportunity to address foraging requirements. Unfortunately, most information available on food habits is based on older research of a small sample size on *pacifica*and *rufa* subspecies (Voth 1968; Allen 1969). Others have documented preferred forage based on availability at feeder holes, which may or may not represent all that the mountain beaver use, and observed clipped vegetation but is usually for limited time period. Using a combination of feeder hole monitoring, clipped vegetation, and remote cameras at feeder holes throughout many seasons may offer a more thorough understanding of forage requirements for the mountain beaver.

Finally, current information on mountain beaver biology and management tools for minimizing damage by mountain beaver is needed for the Pacific Northwest subspecies. Feldhamer et al. (2003) noted that information on the response of mountain beaver populations to sivilcultural activities and a better understanding of populations in association with other habitat characteristics can provide additional direction to management strategies to minimize damage. Additional information on mountain beaver demographics under varying management strategies (e.g., broadcast burning, herbicide treatments, and brush piling) may assist managers in developing integrated pest management strategies that reduce mountain beaver damage.

References

Allen LO (1969) Preferential food habits of *Aplodontia rufa*. MSc Thesis. Central Washington State College, Ellensburg, WA

Arjo WM, Nolte DL (2006) Boomer or bust: managing a Pacific Northwest pest species. Proc Vertebr Pest Conf 22 (in press)

Arjo WM, Nolte DL, Harper J, Kimball B (2004) The effects of lactation on seedling damage by mountain beaver. Proc Vertebr Pest Conf 21:163–168

Beier P (1989) Use of habitat by mountain beaver in the Sierra Nevada. J Wildl Manag 53:649–654

Borrecco JE, Anderson RJ (1980) Mountain beaver problems in the forests of California, Oregon, and Washington. Proc Vertebr Pest Conf 9:135–142

Borrecco JE, Anderson HW, Black HC, Evans J, Guenther KS, Lindsey GD, Mathews RP, Moore TK (1979) Survey of mountain beaver damage to forests in the Pacific Northwest, 1977. State of Washington Department of Natural Resources Note 26, p 16

Caffereta S (1992) Silvicultural methods in relation to selected wildlife species. In: Black HG (ed) Silvicultural approaches to animal damage management in Pacific Northwest Forests. USDA Forest Serv Pacific Northwest Res Stat Tech Rep, Portland, OR. PNW-GTR-287, pp 231–251

Camp DL (1918) Excavations of burrows of the rodent *Aplodontia,* with observations on habitat of the animals. Univ Calif Publ Zool 17:517–536

Campbell DL Evans J (1988) Recent approaches to controlling mountain beavers (*Aplodontia rufa*) in Pacific Northwest forests. Proc Vertebr Pest Conf 13:183–187

Carraway LN, Verts BJ (1993) *Aplodontia rufa*. Mamm Species 431:1–10

Crouch GL (1968) Clipping of woody plants by mountain beaver. J Mammal 49:151–152

Dalquest WW, Scheffer VB (1945) The systematic status of the races of the mountain beaver (*Aplodontia rufa*) in Washington. Murrelet 26:34–37

Engeman RM, Campbell DL, Evans J (1991) An evaluation of 2 activity indicators for use in mountain beaver burrow systems. Wildl Soc Bull 19:413–416

Feldhamer GA, Rochelle JA, Rushton CD (2003) Mountain beaver. In: Feldhamer GA, Thompson BC, Chapman JA (eds) Wild mammals of North America: biology, management, and economics. John Hopkins Univ Press, Baltimore, MD pp 179–187

Fellers GM, Pratt D, Griffin JL (2004) Fire effects on the Point Reyes mountain beaver at Point Reyes National Seashore, California. J Wildl Manage 68:502–508

Fitts KM (1996) Observations on use of two non-native plants by the Point Arena mountain beaver. Calif Fish Game 82:59–60

Fitts KS, Flowers S, Jackson R, Marshall D, Meetenmeyer R, Northern P (2002) Point Arena mountain beaver habitat protection and restoration plan for Manchester State Park. California State Parks

Gyug LW (2000) Status, distribution, and biology of the mountain beaver, *Aplodontia rufa*, in Canada. Can Field Nat 114:476–490
Hacker AL (1992) Population attributes and habitat selection of recolonizing mountain beaver (*Aplodontia rufa*). MSc Thesis, Oregon State Univ, Corvallis, OR
Hacker AL, Coblenz BE (1993) Habitat selection by mountain beavers recolonizing Oregon Coast Range clearcuts. J Wildl Manag 57:847–853
Hall ER (1981) The mammals of North America, 2nd edn. Wiley, New York
Hooven EF (1977) The mountain beaver in Oregon: its life history and control. Oregon State Univ Forest Res Lab, Corvallis, OR. Res paper 30
Hubbard CA (1922) Some data upon the rodent *Aplodontia*. Murrelet 3:14–16
Ingles LG (1959) A quantitative study of mountain beaver activity. Am Mid Nat 61:419–423
Johnson SR (1971) The thermal regulation, microclimate and distribution of the mountain beaver, *Aplodontia rufa pacific Merriam*. PhD Thesis, Oregon State Univ, Corvallis, OR
Lawrence WH, Kverno NB, Hartwell HD (1961) Guide to wildlife feeding injuries on conifers in the Pacific Northwest. Western Forestry Conserv Assoc, Portland, OR
Lovejoy BP, Black HC (1974) Growth and weight of the mountain beaver, *Aplodontia rufa pacifica*. J Mammal 55:364–369
Lovejoy BP, Black HC (1979a) Movements and home range of the Pacific mountain beaver (*Aplodontia rufa pacifica*). Am Mid Nat 101:393–402
Lovejoy BP, Black HC (1979b) Population analysis of the mountain beaver *Aplodontia rufa pacifica*, in western Oregon. Northwest Sci 53:82–89
Lovejoy BP, Black HC, Hooven EF (1978) Reproduction, growth, and development of the mountain beaver (*Aplodontia rufa pacifica*). Northwest Sci 52:323–328
Martin P (1971) Movements and activities of the mountain beaver (*Aplodontia rufa*). J Mammal 52:717–723
Maser C, Mate BR, Franklin JF, Dryness CT (1981) Natural history of Oregon coastal mammals. US Department of Agriculture, Forest Serv, Gen Tech Rep, PNW-133:1–496
McNab BK (1979) The influence of body size on the energetic and distribution of fossorial and burrowing mammals. Ecology 60:1010–1021
Neal FD, Borrecco JE (1981) Distribution and relationship of mountain beaver to openings in sapling stands. Northwest Sci 55:79–86
Nolte DL, Epple G, Campbell DL, Russell JR (1993) Response of mountain beaver (*Aplodontia rufa*) to conspecifics in their burrow system. Northwest Sci 67:251–255
Nungesser WC, Pfeiffer EW (1965) Water balance and maximum concentrating capacity in the primitive rodent, *Aplodontia rufa*. Comp Biochem Physiol 14:289–297
O'Brien JP (1988) Seasonal selection of coniferous trees by the sewellel, *Aplodontia rufa*. Mammalia 52:325–330
Pfeiffer EW (1956) The male reproductive tract of a primitive rodent, *Aplodontia rufa*. Anat Rec 124:629–635
Pfeiffer EW (1958) The reproductive cycle of the female mountain beaver. J Mammal 39: 223–235
Pfeiffer EW, Nungesser NC, Iverson DA, Wallerius JR (1960) The renal anatomy of the primitive rodent, *Aplodontia rufa*, and a consideration of its functional significance. Anat Rec 137:227–235
Scheffer TH (1929) Mountain beavers in the Pacific Northwest: their habits, economic status, and control. Farmer's Bulletin 1958. U.S. Department of Agriculture, Washington D.C.
Steele DT (1986) The mountain beaver in California. MSc thesis, UC Davis, Davis, California
Taylor WP (1918) Revision of the rodent genus *Aplodontia*. Univ Calif Publ Zool 17: 435–504
Thrailkill JA, Andrews LS, Claremont RM (2000) Diet of breeding northern goshawks in the Coast Range of Oregon. J Rapt Res 34:339–340

Todd PA (1992) Mountain beaver habitat use and management implications in Yosemite National Park. Nat Areas J 12:26–31
Toweill DE, Maser C (1985) Food of cougars in the Cascade Range of Oregon. Great Bas Nat 45:77–80
USFWS (1998) Point Arena mountain beaver (*Aplodontia rufa nigra* [Rafinesque]) recovery plan. Region 1, U.S. Fish and Wildlife Service, Portland, Oregon
Voth E (1968) Food habits of the Pacific mountain beaver, *Aplodontia rufa pacifica.* PhD Thesis, Oregon State Univ, Corvallis, OR
Wake TA (2006) Archaeological sewellel (*Aplodontia rufa*) remains from Duncan's Point Cave, Sonoma County, California. J Mammal 87:139–147
Witmer GW, deCalesta DS (1986) Resource use by unexploited sympatric bobcats and coyotes in Oregon. Can J Zool 64:2333–2338
Zielinski WJ, Mazurek MJ (2006) The reproductive ecology and home range of the Point Arena mountain beaver (*Aplodontia rufa nigra*). Final Rep USFWS, USDA Forest Serv, Pacific Southwest Res Stat, Arcata, California

Part V
Molecular Ecology and Evolution

24 Molecular Ecology and Evolution

Eileen A. Lacey

The use of molecular technology is now widespread in evolutionary research, with applications for genetic data ranging from quantification of the adaptive significance of specific traits to characterization of the phylogenetic relationships among disparate lineages of organisms. Studies of subterranean rodents are no exception; an ever-increasing number of publications on these animals incorporate molecular information to address questions regarding the causes and consequences of life underground. Given the now integral role that genetic data play in evolutionary research, it seems only appropriate that one section of this volume should focus on studies that use molecular information to expand our understanding of the biology of subterranean species.

Methodologically, the history of genetic studies of subterranean rodents closely tracks the development of molecular tools for evolutionary research. Among the first studies to apply genetic data to subterranean species were detailed analyses of karyotypic and allozymic variation in geomyid pocket gophers (e.g., Patton 1973; Patton and Yang 1977) and spalacine mole-rats (e.g., Wahrman et al. 1969; Nevo et al. 1982). As sequencing of mitochondrial DNA became routine, this technology was applied not only to geomyids and spalacines (e.g., Nevo et al. 1993; Patton and Smith 1994), but also to ctenomyids (e.g., D'Elia et al. 1998; Gimenez et al. 2002) and bathyergids (e.g., Honeycutt et al. 1987; Faulkes et al. 1997). Microsatellite analyses followed, with these markers being used to quantify genotypic variability in geomyids (Steinberg 1999), ctenomyids (e.g., Lacey 2001; Wlasiuk et al. 2003), and bathyergids (Burland et al. 2002, 2004). Now, as we move into the age of genomics and gene expression, analyses of sequence variation at functional nuclear loci are beginning to appear. Studies of multiple nuclear loci have already been completed for spalacine mole-rats (e.g., Hendriks et al. 1987; Hough et al. 2002) and similar work is underway for bathyergids and ctenomyids (e.g., Kundu and Faulkes 2004; Opazo et al. 2005; Cutrera and Lacey, 2006). This expansionist tendency shows no signs of abating; as the genetic toolkit available to researchers continues to grow, the number and diversity of studies of

Eileen A. Lacey: Museum of Vertebrate Zoology and Department of Integrative Biology, University of California, Berkeley, CA 94720, USA, E-mail: ealacey@berkeley.edu

Subterranean Rodents: News from Underground
S. Begall, H. Burda, C.E. Schleich (Eds.)
© Springer-Verlag Berlin Heidelberg 2007

subterranean species that employ molecular procedures only continues to increase.

Conceptually, the application of molecular techniques to studies of subterranean rodents has generated important insights into numerous aspects of evolutionary diversification (reviewed by Steinberg and Patton 2000). For example, surveys of genetic differentiation among geomyid pocket gophers (Patton 1993; Patton and Smith 1990, 1993) have contributed substantially to the understanding of species limits, hybridization, and processes of speciation. At the same time, geographic and environmental variation among spalacine mole-rats (reviewed by Nevo 1999) have been used to explore the roles of selection and local adaptation in shaping genetic diversity. The contrasting interpretations of genetic structure offered by these research programs have, in turn, drawn attention to another fundamental issue, namely the relative contributions of selection vs selectively neutral forces in generating diversity. Efforts to reconcile these viewpoints have contributed to an integrated, more balanced understanding of evolutionary processes that is applicable to a wide range of taxa (Lessa 2000).

As this synopsis suggests, genetic studies of subterranean rodents have had a substantial impact on evolutionary biology. Given the ever-increasing number of subterranean species under study and the ever-expanding conceptual scope of research on these animals, the potential for such analyses to continue generating insights into organismal diversity seems almost unlimited. Molecular data alone, however, are not sufficient; the ability of this technology to inform biologists can only fully be realized by combining analyses of genetic variability with detailed studies of free-living organisms. The power of this integrated approach is particularly apparent in this final section of the volume, which highlights studies that combine long-term field data with molecular analyses to explore new aspects of the biology of subterranean rodents. While some of the organisms are familiar, the questions addressed are novel and have implications for population and evolutionary biology that extend beyond the limits of underground burrows.

In the first chapter in this section, Braude explores how interpretations of the term "inbreeding" have contributed to our understanding and, in some cases, misunderstanding of the biology of subterranean species. Building upon his more than 20 years of research on social bathyergids, Braude contrasts field evidence for reproduction among closely related individuals with molecular estimates of genetic similarity among potential reproductive partners. The result is revealing with regard to both the language of population genetics and the biology of social mole-rats. In particular, Braude's analyses underscore the complexity of the interactions between behavior and demography that shape genetic structure in free-living organisms.

The second chapter in this section examines the effects of behavior and demography on a different aspect of genetic structure, namely selection for variability at Major Histocompatibility Complex (MHC) genes. Again drawing upon long-term field studies, Lacey and Cutrera use inter- and intra-specific comparisons of ctenomyids to determine whether differences in population structure are associated with predictable differences in selection for and variability at Class II MHC loci. These analyses suggest that social structure and demography are important contributors to selection on MHC genes. At the same time, they reveal previously unsuspected differences in variability across MHC loci. This work is among the first to link variation at MHC loci to specific aspects of population structure and, hence, studies of subterranean rodents promise to contribute significantly to our understanding of the effects of demography on selection at functional nuclear genes.

In the third chapter in this section, Van Daele, Faulkes and Verheyen apply molecular data to a somewhat different set of problems – the systematics and phylogeography of the Bathyergidae. Particular attention is given to central African taxa; the recent formal description of *Fukomys* as a distinct genus (Ingram et al. 2004; Kock et al. 2006) has generated considerable interest in members of *Cryptomys-Fukomys* clade, which extends from the Sudan to South Africa. Evidence for both generic- and species-level divergence among members of this clade is reviewed, revealing a surprising degree of differentiation among taxa that were previously all considered to be *Cryptomys*. These forms include species (e.g., *C. hottentotus*, *F. damarensis*) that are the subjects of intensive, long-term behavioral research programs (Bennett and Faulkes 2000). As a result, in addition to generating intriguing new insights into the phylogeographic history of bathyergids, the analyses by Van Daele et al. have important implications for comparative studies of behavioral and other aspects of phenotypic differentiation among these animals.

The final chapter in this section is a brief tour de force by Nevo, who uses his decades-long research program on spalacine mole-rats to illustrate how multiple data sets can be brought to bear upon studies of adaptation to the subterranean niche. With regard to the theme of this section of the volume, studies of spalacines offer some of the most sophisticated genetic analyses to have been applied to subterranean rodents, as exemplified by Nevo's review of the function and expression of the αA-crystallin gene (associated with optic lens structure) in blind mole-rats. At the same time, the paper underscores the integrative nature of Nevo's work by including information drawn from studies of the morphology, physiology, and ecology of spalacine mole-rats. Given Nevo's long and productive career and his seminal role in promoting studies of subterranean species, it is highly appropriate that the volume concludes with this overview, which serves as an exemplar of the breadth and depth of knowledge that can be derived from studies of these animals.

References

Bennett NC, Faulkes CG (2000) African mole-rats: ecology and eusociality. Cambridge Univ Press, Cambridge

Burland TM, Bennett NC, Jarvis JUM, Faulkes CG (2002) Eusociality in African mole-rats: new insights from patterns of genetic relatedness in the Damaraland mole-rat (*Cryptomys damarensis*). Proc R Soc Lond B 269:1025–1030

Burland TM, Bennett NC, Jarvis JUM, Faulkes CG (2004) Colony structure and parentage in wild colonies of cooperatively breeding Damaraland mole-rats suggest incest avoidance alone may not maintain reproductive skew. Mol Ecol 13:2371–2379

Cutrera AP, Lacey EA (2006) Major histocompatibility complex variation in talas tuco-tucos: the influence of demography on selection. J Mammal 87:706–716

D'Elia G, Lessa EP, Cook JA (1998) Geographic structure, gene flow, and maintenance of melanism in *Ctenomys rionegrensis* (Rodentia: Octodontidae). Z Säugetierk 63:285–295

Faulkes CG, Abbott DH, O'Brien HP, Lau L, Roy MR, Wayne RK, Bruford MW (1997) Micro- and macrogeographical genetic structure of colonies of naked mole-rats *Heterocephalus glaber*. Mol Ecol 6:615–628

Gimenez MD, Mirol PM, Bidau CJ, Searle JB (2002) Molecular analysis of populations of *Ctenomys* (Caviomorpha, Rodentia) with high karyotypic variability. Cyt Gen Res 96:130–136

Hendricks W, Luenisson J, Bloemendal J, Nevo E, De Jong WW (1987) The lens protein αA-crystallin in the blind mole-rat *Spalax ehrenbergi*: evolutionary change and functional constraints. Proc Natl Acad Sci USA 84:5320–5324

Honeycutt RL, Edwards SV, Nelson K, Nevo E (1987) Mitochondrial DNA variation and the phylogeny of African mole-rats (Rodentia: Bathyergidae). Syst Zool 36:280–292

Hough RB, Avivi A, Joel A, Davis J, Nevo E, Piatigorsky J (2002) Adaptive evolution of small heat shock protein/αB-crystallin promoter activity of the blind subterranean mole-rat, *Spalax ehrenbergi*. Proc Natl Acad Sci USA 99:8145–8150

Ingram CM, Burda H, Honeycutt RL (2004) Molecular phylogenetics and taxonomy of the African mole-rats, genus *Cryptomys* and the new genus *Coetomys* Gray, 1864. Molec Phyl Evol 31:997–1014

Kock D, Ingram CM, Frabotta LJ, Honeycutt RL, Burda H (2006) On the nomenclature of Bathyergidae and *Fukomys* n. gen. (Mammalia: Rodentia). Zootaxa 51–55

Kundu S, Faulkes CG (2004) Patterns of MHC selection in African mole-rats, family Bathyergidae: the effects of sociality and habitat. Proc Soc Lond B 271:273–278

Lacey EA (2001) Microsatellite variation in solitary and social tuco-tucos: molecular properties and population dynamics. Heredity 86:628–637

Lessa EP (2000) The evolution of subterranean rodents: a synthesis. In: Lacey EA, Patton JL, Cameron GN (eds) Life underground: the biology of subterranean rodents. Univ Chicago Press, Chicago, pp 390–420

Nevo E (1999) Mosaic evolution of subterranean mammals: regression, progression and global convergence. Oxford Univ Press, Oxford

Nevo E, Heth G, Beiles A (1982) Population structure and evolution in subterranean mole-rats. Evolution 36:1283–1289

Nevo E, Honeycutt RL, Yonekawa H, Nelson K, Hanzawa N (1993) Mitochondrial DNA polymorphisms in subterranean mole-rats of the *Spalax ehrenbergi* superspecies in Israel, and its peripheral isolates. Mol Biol Evol 10:590–604

Opazo JC, Palma RE, Melo F, Lessa EP (2005) Adaptive evolution of the insulin gene in caviomorph rodents. Mol Biol Evol 22:1290–1298

Patton JL (1973) Patterns of geographic variation in karyotype in the pocket gopher, *Thomomys bottae* (Eydoux and Gervais). Evolution 26:574–586

Patton JL (1993) Hybridization and hybrid zones in pocket gophers (Rodentia, Geomyidae). In: Harrison RG (ed) Hybrid zones and the evolutionary proccess. Oxford Univ Press. Oxford, pp 290–308

Patton JL, Smith MF (1990) The evolutionary dynamics of the pocket gopher *Thomomys bottae*, with emphasis on California populations. Univ Calif Publ Zool 123:1–161

Patton JL, Smith MF (1993) Molecular evidence for mating asymmetry and female choice in a pocket gopher (*Thomomys*) hybrid zone. Mol Ecol 2:3–8

Patton JL, Smith MF (1994) Paraphyly, polyphyly, and the nature of species boundaries in pocket gophers (genus *Thomomys*). Syst Biol 43:11–26

Patton JL, Yang SY (1977) Genetic variation in *Thomomys bottae* pocket gophers: macrogeographic patterns. Evolution 31:697–720

Steinberg EK (1999) Characterization of polymorphic microsatellites from current and historic populations of North American pocket gophers (genus *Thomomys*). Mol Ecol 8:1075–1076

Steinberg EK, Patton JL (2000) Genetic structure and the geography of speciation in subterranean rodents: opportunities and constraints for evolutionary diversification. In: Lacey EA, Patton JL, Cameron GN (eds) Life underground: the biology of subterranean rodents. Univ Chicago Press, Chicago, pp 301–331

Wahrman J, Goitein R, Nevo E (1969) Mole rat *Spalax*: evolutionary significance of chromosome variation. Science 164:82–84

Wlasiuk G, Garza JC, Lessa EP (2003) Genetic and geographic differentiation in the Rio Negro tuco-tuco (*Ctenomys rionegrensis*): inferring roles of migration and drift from multiple markers. Evolution 57:913–926

25 The Multiple Meanings of Inbreeding: The Key to Understanding the Social and Genetic Structure of Subterranean Rodent Populations

Stanton Braude

25.1 Introduction

Elements of social structure such as mating system or differential dispersal can contribute to genetic structuring of a population. Genetic structure can, in turn, influence the evolution of social structure in a population. While kin selection is the most obvious mechanism for this, lack of dispersal and inbreeding have also been suggested as contributing factors. For example, inbreeding is frequently cited as the explanation for the evolution of eusociality in s based on a number of early assumptions about naked mole-rat natural history, dispersal, and demography. The case of the naked mole-rat illustrates that the term "inbreeding" has a number of biologically and mathematically distinct definitions and is used to describe a range of very different phenomena at the individual and population levels. Examining the differences among these terms allows us to understand how populations of naked mole-rats, and a wide variety of other subterranean mammals, can be 'inbred' in the drift or pedigree sense, at the same time that individuals avoid 'inbreeding' as their system of mating. It is critical to understand these distinct meanings of inbreeding because only system of mating inbreeding would necessarily foster evolution of eusociality.

25.2 Determinants of Genetic Structure: Mutation, Selection, or Drift

The current discussion regarding the role of genetic forces in the evolution of social structure in rodents has its roots in the early years of the evolutionary synthesis. Fisher (1930) and Haldane (1932) focused their attention on mutation and selection as the primary forces behind evolutionary change. Wright (1932), on the other hand, focused on evolution driven by migration and drift in genetically structured populations.

Stanton Braude: International Center for Tropical Ecology at UMSL and Washington University in St. Louis, Box 1229, St. Louis, MO 63130, USA, E-mail: braude@wustl.edu

Subterranean Rodents: News from Underground
S. Begall, H. Burda, C.E. Schleich (Eds.)
© Springer-Verlag Berlin Heidelberg 2007

The concept of genetic structure within populations was essential for Wright's shifting balance theory of evolution. He focused on geographic barriers limiting gene flow and resulting in genetic structure. Consequently, we expect the dispersal ability of a species to have a strong impact on its genetic structure. Alexander's early model for the evolution of a eusocial rodent was fundamentally Fisherian (Alexander et al. 1991; Braude 1998), as is the more recent aridity food distribution model (Lovegrove and Wissel 1988; Bennett and Faulkes 2000). On the other hand, the inbreeding model of Reeve et al. (1990) for naked mole-rat eusociality is based on genetic structure but posits that differential genetic relatedness, rather than genetic drift, mediated the evolution of eusociality in this case.

More recently, Nevo (1999) has suggested that the low genetic variation in fossorial mammals must be the result of selection rather than the result of isolation and drift. He argued that the narrow subterranean niche drives stabilizing selection and loss of variation. This assumption was based on the low allozyme variation found in *Spalax* over much of its range (Nevo and Shaw 1972; Nevo et al. 1982), and has been supported by Penney and Zimmerman (1976), who reported low allozyme variation in *Geomys*.

In contrast, Patton and Yang (1977) argued that the high levels of genetic variation in the fossorial pocket gopher, *Thomomys bottae*, are evidence against Nevo's niche breadth hypothesis. Pocket gophers occupy as narrow an ecological niche as the blind mole-rats but are found over a larger geographic range. Instead, Patton and Yang suggested that low effective population size and restricted gene flow must be responsible for the typically low genetic variation in fossorial rodent populations. The drift hypothesis is further supported by Daly and Patton (1990) who examined 5 allozymes in *Thomomys*, by Zimmerman and Gayden (1981; 16 allozymes in *Geomys*), by Gallardo et al. (1995; 14 allozymes in *Ctenomys*), and by Bennett and Faulkes (2000) who reviewed population genetic studies in bathyergid mole-rats.

There is a variety of processes that lead to population structure and Patton's classic work on Amazonian rodents at 16 paired sites along the Jurua River has shown how these processes lead to identifiable differences in population structure. Population size, history, migration, cycles of habitat expansion and contraction, and social system each result in different forms of population structure and leave telltale patterns in the genetic structure (Patton et al. 1994, 1996; Matocq et al. 2000). These are far more informative than earlier discussions based on greater or lower levels of "genetic variation" (i.e. polymorphism).

25.3 The Influence of Social Structure on Genetic Structure and Genetic Structure on Social Structure

We have come to recognize that social structure as well as geography can influence genetic structure in a wide range of mammalian species (Chepko-Sade and Halpin 1987; Dobson 1998; Dobson et al. 1998; Pope 1998; Storz 1999). Social structure is the non-random association of animals resulting from cooperative interactions, philopatry, or a variety of mating systems. Further, the relatively high reproductive rate, small body size and mortality due to predation all contribute to population structuring in rodents (Shields 1987). Wasser and Jones (1983) pointed out that these characteristics would not be independent of social structure. Small, short-lived species are not expected to disperse far and will therefore be surrounded by close relatives. While the effect of social structure on genetic structure has been well studied, our understanding of the effect of genetic structure on social structure is based mostly on familial relatedness that follows from the predictions of basic kin-selection theory (reviewed in Frank 1998). Chesser (1998) has modeled the effects of genetic structure on social structure at the population level, but it is not clear whether this represents a phenomenon that is distinct from kin-selection and he was careful to specify that he was looking at the effects of family groups compared to broader subpopulations.

25.4 Genetic Structure and Social Structure in Naked Mole-rats: A Classic Example of the Importance of Understanding the Multiple Meanings of Inbreeding

Naked mole-rats are cited as a classic example of genetic structure leading to the evolution of social structure. Naked mole-rats have extremely high reproductive skew for a vertebrate, living in colonies with up to 290 workers and a single queen (Sherman et al. 1995). Although the discovery of eusociality in this species was based on an ecological constraints model (Alexander et al. 1991; Braude 1998), inbreeding is often cited in explaining naked mole-rat social evolution (Honeycutt 1992; Sherman et al. 1992).

The inbreeding hypothesis followed from a number of basic misunderstandings about the natural history of this species (Braude 2000). Because early field work on this species was primarily focused on capturing animals for laboratory study, small colonies were rarely trapped or recognized. In addition, their blind hairless morphology suggested that above ground dispersal would be difficult, if not impossible. These assumptions led to the conclusion that naked mole-rats would not disperse and could not survive in small groups. It was therefore concluded that new colonies must result from fissioning of existing colonies. This would mean that all breeding pairs would be either parent-offspring or sibling pairs and thus naked mole-rats would be extremely inbred and closely related. Since relatedness is central to evolution of cooperation by kin-selection, this would be a tidy example of genetic structure promoting the evolution of eusocial structure.

Much of the misunderstanding began with the first discussion of inbreeding in naked mole-rats which estimated an F statistic but conflated the meanings of three distinct inbreeding coefficients (Reeve et al. 1990). Population structure is quantified by a series of hierarchical F statistics which describe the genetic similarity within and between groups (e.g. F_{ST}, F_{IT}, F_{IS}, F_{LS}, F_{IL}). The algorithms for calculating these F statistics have been borrowed and modified to estimate inbreeding and relatedness, but non-population geneticists are often unclear about the particular algorithm which has been borrowed (Balloux et al. 1998) and this has led to considerable misunderstanding.

Perhaps more important than the estimator of inbreeding from population structure F statistics is the misunderstanding of the multiple meanings of inbreeding, also represented by various F values. Before considering the naked mole-rat example, we must understand a few of the biologically distinct meanings of inbreeding. Templeton and Read (1994, 1998) point out that inbreeding as a measure of shared ancestry (F), inbreeding as a measure of genetic drift ($\bar{F}$), and inbreeding as a measure of the system of mating (f) are often confused because the generic term "inbreeding" is used to describe all of them. Inbreeding as a measure of shared ancestry, F (Pedigree inbreeding) is a characteristic of individuals, not populations. Therefore it cannot measure "genetic diversity" in a population, nor can it measure avoidance of inbreeding because both are population characteristics. Another way to think of this is that the F values can only be greater or equal to zero but avoidance of inbreeding (outbreeding) is reflected by a negative value of f. Inbreeding as a measure of drift ($\bar{F}$) is a population parameter, and measures deviation from Hardy-Weinberg but does not measure "genetic diversity" in a population. Furthermore, $\bar{F}$ cannot measure avoidance of inbreeding because $\bar{F}$ values can only be greater or equal to zero. Inbreeding as a measure of the system of mating (f) is an index of deviation from random mating, and can be greater or less than zero. While f cannot measure "genetic diversity" in the population either, it can measure avoidance of inbreeding. Only inbreeding as the system of mating is relevant to the evolution of eusociality because it implies a mechanism which could enhance the spread of "altruism" genes. Pedigree inbreeding and inbreeding in the drift sense reflect the consequences of events that have already happened and would not necessarily foster the spread of new "altruism" genes.

In the case of the naked mole-rats, Reeve et al. (1990) calculated an F statistic from relatedness estimates. The F statistic is described as "the probability that two alleles randomly drawn from an individual are identical by descent" (Reeve et al. 1990). This would be an estimate of pedigree inbreeding (F) but they had no pedigree data. It is then discussed as a measure of "genetic diversity", and later interpreted as an indication of the system of mating in the species. Their algorithm for estimating inbreeding was used to estimate relatedness from pedigree inbreeding, $r = (1+3F)/2(1+F)$. However, the issue is further confused by substituting average estimated relatedness for a population and solving for F. At best this is an estimate of $\bar{F}$.

Reeve et al. (1990) also compared M13 and Jeffrey's Probe fingerprint band sharing in naked mole-rats with 14 other species from house sparrows to domestic dogs. They noted that within colonies naked mole-rats share a greater proportion of bands than any of the other species and conclude that close pedigree inbreeding must be typical, and extreme, for naked mole-rats. However, cross species comparisons of DNA fingerprint estimates of band sharing similarities are not a strong indicator of pedigree inbreeding regardless of the loci. While there is actually no direct evidence for pedigree inbreeding in wild populations of naked mole-rats, pedigree inbreeding has been reported in a wide variety of non-eusocial species including feral horses (Berger 1987), wolves (Packard et al. 1985; Mech 1987), dwarf mongoose (Rood 1987), prairie dogs (Chesser 1983; Halpin 1987), pikas (Smith 1987), and killer whales (*Orcinus orca*) (Hoelzel et al. 1998). Ralls et al. (1986) also report pedigree inbreeding in 1.2% to 4.4% of matings in wild populations of yellow bellied marmot, beaver, chimpanzee, mountain gorilla, and lion. In fact, Shields (1982) suggested that natal philopatry across mammals typically leads to some pedigree inbreeding. Hence, even if we find that naked mole-rats occasionally mate with close relatives, it would not be unusual for a mammal and it does not necessarily follow that this would lead to the evolution of eusociality.

Confusion among the various meanings of inbreeding along with a lack of basic natural history data on the species led to the suggestion that inbreeding was the system of mating in naked mole-rats. However, the roughly estimated $\bar{F}$ reported for the wild population of Reeve et al. (1990) actually suggests a bottleneck event. In fact they also note that their data indicate that the study population went through a bottleneck event and descends from a single female. In addition, the vast majority of naked mole-rat samples used in published genetic studies (and many of the captive naked mole-rat colonies in zoos and research institutions) have come from the same recent bottleneck population south of the Athi River in Kenya (Jarvis 1985).

Drift has likely had a major impact on the small founder population that crossed the Athi river. Inbreeding due to drift should also play a role in older populations of naked mole-rats that are geographically isolated. On the other hand, Wasser (1987) suggested that in species with long lived breeders (such as naked mole-rats), dispersers are forced to move long distances, thus increasing the population effective size ($Ne_{migration}$) and reducing the effect of drift on population structure. We expect this result in naked mole-rat populations in which queens can live more than 20 years and dispersers move great distances (> 2 km). This combination of forces may help explain the complex pattern of genetic structure observed in naked mole-rats across East Africa (Faulkes et al. 1997).

It was not until recently that a comprehensive population genetic analysis was completed for naked mole-rats (Hess 2004). Hess genotyped 1144 animals at seven variable microsatellite loci. His extensive sampling included animals from 126 colonies of naked mole-rats over an 870 km^2 area. Once he took historic population subdivision by rivers into account, he estimated an average within group relatedness of 0.406. He suggests that this low value could result from the fact that breeding females can have more than one mate who may be unrelated males. This is also consistent with his finding of significantly negative F_{IS} values which suggest avoidance of inbreeding as the system of mating. Ciszek (2000) also found that captive naked mole-rats avoid inbreeding when given a choice of mates.

25.5 Dispersal

In their discussion of small rodent populations, Lidicker and Patton (1987) identified three different types of dispersal: seasonal, ontogenetic, and saturation. These will have different effects on gene flow and resulting population structure because seasonal and saturation dispersal may not be related to reproduction. In naked mole-rats we expect genetic structure to indicate isolation by distance and isolation by major dispersal barriers such as rivers. We do not expect colonies or populations to remain isolated due to ecologically impossible dispersal, as has previously been assumed. Naked mole-rat dispersal may be seasonal with dispersers moving above ground at the end of both the long and the short rainy seasons. It is well established that dispersal in naked mole-rats is also ontogenetic since dispersers must develop the appropriate morphology before leaving the burrow (O'Riain et al. 1996; O'Riain and Braude 2001). At this point, we do not have the data to determine whether dispersal is also correlated with habitat saturation.

25.6 Conclusions

It is clear that social structure can influence genetic structure and that genetic relatedness influences social structure in rodents, as well as in other mammals. An inbreeding system of mating, however, does not appear to contribute to this relationship in naked mole-rats. In fact, as Craig (1982) pointed out, since pedigree inbreeding would increase relatedness between parents and offspring just as much as between siblings, we should expect it to have little effect on the evolution of social behavior. Strong evidence of outbreeding as the system of mating ($f < 0$) in another eusocial bathyergid, *Fukomys damarensis*, suggests that we must look elsewhere to understand eusociality in fossorial mammals. A full understanding of the different meanings and uses of the terms "inbreeding" and "dispersal" will be necessary for the development of any meaningful model of the interaction between genetic and social structure in bathyergids and rodents in general.

Acknowledgements. I am grateful to Sabine Begall for inviting me to the subterranean rodent symposium at the Sapporo IMC. Thanks also to Bobbi Low for her valuable feedback and to Eileen Lacey for comments on an early draft of this manuscript. This work was supported by the St. Louis Zoo's FRC program.

References

Alexander R, Noonan K, Crespi BJ (1991) The evolution of eusociality. In: Sherman P, Jarvis J, Alexander RD (eds) The biology of the naked mole-rat. Princeton, Princeton, pp 3-44

Balloux F, Goudet J, Perrin N (1998) Breeding system and genetic variance in the monogamous, semi-social shrew, *Crocidura russula*. Evolution 52:1230–1235

Bennett N, Faulkes C (2000) African mole-rats: ecology and eusociality. Cambridge Univ Press, Cambridge, p 273

Berger J (1987) Reproductive fates of dispersers in a harem-dwelling ungulate: the wild horse. In: Chepko-Sade BD, Halpin Z (eds) Mammalian dispersal patterns: the effects of social structure on population genetics. Univ Chicago Press, Chicago, pp 41–54

Braude S (1998) The predictive power of evolutionary biology and the discovery of eusociality in the naked mole-rat. NCSE Reports 17(4):12–15

Braude S (2000) Dispersal and new colony formation in wild naked mole-rats: evidence against inbreeding as the system of mating. Behav Ecol 11:7–12

Chepko-Sade BD, Halpin Z (1987) Mammalian dispersal patterns: the effects of social structure on population genetics. Univ Chicago Press, Chicago

Chesser R (1983) Genetic variability within and among populations of the black tailed prairie dog. Evolution 37:320–331

Chesser R (1998) Relativity of behavioral interactions in socially structured populations. J Mammal 79:713–724
Ciszek D (2000) New colony formation in the "highly inbred" eusocial naked mole-rat: outbreeding is preferred. Behav Ecol 11:1–6
Craig R (1982) Evolution of eusociality by kin selection: the effect of inbreeding between siblings. J Theor Biol 94:119–128
Daly J, Patton JL (1990) Dispersal, gene flow, and allelic diversity between local populations of *Thomomys bottae* pocket gophers in the coastal ranges of California. Evolution 44: 1283–1294
Dobson S (1998) Social structure and gene dynamics in mammals. J Mammal 79:667–670
Dobson S, Chesser R, Hoogland J, Sugg D, Foltz D (1998) Breeding groups and gene dynamics in a socially structured population of prairie dogs. J Mammal 79:671–680
Faulkes C, Abbott D, O'Brien H, Lau L, Roy M, Wayne R, Bruford M (1997) Micro- and macrogeographical genetic structure of colonies of naked mole-rats *Heterocephalus glaber*. Mol Ecol 6:615–628
Fisher RA (1930) [revised edition 1958] The genetical theory of natural selection. Dover, New York
Frank S (1998) Foundations of social evolution. Princeton Univ Press, Princeton
Gallardo M, Köhler N, Araneda C (1995) Bottleneck effects in local populations of fossorial *Ctenomys* (Rodentia, Ctenomyidae) affected by vulcanism. Heredity 74:638–646
Haldane J (1932) The causes of evolution. Longmans Green, London
Halpin Z (1987) Natal dispersal and formation of new social groups in a newly established town of black tailed prairie dogs (*Cynomys ludovicianus*). In: Chepko-Sade BD, Halpin Z (eds) Mammalian dispersal patterns: the effects of social structure on population genetics. Univ Chicago Press, Chicago
Hess J (2004) A population genetic study of the eusocial naked mole-rat (*Heterocephalus glaber*). PhD Dissertation, Washington Univ, St. Louis
Hoelzel A, Dahlheim M, Stern S (1998) Low genetic variation among killer whales (*Orcinus orca*) in the eastern North Pacific and genetic differentiation between foraging specialists. J Hered 89:121–128
Honeycutt RL (1992) Naked mole-rats. Am Sci 80:43–53
Jarvis JUM (1985) Ecological studies on *Heterocephalus glaber*, the naked mole rat, in Kenya. Natl Geog Soc Res Rep 20:429–437
Lidicker WZ, Patton JL (1987) Patterns of dispersal and genetic structure in populations of small rodents. In: Chepko-Sade BD, Halpin Z (eds) Mammalian dispersal patterns: the effects of social structure on population genetics. Univ Chicago Press, Chicago
Lovegrove BG, Wissel C (1988) Sociality in mole-rats: metabolic scaling and the role of risk sensitivity. Oecology 74:600–606
Matocq M, Patton J, DaSilva M (2000) Population genetic structure of two ecologically distinct Amazonian spiny rats: separating history and current ecology. Evolution 54: 1423–1432
Mech LD (1987) Age, season, distance, direction and social aspects of wolf dispersal from a Minnesota pack. In: Chepko-Sade BD, Halpin Z (eds) Mammalian dispersal patterns: the effects of social structure on population genetics. Univ Chicago Press, Chicago
Nevo E (1999) Mosaic evolution of subterranean mammals: regression, progression and global convergence. Oxford Univ Press, Oxford
Nevo E, Shaw C (1972) Genetic variation in a subterranean mammal, *Spalax ehrenbergi*. Biochem Genet 7:235–241

Nevo E, Heth G, Beiles A (1982) Population structure and evolution in subterranean mole-rats. Evolution 36:1283–1289
O'Riain MJ, Braude S (2001) Inbreeding versus outbreeding in captive and wild populations of naked mole-rats. In: Clobert J, Danchin E, Dhondt A, Nichols J (eds) Dispersal. Oxford Univ Press, Oxford
O'Riain MJ, Jarvis JUM, Faulkes C (1996) A dispersive morph in the naked mole-rat. Nature 380:619–621
Packard J, Seal U, Mech LD (1985) Social influences on reproduction in wolves. In: Carbyn LN (ed) Wolves in Canada and Alaska. Canadian Wild Serv Rep Ser 45, Ottawa, pp 78–85
Patton J, Yang S (1977) Genetic variation in *Thomomys bottae* pocket gophers: macrogeographic patterns. Evolution 31:697–720
Patton J, Da Silva M, Malcolm J (1994) Gene genealogy and differentiation among arborial spiny rats (Rodentia: Echimyidae) of the Amazon basin: a test of the riverine barrier hypothesis. Evolution 48:1314–1323
Patton J, Da Silva M, Malcolm J (1996) Hierarchical genetic structure and gene flow in three sympatric species of Amazonian rodents. Mol Ecol 5:229–238
Penney D, Zimmerman E (1976) Genic divergence and local population differentiation by random drift in the pocket gopher genus *Geomys*. Evolution 30:473–483
Pope T (1998) Effects of demographic change on group kin structure and gene dynamics of populations of red howling monkeys. J Mammal 79:692–712
Ralls K, Harvey P, Lyles A (1986) Inbreeding in natural populations of birds and mammals. In: Soule M (ed) Conservation biology. Sinauer, Sunderland
Reeve HK, Westneat D, Noon W, Sherman P, Aquadro C (1990) DNA "fingerprinting" reveals high levels of inbreeding in colonies of the eusocial naked mole-rat. Proc Natl Acad Sci USA 87:2496–2500
Rood JP (1987) Dispersal and intergroup transfer in the dwarf mongoose. In: Chepko-Sade BD, Halpin Z (eds) Mammalian dispersal patterns: the effects of social structure on population genetics. Univ Chicago Press, Chicago
Sherman PW, Jarvis JUM, Braude S (1992) Naked mole-rats. Sci Am 267:72–78
Sherman PW, Lacey EA, Reeve HK, Keller L (1995) The eusociality continuum. Behav Ecol 6:102–108
Shields W (1982) Philopatry, inbreeding and the evolution of sex. State Univ New York Press, Albany
Shields W (1987) Dispersal and mating systems: investigating their causal connections. In: Chepko-Sade BD, Halpin Z (eds) Mammalian dispersal patterns: the effects of social structure on population genetics. Univ Chicago Press, Chicago
Smith A (1987) Population structure of pikas: dispersal versus philopatry. In: Chepko-Sade BD, Halpin Z (eds) Mammalian dispersal patterns: the effects of social structure on population genetics. Univ Chicago Press, Chicago
Storz J (1999) Genetic consequences of mammalian social structure. J Mammal 80:553–569
Templeton AR, Read B (1994) Inbreeding: one word, several meanings, much confusion. In: Loeschcke V, Tomiuk J, Jain S (eds) Conservation genetics. Birkhauser, Basel
Templeton AR, Read B (1998) Elimination of inbreeding depression from a captive population of Speke's gazelle: validity of the original statistical analysis and confirmation by permutation testing. Zoo Biol 17:77–94
Wasser P (1987) A model predicting dispersal distance distributions. In: Chepko-Sade BD, Halpin Z (eds) Mammalian dispersal patterns: the effects of social structure on population genetics. Univ Chicago Press, Chicago

Wasser P, Jones A (1983) Natal philopatry among solitary mammals. Quart Rev Biol 50:355–390

Wright S (1932) The roles of mutation, inbreeding, crossbreeding, and selection in evolution. Proc 6th Int Cong Genetics, vol 1, pp 356–366

Zimmerman E, Gayden N (1981) Analysis of genic heterogeneity among local populations of the pocket gopher, *Geomys bursarius*. In: Smith M, Joule J (eds) Mammalian population genetics. Univ Georgia Press, Athens, pp 273–287

26 Behavior, Demography, and Immunogenetic Variation: New Insights from Subterranean Rodents

Eileen A. Lacey, Ana Paula Cutrera

26.1 Introduction

Identifying the factors that shape genetic variation is a fundamental goal of evolutionary research (Lewontin 1974; Gillespie 1991; Ford 2002). Although the effects of demography on neutral genetic variation have been characterized for numerous species (reviewed by Endler 1992; Avise 2004), relationships among demography, selection, and variability at functional loci remain largely unstudied (Meyers and Bull 2002). The demographic structure of a population is expected to influence the nature and intensity of selection (Wright 1949) and, hence, determining how demography and selection interact represents a critical step toward understanding genetic structure at functional portions of the genome.

Subterranean rodents have played a prominent role in studies of neutral genetic structure in natural populations of vertebrates (Honeycutt et al. 1991; Steinberg and Patton 2000). As the scope of these analyses has expanded to include variation at Major Histocompatibility Complex (MHC) genes, studies of subterranean rodents have also begun to generate important insights into determinants of genetic structure at functional loci (Nevo 1999; Hambuch and Lacey 2002; Kundu and Faulkes 2004; Cutrera and Lacey, 2006). Here, we review analyses of immunogenetic variation in these animals, with emphasis upon the distinctive attributes of subterranean rodents and MHC loci that make these systems particularly appropriate for elucidating relationships among demography, selection, and genetic structure. Because much of the information on MHC diversity is drawn from our ongoing studies of ctenomyid rodents, we pay particular attention to this lineage and to the role that it is playing in efforts to understand variability at these genes. We conclude by suggesting several directions for future

Eileen A. Lacey: Museum of Vertebrate Zoology and Department of Integrative Biology, University of California, Berkeley, CA 94720, USA, E-mail: ealacey@berkeley.edu

Ana Paula Cutrera: Laboratorio de Ecofisiología, Facultad de Ciencias Exactas y Naturales, Universidad Nacional de Mar del Plata, CC 1245 Mar del Plata, Argentina

Subterranean Rodents: News from Underground
S. Begall, H. Burda, C.E. Schleich (Eds.)
© Springer-Verlag Berlin Heidelberg 2007

research that we hope will stimulate further investigation of interactions between demography and immunogenetic structure in subterranean and other rodent species.

26.2 Genetic Structure in Subterranean Rodents

Historically, studies of subterranean rodents have been instrumental in revealing patterns and processes of genetic diversification (reviewed by Honeycutt et al. 1991; Nevo 1999; Steinberg and Patton 2000). Several factors have contributed to the popularity of these animals as subjects of genetic research, including their tendency toward limited individual vagility and patchily distributed local populations (Nevo 1979; Reig et al. 1990; Busch et al. 2000). These attributes are expected to increase the extent to which genetic variation is apportioned among vs within populations (Steinberg and Patton 2000), thereby enhancing opportunities to explore the interplay between drift and gene flow. Accordingly, numerous studies have examined genetic structure in subterranean rodents, with an emphasis upon the geographic structure of genetic variation in geomyid pocket gophers (reviewed by Patton 1990; Smith 1998) and spalacine mole-rats (reviewed by Nevo 1999).

Among ctenomyids, genetic variation has been characterized for at least 11 species (Table 26.1). These studies have explored aspects of genetic differentiation ranging from phylogenetic relationships and rates of intrageneric diversification (e.g. Cook and Lessa 1998; Lessa and Cook 1998; Slamovits et al. 2001; Honeycutt et al. 2003; Castillo et al. 2005) to the impact of specific environmental events on local genotypic variability (Gallardo et al. 1995). Accordingly, the spatial scales over which genetic differentiation has been assessed vary from taxon-wide or regional analyses (e.g., Castillo et al. 2005; Tomasco and Lessa, in press) to analyses of variation within single populations of conspecifics (e.g., Gallardo and Köhler 1994; Lacey 2001). While these studies reveal a general tendency toward strong spatial structuring of genetic diversity within ctenomyids, the details of this structure clearly differ among species, suggesting that comparative analyses of these animals can be used to generate important insights into relationships between observed patterns of genetic variability and the underlying processes of genotypic diversification. Given that genetic variation has been characterized for only ca. 20% of the 60 currently recognized species of ctenomyids (Woods and Kilpatrick 2005), this family of subterranean rodents appears to hold considerable potential for studies aimed at improving understanding of the determinants of genetic structure.

Table 26.1. Studies of genetic structure in ctenomyid rodents. Analyses of the molecular systematics of these animals are not included because such studies do not typically consider intraspecific genetic variation. Because studies of immunogenetic variation typically report nucleotide or allelic variation at MHC loci, only studies employing molecular markers that capture variation in DNA sequences or gene products (e.g., allozymes) are considered here; for this reason, the extensive literature on chromosomal variation in ctenomyids is not included in this review

Species	Genetic marker(s)	Research question(s)	References
Multiple	Allozymes	Comparative levels of variability	Sage et al. (1986)
Multiple	Allozymes	Comparative levels of variability, patterns of genetic structure	Gallardo and Palma (1992)
Multiple	mtDNA	Evolutionary processes underlying chromosomal differentiation	Gimenez et al. (2002)
Multiple	mtDNA, nuclear introns	Evolutionary history of genus *Ctenomys*	Castillo et al. (2005)
C. australis	Allozymes	Levels of variability, local genetic structure	Apfelbaum et al. (1991)
C. australis	mtDNA	Phylogeography, impact of historical events on genetic structure	Mora et al., 2006
C. flamarioni	Microsatellites	Population structure, microgeographic processes of genetic differentiation	Fernández et al., in press
C. haigi	Microsatellites	Population structure, impact of behavior and demography on genetic variation	Lacey (2001)
C. lami	Microsatellites	Population genetic structure, impact of demography on genetic variation	El Jundi and de Freitas (2004)
C. maulinus	Allozymes	Levels of variability, genetic structure	Gallardo and Köhler (1992)
C. maulinus	Allozymes	Effects of recent bottleneck on genetic variability	Gallardo and Köhler (1994); Gallardo et al. (1995)
C. minutus	Microsatellites	Microgeographic genetic structure, hybrid zone dynamics	Gava and de Freitas (2004)
C. pearsoni	mtDNA	Phylogeography, evolutionary processes underlying chromosomal differentiation	Tomasco (2003); Tomasco and Lessa, in press
C. porteousi	Allozymes	Levels of variability	Apfelbaum et al. (1991)
C. porteousi	Allozymes	Genetic variability, population structure	Massarini et al. (1992)
C. rionegrensis	Allozymes, mtDNA, microsatellites	Evolutionary processes underlying maintenance of pelage color dimorphism	D'Elia et al. (1998); Wlasiuk et al. (2003); Lessa et al. (2005)

Table 26.1. (continued)

Species	Genetic marker(s)	Research question(s)	References
C. sociabilis	Microsatellites	Population structure, impact of behavior and demography on genetic variation	Lacey (2001)
C. sociabilis	Microsatellites, MHC exon	Comparative genetic structure at functional and non-functional loci	Hambuch (2000); Hambuch and Lacey (2002)
C. sociabilis	mtDNA	Historical genetic diversity, genetic response to climate change	Hadly et al. (2003); Chan et al. (2005)
C. talarum	Microsatellites	Genetic structure in relation to intraspecific variation in demography	Cutrera et al. (2005, 2006)
C. talarum	Microsatellites, MHC intron and exons	Genetic structure at functional and non-functional loci in relation to intraspecfic variation in demography	Cutrera and Lacey, 2006

26.2.1 Neutrality vs Selection in Genetic Differentiation

One point of contention that has emerged from studies of genetic structure in subterranean rodents concerns the importance of natural selection vs selectively neutral processes (e.g., drift) in shaping genetic variation (Lessa 2000). For example, studies of spalacine mole-rats have emphasized selective pressures imposed by the subterranean niche as determinants of genetic structure (Nevo et al. 1974; Nevo 1999). In contrast, studies of geomyids have focused on the role of neutral evolutionary processes in promoting genetic diversity (Patton and Yang 1977; Patton and Smith 1990). We concur with Lessa (2000) that it is the *balance* of selective and neutral processes acting on genotypes that determines patterns of variability. Accordingly, our research program explicitly compares variation at functional and presumptively non-functional (neutral) loci in order to assess the relative contributions of these processes to genetic structure in subterranean rodents.

26.2.2
Genetic Structure at Major Histocompatibility Complex (MHC) Loci

The genes of the Major Histocompatibility Complex (MHC) provide a particularly appropriate system for exploring the impact of selection on genetic structure. In addition to being highly variable, these genes are often subject to strong selection (She et al. 1990; Edwards and Hedrick 1998) and, hence, they offer an important contrast to presumably non-functional loci. The primary selective forces thought to operate on MHC loci are pathogen exposure and inbreeding depression (reviewed by Penn and Potts 1999; Bernatchez and Landry 2003). Both selective pressures are expected to vary as a function of demography (Anderson and May 1979; Coté and Poulin 1995; Arneberg 2002; Solomon 2003; Matocq and Lacey 2004), suggesting that population structure should be an important determinant of genetic structure at MHC loci (Piertney and Oliver 2006; Cutrera and Lacey, 2006). While numerous studies - including multiple studies of subterranean rodents - have examined interactions between demography and patterns of neutral genetic structure (Storz et al. 2002; Matocq and Lacey 2004; El Jundi and Freitas 2004; Cutrera et al. 2005, 2006), the effects of demography on selection at MHC loci remain largely unexplored.

26.3
MHC Variation in Subterranean Rodents

The first studies to examine MHC diversity in subterranean rodents were conducted on the blind mole-rat, *Spalax ehrenbergi* (Nevo and Klein 1990; reviewed in Nevo 1999). In a methodologically diverse series of analyses, Nevo and colleagues characterized the structure of the Major Histocompatibility Complex in *S. ehrenbergi* (Ben-Shlomo et al. 1988), quantified variability at multiple MHC loci (Nizetic et al. 1985), and examined evolutionary relationships among MHC alleles from diverse mammalian species (Schopfer et al. 1987). Particularly relevant to this review is a multi-population survey of *S. ehrenbergi* (Nevo and Beiles 1992), which revealed that parasite infectivity is positively associated with heterozygosity at two Class II MHC loci. While the authors of this study emphasize the role of environmental variation (ambient humidity and temperature) in driving this relationship (Nevo and Beiles 1992), Nevo (1999) indicates that population structure and dynamics are also likely to be important in shaping variation at MHC loci.

Demography may also contribute to MHC variation in the Balkan mole-rat, *S. leucodon*. According to Nevo (1999), low levels of variability at Class I and II genes in this species (Nizetic et al. 1988) reflect the impact of population dynamics (e.g., founder effects, bottlenecks) and ecological conditions on MHC polymorphism and heterozygosity. Expanding the taxonomic scope of these analyses, Kundu and Faulkes (2004) explored the effects of habitat type and social structure on immunogenetic variation in four species of bathyergid mole-rats. Although these authors report a positive relationship between sociality and risk of parasite transmission, the intensity of selection on MHC loci in the two most social species examined (*Heterocephalus glaber* and *Fukomys damarensis*) was relatively weak, leading Kundu and Faulkes (2004) to conclude that social structure is not the only demographic parameter that influences immunogenetic variation in these animals. Collectively, these analyses suggest that further studies of subterranean rodents should yield significant new insights into the role of demography in shaping genetic structure at functional loci.

To this end, we are exploring interactions among demography, selection, and MHC variation in three species of ctenomyids. These analyses were prompted by data indicating that the variable demographic structures of these animals are associated with pronounced differences in patterns of neutral genetic variation (Lacey 2001; Cutrera et al. 2005, 2006). As a logical conceptual extension of these studies, we are characterizing MHC variation in these species in order to assess the impact of demography on selection at these genes. Here, we provide a brief overview of these studies and the insights that they have generated into selection on MHC loci in ctenomyids:

(1) Interspecific differences in social structure and selection on MHC genes

Differences in social structure, in particular the tendency for members of a species to live alone or in groups, may significantly impact patterns of selection on MHC loci. Among mammals, sociality (defined as group living; Alexander 1974) typically arises due to natal philopatry. As a result, sociality has important implications for a number of demographic parameters, including movement of individuals and their genes among populations, as well the spatial distribution of close kin within populations. While these demographic correlates of group living are expected to impact variation at neutral as well as functional loci, sociality may also have effects that are specific to MHC genes. In particular, increased exposure to pathogens is thought to represent an unavoidable cost of sociality that arises due to the extensive contact among conspecifics associated with group living (Alexander 1974). Given the apparent functional link between pathogen exposure and immunogenetic variation (Nevo and Beiles 1992; Wegner et al. 2003), this aspect of sociality leads to the prediction that, all else being equal, selection for variation at MHC loci should be greater in social versus solitary species.

The ctenomyids of the Limay Valley region of southwestern Argentina provide an important opportunity to test this prediction. The western Limay Valley and surrounding hills are occupied by the colonial tuco-tuco (*Ctenomys sociabilis*), which is the only species of ctenomyid that has been demonstrated to be group living (Lacey et al. 1997). In contrast, the eastern Limay Valley and adjacent areas are home to the Patagonian tuco-tuco (*C. haigi*), which is solitary (Lacey et al. 1998). Although these species are not sister taxa (Chan et al. 2005; Lacey, unpublished data), they occur in effectively identical habitats that are separated only by the width of the Río Limay (ca 300 m; Lacey and Wieczorek 2004). Given their close physical proximity and apparent ecological similarity, the striking differences in social structure between *C. sociabilis* and *C. haigi* represent an ideal system for exploring relationships among behavior, demography, and selection on MHC loci.

Genetic analyses of the population of each species that has been monitored since 1992 have revealed that microsatellite variation is significantly reduced in *C. sociabilis* relative to *C. haigi* (Lacey 2001). In marked contrast, however, allelic variation and heterozygosity at the MHC Class II DQB locus are comparable in these populations (Hambuch 2000; Hambuch and Lacey 2002; Fig. 26.3). Although population-level evidence of departure from neutrality at the DQB locus is limited to *C. sociabilis*, the results of molecular-level analyses are consistent with selection acting on this gene in both species (Hambuch and Lacey 2002). Estimates of the intensity of selection (corrected for differences in effective population size) indicate that this parameter is consistently greater for *C. sociabilis* (Hambuch and Lacey 2002; Table 26.2), suggesting that behavioral and demographic differences between the study populations influence selection on this locus. Work currently in progress is expanding these analyses to include multiple populations per species as well as multiple MHC Class II loci. Collectively, these studies will yield a more comprehensive picture of the effects of social structure on patterns and processes of diversification at MHC genes.

(2) Intraspecific differences in demography and selection on MHC genes

Studies that exploit intraspecific variation in population structure provide a particularly powerful means of assessing the effects of demographic variables on selection at MHC loci. Long-term field studies of the talas tuco-tuco (*C. talarum*) by Cristina Busch and colleagues have revealed pronounced differences in demography and behavior between populations of this species at Mar de Cobo and Necochea, Buenos Aires Province, Argentina. For example, population density at Mar de Cobo is significantly greater than that at Necochea (Busch et al. 1989; Malizia et al. 1991, 1995; Cutrera and Zenuto, unpublished data); if pathogen exposure and trans-

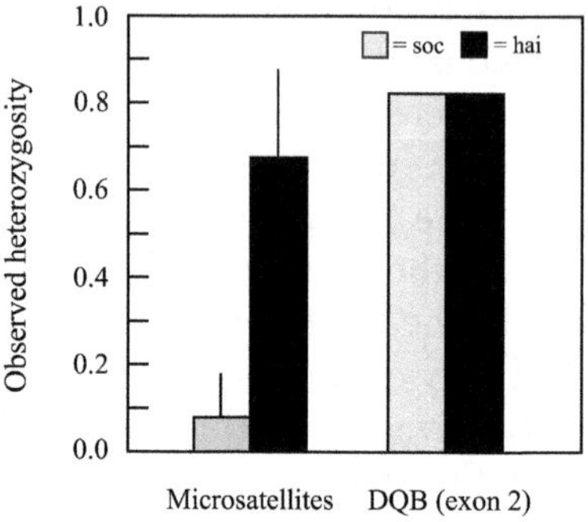

Fig. 26.1. Comparison of observed heterozygosity at microsatellite and MHC loci in the group-living colonial tuco-tuco (*C. sociabilis*) and the solitary Patagonian tuco-tuco (*C. haigi*). Data are based on genetic variation detected for 35 adults per species; all conspecifics were members of the same local population located in the Limay Valley of southwestern Argentina. Mean (±1 SD) heterozygosity is presented for 15 microsatellite loci screened by Lacey (2001). Data on observed heterozygosity at exon 2 of the MHC Class II DQB locus are from Hambuch and Lacey (2002)

mission increase with density (Anderson and May 1978; May and Anderson 1979), this interpopulation variation may translate into differences in the intensity of selection for immunogenetic variability. At the same time, the coefficient of inbreeding among potential reproductive partners is greater for Mar de Cobo (Cutrera et al. 2005), suggesting that if inbreeding is detrimental, selection for MHC diversity as a mechanism of kin recognition may also vary between populations. In short, the well-documented demographic differences between these populations lead us to predict that selection for immunogenetic variability should be greater at Mar de Cobo, which should lead to greater MHC diversity in this population relative to the one at Necochea.

Table 26.2. Estimated intensities of selection (s) on MHC Class II loci from three species of ctenomyids

Species	Locus	d_N/d_S	$S_{Ne\ (\%)}$	Reference
C. sociabilis	DQB	1.97	0.04	Hambuch and Lacey (2002)
C. haigi	DQB	1.68	0.001	Hambuch and Lacey (2002)
C. talarum				
Mar de Cobo	DRB	1.68	0.60	Cutrera and Lacey, 2006
Necochea	DRB	0.50	0.04	Cutrera and Lacey, 2006
C. talarum				
Mar de Cobo	DQA	0.64	0.06	Cutrera and Lacey, 2006
Necochea	DQA	0.54	0.03	Cutrera and Lacey, 2006

DQB = 231 bp of exon 2 of the DQB locus, *DRB* = 273 bp of exon 2 of the DRB locus, and *DQA* = 177 bp of exon 2 of the DQA locus. Estimates of *s* were generated according to Satta et al. (1994). d_N/d_S values were obtained from the references cited in the table and reflect rates of non-synonymous to synonymous changes over the entire length of each exon examined. Estimates of *s* were corrected for differences in effective population size by dividing these values by $2N_e$ (Satta et al. 1994)

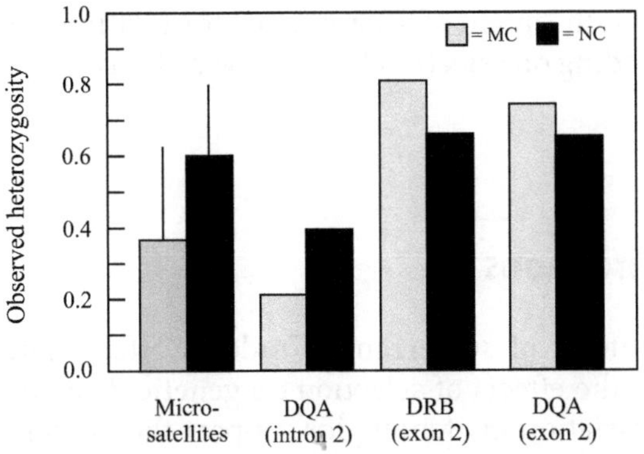

Fig. 26.2. Comparison of observed heterozygosity at microsatellite and MHC loci in two demographically distinct populations of the talar tuco-tuco (*C. talarum*). Data are based on genetic variation detected for 30 adults per population for populations of *C. talarum* located at Mar de Cobo and Necochea, Buenos Aires Province, Argentina. Mean (±1 SD) heterozygosity is presented for eight microsatellite loci screened by Cutrera et al. (2006). Data on observed heterozygosity at exon 2 of the MHC Class II DRB locus, exon 2 of the MHC Class II DQA locus, and intron 2 of the MHC Class II DQA locus are from Cutrera and Lacey (2006). Redrawn with permission from Cutrera and Lacey (2006)

Comparisons of levels of allelism and heterozygosity at the MHC Class II DRB and DQA loci have revealed that variability at these genes is greater among animals from Mar de Cobo (Cutrera and Lacey, 2006; Fig. 26.3). In contrast, microsatellite variation and variation at the apparently non-functional DQA intron 2 are greater for Necochea (Cutrera and Lacey, 2006; Fig. 26.3), suggesting that the forces maintaining variability at MHC exons differ from those acting on presumably neutral genetic markers. For DQA, significant population-level evidence of selection occurs in both study populations; in contrast, for DRB, significant evidence of selection is limited to molecular-level analyses of data from Mar de Cobo (Cutrera and Lacey, 2006). Estimates of the intensity of selection (corrected for differences in effective population size) are consistently greater for Mar de Cobo (Cutrera and Lacey, 2006; Table 26.2), as predicted based on the greater density and potential for inbreeding in this population. At the same time, interpopulation differences in the intensity of selection are more pronounced at the DRB locus (Cutrera and Lacey, 2006), raising intriguing questions concerning potential functional differences between the MHC genes surveyed. Thus, while comparisons of the Mar de Cobo and Necochea populations support the hypothesis that demography is an important determinant of selection on MHC genes, these analyses suggest that the impact of demographic variables may differ among loci in this gene complex. This work is currently being expanded to include multiple populations of *C. talarum*, with the intent of teasing apart the relative contributions of density vs inbreeding on selection for MHC variability in this species.

26.4 Overview and Future Directions

As this synopsis suggests, studies of subterranean rodents offer multiple opportunities to examine the effects of selection on genetic diversity. In general, studies of MHC variation in ctenomyids support the expectation that demography is an important determinant of selection on these genes. Despite the insights offered by these analyses, however, we have barely scratched the surface of this aspect of population genetics. Although studies of MHC variation in natural populations of vertebrates are increasing rapidly, few have incorporated the data necessary to relate evidence of selection to specific pressures such as pathogen exposure or inbreeding depression. Even fewer have attempted to link those selective pressures to relevant demographic parameters such as population density or kin structure. Thus, despite compelling conceptual models and clear, testable predictions, the role of population structure in shaping selection on MHC loci remains largely unexplored. Future studies of MHC genes will benefit from incorporating multiple loci (functional and non-functional), multiple populations of conspecifics, and multiple statistical estimators of selection to explore interactions among demography, selection, and genetic structure.

Future studies of subterranean rodents should prove especially useful in furthering our understanding of the factors contributing to selection on MHC loci. In particular, analyses of other ctenomyids will add to the growing comparative picture of genetic structure in this lineage while at the same time facilitating studies of MHC evolution over larger geographic distances and deeper time scales. Concomitantly, analyses of MHC variation in other, independently evolved examples of life underground may yield important insights into the role that the shared subterranean *habitus* plays in generating selection on these loci. In short, given their long history as subjects of population genetic research and the existing foundation of studies of MHC variability in these animals, we expect that subterranean rodents will play a significant role in the continuing effort to untangle relationships among demography, selection, and variation at functional genes.

Acknowledgements. We particularly wish to thank Tina Hambuch for her pioneering studies of MHC variation in ctenomyids. For assistance with the fieldwork that underlies our genetic analyses of tuco-tucos, we thank

John Wieczorek (*C. sociabilis* and *C. haigi*) and M. Mora (*C. talarum*). Many members of the Laboratorio de Ecofisiología have contributed to our understanding of the biology of *C. talarum*, notably R. Zenuto, C.D. Antinuchi, M. Mora, C. Schleich, F. Luna, M.S. Fanjul, A. Vassallo, M. Kittlein, and C. Busch. Financial support for the studies summarized here was provided by the Universidad Nacional de Mar del Plata, the Agencia de Promoción Científica y Tecnológica (FONCYT-PICT 01-09846), the National Science Foundation (DEB-9704462 and DEB-0128857), and CONICET (Beca Doctoral Mixta to APC). Finally, we thank the editors of this volume for inviting us to participate in this endeavor.

References

Alexander RD (1974) The evolution of social behavior Ann Rev Ecol Syst 5:325–383

Anderson RM, May RM (1978) Regulation and stability of host-parasite population interactions. I. Regulatory processes. J Anim Ecol 47:219–247

Anderson RM, May RM (1979) Population biology of infectious disease: part I. Nature 280:361–367

Apfelbaum LI, Massarini AI, Daleffe LE, Reig OA (1991) Genetic variability in the subterranean rodents *Ctenomys australis* and *Ctenomys porteousi* (Rodentia: Octodontidae). Biochem Syst Ecol 19:467–476

Arneberg P (2002) Host population density and body mass as determinants of species richness in parasite communities: a comparative analysis of directly transmitted nematodes of mammals. Ecography 25:88–94

Avise JC (2004) Molecular markers, natural history, and evolution. Sinauer, Sunderland, MA

Ben-Shlomo RF, Figueroa J, Klein J, Nevo E (1988) MHC class II DNA polymorphisms within and between chromosomal species of the *Spalax ehrenbergi* in Israel. Genetics 119:141–149

Bernatchez L, Landry C (2003) MHC studies in nonmodel vertebrates: what have we learned about natural selection in 15 years? J Evol Biol 16:363–377

Busch C, Malizia AI, Scaglia OA, Reig OA (1989) Spatial distribution and attributes of a population of *Ctenomys talarum* (Rodentia: Octodontidae). J Mammal 70:204–208

Busch C, Antinuchi CD, del Valle JC, Kittlein MJ, Malizia AI, Vassallo AI, Zenuto RR (2000) Population ecology of subterranean rodents. In: Lacey EA, Patton JL, Cameron GN (eds) Life underground: the biology of subterranean rodents. Univ Chicago Press, Chicago, IL, pp 183–226

Castillo AH, Cortinas MN, Lessa EP (2005) Rapid diversification of South American tucotucos (*Ctenomys*; Rodentia, Ctenomyidae): contrasting mitochondrial and nuclear intron sequences. J Mammal 86:170–179

Chan YL, Lacey EA, Pearson OP, Hadly EA (2005) Ancient DNA reveals Holocene loss of genetic diversity in a South American rodent. Proc R Soc Lond B 1:423–426

Cook JA, Lessa EP (1998) Macroevolutionary patterns of diversification in subterranean rodents: the case of the South American tuco-tucos (genus *Ctenomys*, Rodentia: Octodontidae). Evolution 52:1521–1527

Coté IM, Poulin R (1995) Parasitism and group-size in social animals: a meta-analysis. Behav Ecol 6:159–165

Cutrera AP, Lacey EA (2006) Major histocompatibility complex variation in talar tuco-tucos: the influence of demography on selection. J Mammal 87:706–716
Cutrera AP, Lacey EA, Busch C (2005) Genetic structure in a solitary rodent (*Ctenomys talarum*): implications for kinship and dispersal. Mol Ecol 14:2511–2523
Cutrera AP, Lacey EA, Busch C (2006) Intraspecific variation in effective population size in tuco-tucos (*Ctenomys talarum*): the role of demography. J Mammal 87:108–116
D'Elia G, Lessa EP, Cook JA (1998) Geographic structure, gene flow and maintenance of melanism in *Ctenomys rionegrensis* (Rodentia: Octodontidae). Z Säugetierk 63:285–295
Edwards SV, Hedrick PW (1998) Evolution and ecology of MHC molecules: from genomics to sexual selection. Trends Ecol Evol 13:305–311
El Jundi TAR, de Freitas TRO (2004) Genetic and demographic structure in a population of *Ctenomys lami* (Rodentia-Ctenomyidae). Hereditas 140:18–23
Endler JA (1992) Genetic heterogeneity and ecology. In: Berry RJ, Crawford TJ, Hewitt GM (eds) Genes in ecology. Blackwell, London, pp 281–312
Fernández-Stolz GP, Stolz JFB, de Freitas TRO (in press) Bottlenecks and dispersal in the tuco-tuco-das-dunas, *Ctenomys flamarioni* (Rodentia: Ctenomyidae) in southern Brazil. J Mammal
Ford MJ (2002) Applications of selective neutrality tests to molecular ecology. Mol Ecol 11:1245–1262
Gallardo MH, Köhler N (1992) Genetic divergence in *Ctenomys* (Rodentia, Ctenomyidae) from the Andes of Chile. J Mammal 73:99–105
Gallardo MH, Köhler N (1994) Demographic changes and genetic losses in populations of a subterranean rodent (*Ctenomys maulinus brunneus*) affected by a natural catastrophe. Z Säugetierk 59:358–365
Gallardo MH, Palma RE (1992) Intra- and interspecific genetic variability in *Ctenomys* (Rodentia: Ctenomyidae). Biochem Syst Ecol 20:523–534
Gallardo MH, Köhler N, Araneda C (1995) Bottleneck effects in local populations of fossorial *Ctenomys* (Rodentia: Ctenomyidae) affected by vulcanism. Heredity 74:638–646
Gava G, de Freitas TRO (2004) Microsatellite analysis of a hybrid zone between chromosomally divergent populations of *Ctenomys minutus* from southern Brazil (Rodentia: Ctenomyidae). J Mammal 85:1201–1206
Gillespie JH (1991) The causes of molecular evolution. Oxford Univ Press, New York
Giménez MD, Mirol PM, Bidau CJ, Searle JB (2002) Molecular analysis of populations of *Ctenomys* (Caviomorpha, Rodentia) with high karyotypic variability. Cyt Geno Res 96:130–136
Hadly EA, Van Tuinen M, Chan YL, Heiman K (2003) Ancient DNA evidence of prolonged population persistence with negligible genetic diversity in an endemic tuco-tuco (*Ctenomys sociabilis*). J Mammal 84:403–417
Hambuch TM (2000) Major histocompatibility complex diversity in species of the genus *Ctenomys*: the role of selection in the evolution of genetic variation. PhD Dissertation, Univ California, Berkeley
Hambuch TM, Lacey EA (2002) Enhanced selection for MHC diversity in social tuco-tucos. Evolution 56:841–845
Honeycutt RL, Allard MW, Edwards SV, Schlitter DA (1991) Systematics and evolution of the Bathyergidae. In: Sherman PW, Jarvis JUM, Alexander RD (eds) The biology of the naked mole-rat. Princeton Univ Press, Princeton, NJ, pp 45–65
Honeycutt RL, Rowe DL, Gallardo MH (2003) Molecular systematics of the South American caviomorph rodents: relationships among species and genera in the family Octodontidae. Mol Phyl Evol 26:476–489
Kundu S, Faulkes CG (2004) Patterns of MHC selection in African mole-rats, family Bathyergidae: the effects of sociality and habitat. Proc R Soc Lond 271:273–278

Lacey EA (2001) Microsatellite variation in solitary and social tuco-tucos: molecular properties and population dynamics. Heredity 86:628–637

Lacey EA, Wieczorek JR (2004) Kinship in colonial tuco-tucos: evidence from group composition and population structure. Behav Ecol 15:988–996

Lacey EA, Braude SH, Wieczorek JR (1997) Burrow sharing by colonial tuco-tucos (*Ctenomys sociabilis*). J Mammal 78:556–562

Lacey EA, Braude SH, Wieczorek JR (1998) Spatial relationships among adult Patagonian tuco-tucos (*Ctenomys haigi*). J Mammal 79:986–991

Lessa EP (2000) The evolution of subterranean rodents: a synthesis. In: Lacey EA, Patton JL, Cameron GN (eds) Life underground: the biology of subterranean rodents. Univ Chicago Press, IL, pp 390–420

Lessa EP, Cook JA (1998) The molecular phylogenetics of tuco-tucos (genus *Ctenomys*, Rodentia: Octodontidae) suggests an early burst of speciation. Mol Phyl Evol 9:88–99

Lessa EP, Wlasiuk G, Garza JC (2005) Dynamics of genetic differentiation in the Rio Negro tuco-tuco (*Ctenomys rionegrensis*) at the local and geographic scale. In: Lacey EA, Myers P (eds) Mammalian diversification: from chromosomes to phylogeography. Univ California Publications in Zoology, vol. 133. Univ California Press, Berkeley, CA, pp 155–174

Lewontin RC (1974) The genetic basis of evolutionary change. Columbia Univ Press, New York, NY

Malizia AI, Vassallo AI, Busch C (1991) Population and habitat characteristics of two sympatric species of *Ctenomys* (Rodentia: Octodontidae). Acta Theriol 36:87–94

Malizia AI, Zenuto RR, Busch C (1995) Demographic and reproductive attributes of dispersers in two populations of the subterranean rodent *Ctenomys talarum* (tuco-tuco). Can J Zool 73:732–738

Massarini A, Dopazo HJ, Bouzat JL, Hasson E, Reig OA (1992) The population genetic structure of *Ctenomys porteousi* (Rodentia: Octodontidae). Biochem Syst Ecol 20: 723–734

Matocq MD, Lacey EA (2004) Philopatry, kin clusters, and genetic relatedness in a population of woodrats (*Neotoma macrotis*). Behav Ecol 15:647–653

May RM, Anderson RM (1979) Population biology of infectious disease. Part II. Nature 280:455–461

Meyers LA, Bull JJ (2002) Fighting change with change: adaptive variation in an uncertain world. Trends Ecol Evol 17:551–557

Mora MS, Lessa EP, Kittlein MJ, Vassallo AI (2006) Phylogeography of the subterranean rodent *Ctenomys australis* (Rodentia: Ctenomyidae) in sand-dune habitats: evidence of recent population expansion. J Mammal 87:1192–1203

Nevo E (1979) Adaptive convergence and divergence of subterranean mammals. Ann Rev Ecol Syst 10:269–308

Nevo E (1999) Mosaic evolution of subterranean mammals: regression, progression and global convergence. Oxford Univ Press, Oxford

Nevo E, Beiles A (1992) Selection for class II MHC heterozygosity by parasites in subterranean mole rats. Cel Mol Life Sci 48:512–515

Nevo E, Klein J (1990) Structure and evolution of MHC in subterranean mammals of the *Spalax ehrenbergi* superspecies in Israel. In: Nevo E, Reig OA (eds) Evolution of subterranean mammals at the organismal and molecular levels. Wiley-Liss, New York, pp 397–411

Nevo E, Kim YJ, Shaw CR, Thaeler CS (1974) Genetic variation, selection and speciation in *Thomomys talpoides* pocket gophers. Evolution 28:1–23

Nizetic D, Figueroa F, Dembic Z, Nevo E, Klein J (1985) Major histocompatibility complex of the mole rat. II. Restriction fragment length polymorphism. Immunogen 22: 55–67

Nizetic D, Stevanovic M, Soldatovic B, Savic I, Crkvenjakov R (1988) Limited polymorphism of both classes of MHC genes in four different species of the Balkan mole rat. Immunogen 28:91–98

Patton JL (1990) Geomyid evolution: the historical, selective, and random basis for divergence patterns within and among species. In: Nevo E, Reig OA (eds) Evolution of subterranean mammals at the organismal and molecular levels. Wiley-Liss, New York, pp 49–69

Patton JL, Smith MF (1990) The evolutionary dynamics of the pocket gopher *Thomomys bottae*, with emphasis on California populations. Univ California Publications in Zoology, vol 123. Univ California Press, Berkeley, CA

Patton JL, Yang SY (1977) Genetic variation in *Thomomys bottae* pocket gophers: macrogeographic patterns. Evolution 31:697–720

Penn DJ, Potts WK (1999) The evolution of mating preferences and major histocompatibility complex genes. Am Nat 153:145–164

Piertney SB, Oliver MK (2006) The evolutionary ecology of MHC. Heredity 96:7–21

Reig OA, Busch C, Ortells MO, Contreras JR (1990) An overview of evolution, systematics, population biology, cytogenetics, molecular biology, and speciation in *Ctenomys*. In: Nevo E, Reig OA (eds) Evolution of subterranean mammals at the organismal and molecular levels. Wiley-Liss, New York, NY, pp 71–96

Sage RD, Contreras JR, Roig VG, Patton JL (1986) Genetic variation in the South American burrowing rodents of the genus *Ctenomys* (Rodentia: Ctenomyidae). Z Säugetierk 51:158–172

Satta YC, O'Huigin C, Takahata N, Klein J (1994) Intensity of natural selection at the major histocompatibility complex loci. Proc Natl Acad Sci USA 91:7184–7188

Schopfer RF, Figueroa D, Nizetic D, Nevo E, Klein J (1987) Evolutionary diversification of class II P loci in the MHC of the mole rat, *Spalax ehrenbergi*. Mol Biol Evol 4:287–299

She JX, Boehme S, Wang TW, Bonhomme F, Wakeland EK (1990) The generation of MHC class II gene polymorphism in the genus *Mus*. Biol J Linn Soc 41:141–161

Slamovits CH, Cook JA, Lessa EP, Rossi MS (2001) Recurrent amplifications and deletions of satellite DNA accompanied chromosomal diversification in South American tuco-tucos (genus *Ctenomys*, Rodentia: Octodontidae): a phylogenetic approach. Mol Biol Evol 18:1708–1719

Smith MF (1998) Phylogenetic relationships and geographic structure in pocket gophers in the genus *Thomomys*. Mol Phyl Evol 9:1–14

Solomon NG (2003) A reexamination of factors influencing philopatry in rodents. J Mammal 84:1182–1197

Steinberg EK, Patton JL (2000) Genetic structure and the geography of speciation in subterranean rodents: opportunities and constraints for evolutionary diversification. In: Lacey EA, Patton JL, Cameron GN (eds) Life underground: the biology of subterranean rodents. Univ Chicago Press, Chicago, IL, pp 301–331

Storz JF, Ramakrishnan U, Alberts SC (2002) Genetic effective size of a wild primate population: influence of current and historical demography. Evolution 56:817–829

Tomasco IH (2003) Filogeografía del tucu-tucu *Ctenomys pearsoni*: variación del ADN mitocondrial y sus implicancias para la diferenciación cromosómica. MSc Thesis. PEDECIBA. Univ República, Uruguay

Tomasco IH, Lessa EP (in press) Phylogeography of the tuco-tuco *Ctenomys pearsoni*: mtDNA variation and its implication for chromosomal differentiation. In: Kelt DA, Lessa EP, Salazar-Bravo J, Patton JL (eds) Essays in honor of Oliver Pearson. Univ California Publications in Zoology Series, Univ California Press, Berkeley, CA

Wegner KM, Reusch TBH, Kalbe M (2003) Multiple parasites are driving major histocompatibility complex polymorphism in the wild. J Evol Biol 16:224–232

Wlasiuk G, Garza JC, Lessa EP (2003) Genetic and geographic differentiation in the Rio Negro tuco-tuco (*Ctenomys rionegrensis*): inferring the roles of migration and drift from multiple genetic markers. Evolution 57:913–926

Wright SA (1949) Adaptation and selection. In: Jepson GL, Simpson GG, Mayr E (eds) Genetics, paleontology, and evolution. Princeton Univ Press, Princeton NJ, pp 365–389

Woods CA, Kilpatrick CW (2005) Infraorder Hystricognathi. In: Wilson DO, Reeder DM (eds) Mammal species of the world: a taxonomic and geographic reference. Johns Hopkins Univ Press, Baltimore, MD, pp 1538–1600

27 African Mole-rats (Bathyergidae): A Complex Radiation in Tropical Soils

Paul A. A. G. Van Daele, Chris G. Faulkes, Erik Verheyen, Dominique Adriaens

27.1 Introduction

African mole-rats are obligatory subterranean hystricognath rodents endemic to the Afrotropical region. The family contains solitary, social and two evolutionary divergent eusocial taxa, the naked mole-rat (*Heterocephalus glaber*) and the Damaraland mole-rat (*Fukomys damarensis*; Jarvis 1981; Jarvis and Bennett 1993; Bennett and Faulkes 2000) (but see Burda et al. 2000 who consider also other species of the genus *Fukomys* to be eusocial). This social diversity has lead to a great deal of recent research interest in the group, although many holotypes had been described by the late nineteenth century (reviewed by Ellerman et al. 1940, 1953). In this chapter it is our intention to review what is known about the phylogenetic relationships in Bathyergidae, with particular emphasis on the central African clade (*Cryptomys/Fukomys*) and present the evidence that shows how taxonomic differentiation has been influenced by past climatic and geomorphological factors. We realise that factors other than the historical geographic setting affect evolution in these subterranean rodents, including evolutionary constraints set by population demography on reproductive isolating mechanisms and enhancement of divergence through hybridisation (Golding and Strobeck 1983; Harrison 1993; Ruedi et al. 1997; Grant and Grant 1997; Steinberg and Patton 2000). At this stage, however, demographic studies and the role of hybridisation as an evolutionary mechanism have not received explicit attention in bathyergids. Data on chromosomal diversification in *Fukomys*, however, suggest that this genus at least may bear the stamp of reticulation events (Van Daele et al. 2004).

Paul A. A. G. Van Daele: Evolutionary Morphology of Vertebrates, Ghent University, K.L.Ledeganckstr. 35, 9000 Ghent, Belgium, E-mail: Paul.VanDaele@ugent.be

Chris G. Faulkes: School of Biological & Chemical Sciences, Queen Mary, University of London, Mile End Road, London E14NS

Erik Verheyen: Royal Belgian Institute of Natural Sciences, Vautierstr. 29, 1000 Brussels, Belgium

Dominique Adriaens: Evolutionary Morphology of Vertebrates, Ghent University, K.L. Ledeganckstr. 35, 9000 Ghent, Belgium

Subterranean Rodents: News from Underground
S. Begall, H. Burda, C.E. Schleich (Eds.)
© Springer-Verlag Berlin Heidelberg 2007

27.2 Intergeneric Relationships

Bathyergids have been the subject of numerous phylogenetic studies incorporating data from morphology (Wood 1985), allozyme data (Nevo et al. 1987; Filippucci et al. 1994, 1997; Janecek et al. 1992), mitochondrial restriction fragment analysis (Honeycutt et al. 1987), nuclear Transthyretin (TTR) sequence data (Walton et al. 2000), the nuclear von Willebrand Factor gene (Huchon and Douzery 2001), mitochondrial 12S rRNA (12S) (Honeycutt et al. 1987; Allard and Honeycutt 1992; Nedbal et al. 1994) and mitochondrial cytochrome b (cyt b) sequence data (Faulkes et al. 1997). Systematic overviews of the Bathyergidae by Jarvis and Bennett (1990) and Honeycutt et al. (1991) provided the basis for more recent studies on intergeneric and intrageneric relationships (Faulkes et al. 2004; Ingram et al. 2004). The Bathyergidae are the most species rich of the four monophyletic families that constitute the Phiomorpha (the African clade of Hystricomorph rodents; Nedbal et al. 1994). At the subfamilial level, groupings based on morphology are not corroborated by molecular evidence (Walton et al. 2000). The consensus now is that the family Bathyergidae consists of six extant genera (Kock et al. 2006), the highest number among subterranean rodents: *Heterocephalus*, *Heliophobius*, *Bathyergus Georychus*, *Cryptomys* and *Fukomys* (Fig. 27.1). Molecular genetic data unambiguously support this treatment at the generic level. All recent studies indicate that the eusocial mole-rat *Heterocephalus glaber* is sister to a group comprising all other genera in the family. At the next dichotomy, *Heliophobius* forms a sister lineage to the remaining genera. Following Faulkes et al. (2004) and Ingram et al. (2004) *Georychus* and *Bathyergus* group in a monophyletic clade, but the relationships with *Cryptomys* and *Fukomys* remain poorly resolved. Several studies have observed that *Cryptomys sensu lato* actually comprised two divergent clades and recommended the recognition of two separate genera (Nevo et al. 1987; Honeycutt et al. 1991; Janecek et al. 1992; Filippucci et al. 1994, 1997; Faulkes et al. 1997, 2004; Walton et al. 2000; Ingram et al. 2004). It is now generally accepted that *Cryptomys sensu lato* should be split into *Cryptomys* and the new genus *Fukomys* based on the reciprocal monophyly in both nuclear and mitochondrial analyses, the high level of sequence divergence between the two taxa and differential patterns of chromosomal evolution. The available data suggest that intrageneric variability, particularly in *Fukomys*, is higher then hitherto accepted and many new species await description (Van Daele et al. 2004). Morphological synapomorphies allow a straightforward diagnosis of *Heterocephalus*, *Bathyergus*, *Georychus* and *Heliophobius*. However, *Cryptomys* and *Fukomys* cannot be clearly separated on morphological and traditional morphometric criteria. Preliminary geometrical morphomet-

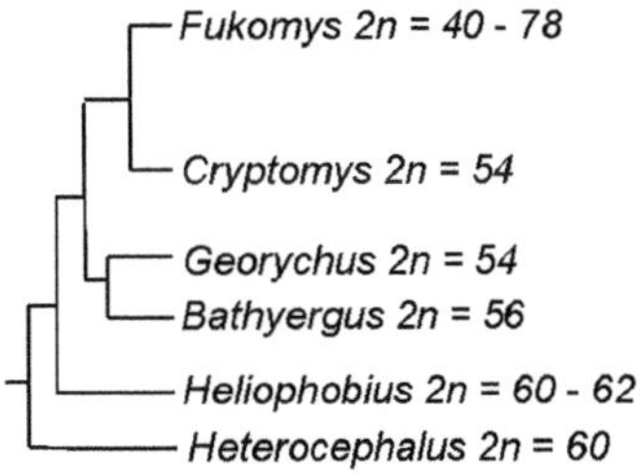

Fig. 27.1. Phylogenetic relationships of African mole-rats

rical data indicate subtle differences in skull morphology between the two genera (Van Daele et al. 2006; Van Daele, Corkery and Murtas, unpublished data).

27.3 Biogeography of the Bathyergidae

Recent molecular phylogenies that have inferred a molecular clock, have consistently suggested an ancient (Eocene) origin for the Bathyergidae (e.g. Huchon and Douzery 2001). Unfortunately, fossil-bearing strata of the Eocene/Oligocene are extremely rare in Africa (Lavocat 1978), and fossil bathyergids first appear later on in the more common early Miocene deposits. Of the extant genera, *Heterocephalus* is the first to appear in the fossil record and the temporal association of *Heterocephalus* fossils with extinct bathyergid ancestors supports the early divergence of *Heterocephalus* within the family that is suggested by molecular phylogenies (for a recent review of fossil bathyergids see Faulkes et al. 2004).

The availability of robust molecular phylogenies now enables us to infer the potential phylogeographic influences that underlie the adaptive radiation of the Bathyergidae. Molecular phylogenies of the Bathyergidae are rooted in East Africa with *Heterocephalus* and *Heliophobius* being the earliest divergences (Fig. 27.2a). Molecular clock-based timing of the major divergences within the family suggest that the initial cladogenesis of the Bathyergidae was sufficiently early to be independent of the rifting and volcanism that characterise this part of Africa today, and a general radiation appears to have occurred from East Africa into southern and south-central Africa. Divergence of the basal lineages in the Bathyergidae, *Heterocephalus* (40–48 mya) and *Heliophobius* (32–40 mya) may have been constrained northwards by the Ethiopian highlands, but was relatively unimpeded south and west (Fig. 27.2b). It has been suggested (Honeycutt et al. 1991) that these early Bathyergids may have spread via a corridor of fluctuating aridity linking east and southwest Africa (Van Zinderen Bakker 1967). This arid corridor has been implicated in the distribution of both modern

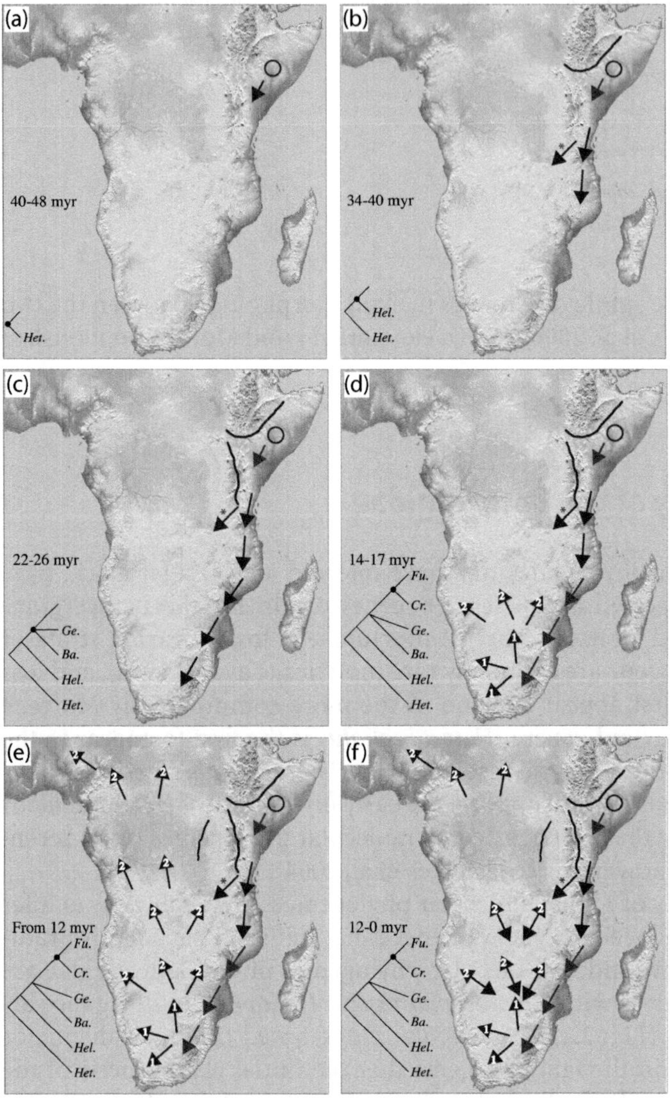

Fig. 27.2a–f. Phylogeographic trends in the Bathyergidae inferred from analysis of mitochondrial 12S rRNA and cytochrome b sequence differences: **a** initial divergence of the *Heterocephalus* (Het.) lineage from the common ancestor of the family in East Africa; **b** radiation of *Heliophobius* (Hel.) and movement from East Africa into Southern Africa, with some populations crossing the Rift Valley (*); **c** *Bathyergus* (Ba.) and *Georychus* (Ge.) lineages diverge in South Africa; **d** *Cryptomys* (Cr.) diverges into two clades, one radiating predominantly in South Africa (1), and *Fukomys* spreading north into Southern, Central and West Africa (2); **e** formation of Lake Tanganyika; **f** later diversification of *Fukomys* clade in South-central Africa leaving disjunct populations in Central and West Africa (data adapted from Faulkes et al. 2004)

and Early Miocene fossil faunas (Van Couvering and Van Couvering 1976). During this time, large-scale physical and climatic changes were also occurring in this part of Africa as a result of continental drift, leading to increased volcanism and the gradual formation of the African Rift Valley. These volcanic uplands and deep valleys, some of which eventually formed the great lakes of Africa, constitute a physical barrier for subterranean rodents. In addition, it is likely that the climatic and vegetative changes that indirectly resulted from the rifting process were of importance in determining the distribution and social evolution of the Bathyergidae. For example, estimates of the divergence of *Georychus/Bathyergus* from their common ancestor with *Cryptomys/Fukomys* (20–26 mya; Fig. 27.2c) coincide with the beginning of volcanism in the Kenya rift, and this may have favoured the expansion of these lineages into southern Africa, rather than to the north and west. At the same time, aridification of the Saharo-Arabian belt was beginning, and this would have further restricted the Bathyergidae to sub-Saharan Africa. The first fossil bathyergids are known from this period (Early Miocene) in both East Africa and Namibia, confirming these molecular-based estimates that southern Africa was also being colonized at this time. During the later Early and Middle Miocene the *Cryptomys* genus diverged into its two subclades (*Cryptomys/Fukomys*; at 12–17 mya; Fig. 27.2d). Again, this coincides with a critical period when rifting was progressing in the Kenya rift and also just beginning in the Western rift. While the *Cryptomys* clade appears to have speciated almost exclusively in South Africa, the *Fukomys* clade underwent a more extensive radiation, particularly in Zambia and Central Africa (Fig. 27.2e), resulting in a wide diversity of genetically divergent chromosomal forms (see below). This initial spread of *Fukomys* has left extant, but disjunct populations in central Africa as far west as Ghana (*F. zechi*), Nigeria (*F. foxi*) and Nigeria/Sudan/Uganda (*F. ochraceocinereus*). These species at the northern periphery of the bathyergid range form a basal clade in the *Fukomys* genus (Ingram et al. 2004; C.G. Faulkes and N.C. Bennett, unpublished data), and may represent relic populations from the initial radiation of *Fukomys*, isolated by the formation of tropical rainforest in the Congo Basin. If molecular timings are correct, then the initial radiation of *Fukomys* is coincident with the onset of volcanic activity in the adjacent Western rift. It is possible that the resulting environmental challenges brought about as climate and vegetation changed, may have influenced patterns of speciation and social elaboration. This period has previously been reported as a time of faunal turnover (Van Couvering and Van Couvering 1976). Further volcanism and formation of the East Africa rift during the Miocene appears to have almost completely isolated the populations of *Heterocephalus* and *Heliophobius* to the east, and restricted *Cryptomys/Fukomys* to the west of the Rift. Exceptions to this are a few populations of *Fukomys whytei* in western Tanzania, and *Heliopho-*

bius found in Malawi, D.R. Congo and Zambia. The latter appear to have diverged before local rifting restricted their movement (Fig. 27.2f; Faulkes et al. 2004).

27.4 Overview of the Six Genera

Bathyergids exhibit considerable geographically structured morphological and genetical variation (see, e.g. Faulkes et al. 2004). In the past this fact triggered the description of many species and subspecies of bathyergids. Although this is not the place to revive the entire discussion about species concepts, it seems that the alternative species concepts cannot be neatly applied to bathyergids and subterranean rodents in general (see, e.g. Lessa 2000). By trying to box dynamic evolutionary processes in a static framework, we will inevitably be oversimplifying the complex relationships within certain taxa (e.g. *Fukomys*). We take the view that species limits should be established using multiple evidence, which takes into account variation at the molecular, cytogenetical and organismal levels. This section therefore highlights known polymorphism and divergence as observed in morphological, chromosomal, and molecular analyses to hint at both the validity of named forms and the possibility of finding additional, unrecognized forms among African mole-rats.

27.4.1 The Eusocial *Heterocephalus* Rüppell, 1842

Heterocephalus forms a monotypic genus. Despite genetic differences of 10.6% in cyt b between populations at the extremes of its range (Faulkes et al. 1997) and life history characteristics that seem to make *Heterocephalus* prone to speciation, *Heterocephalus* shows no morphological variation nor variation in chromosome number ($2n = 60$; George 1979; Capanna and Merani 1980).

27.4.2 The Three Solitary Genera

Heliophobius Peters, 1846

Currently only *H. argenteocinereus* is recognised as a valid species in this genus. Karyotypic variation ($2n = 60$–62; George 1979; Scharff et al. 2001), high levels of cyt b sequence divergence (up to 13.3%; Faulkes et al. 2004) and 12S sequence divergence (7.3%; Ingram et al. 2004) and morphological differences suggest that *Heliophobius* is actually polytypic.

Bathyergus Illiger, 1811

Two species are currently recognized with different karyotypes (Nevo et al. 1986): *B. suillus* ($2n = 56$) and *B. janetta* ($2n = 54$). However two recent independent studies have suggested that both species have a $2n = 54$ (C.G. Faulkes, unpublished data; J. Deuve and T.J. Robinson, unpublished data). Faulkes et al. 2004 reported relatively low levels of cyt *b* sequence divergence (up to 5.6%) and suggested the existence of a hybrid zone in an area where the species are sympatric. While Ingram et al. (2004) were unable to resolve interspecific relationships they suggested *Bathyergus suillus* might be more variable, warranting recognition of additional taxa.

Georychus Illiger, 1811

Although the accepted view is that *Georychus* just comprises one species: *G. capensis*, allozyme data (Nevo et al. 1987), mitochondrial RFLP data (Honeycutt et al. 1987) and high levels of mitochondrial sequence divergence (cyt b: up to 13.7%; Faulkes et al. 2004; Ingram et al. 2004) all suggest the possibility of two species.

27.5 Diversification in Social Mole-rats of the Genera *Cryptomys* Gray, 1864 and *Fukomys* (Kock, Ingram, Frabotta, Honeycutt and Burda 2006)

Cryptomys sensu lato

Cryptomys sensu lato forms stable colonies with one reproductive female, one or more reproductive males and several non-reproductive animals that help with the care for the off-spring and maintenance of the colony. They are found in a variety of habitats, all of which are characterised by the presence of geophytes (staple food). Roots, bulbs and corms are harvested along extensive, anastomosing tunnel systems. The reproductive skew and other life history characteristics as well as differences in habitat are at the basis of intra- and intercolonial polymorphism in quantifiable traits (Rosevear 1969; Honeycutt et al. 1987; Burda et al. 1999; Bennett and Faulkes 2000; Janse Van Rensburg et al. 2004).

Therefore, it is no surprise that classifications of *Cryptomys sensu lato* based on morphological data (often subtle differences in a size variable, pelage colour or the extent of the headspot) show several inconsistencies. Accordingly there has been a great discrepancy in the number of

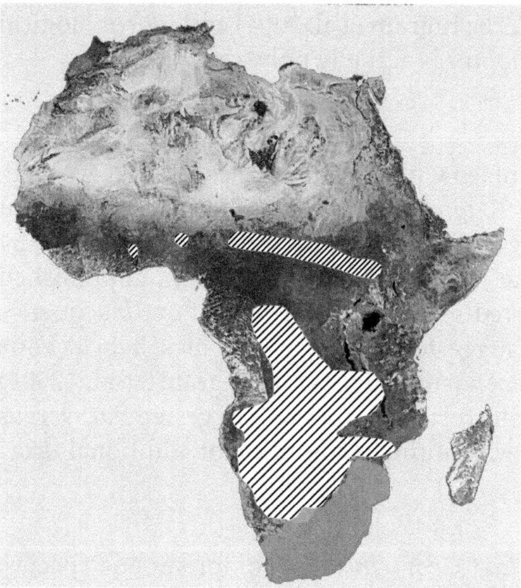

Fig. 27.3. Distribution map of *Fukomys* (*hatched*) and *Cryptomys* (*grey*). The main distribution of extant *Fukomys* is closely associated with mesic savannas North and South of the tropical rain forest belt

species recognised by different authors (Allen 1939; Ellerman et al. 1940; Roberts 1951; Ansell 1978; De Graaff 1981; Nowak 1991; Wilson and Reeder 2005).

Cryptomys sensu stricto is a South African radiation while *Fukomys* (synonym *Coetomys*; Kock et al. 2006) is widely distributed throughout the mesic and arid Savannahs (Fig. 27.3). The two genera show remarkably different evolutionary pathways. On the one hand *Cryptomys* sensu stricto is chromosomally conservative while showing deep DNA sequence divergences. *Fukomys* on the other hand, being chromosomally hypervariable, shows varying levels of sequence divergence (Van Daele et al., accepted; cf. below).

Cryptomys Gray, 1864

Cryptomys is a polytypic genus with at least six species. While the available data indicate it is chromosomally conservative ($2n = 54$; Nevo et al. 1986; Faulkes et al. 2004), the species richness of the group is supported by recent genetic studies. Ingram et al. (2004) recovered five monophyletic lineages, which they assigned to *C. hottentotus*, *C. natalensis*, *C. nimrodi*, *C. holosericeus* and *C. anomalus*, the latter from Pretoria. However, Faulkes et al. (2004) found two highly divergent populations around Pretoria, separated by only 30 km (using a subspecific nomenclature for all taxa, the Pretoria samples were assigned to *C. hottentotus pretoriae* and *C.h. mahali*). Several formally described races have not been thoroughly studied and therefore diversity could be much higher (Roberts 1951). However, one should take great care in assigning specimens or naming species as it has been hampered by the existence of many synonyms (there are 34 synonyms available for Southern African races of the *C. hottentotus* "superspecies" (De Graaff 1981) and vague type locality descriptions. Therefore a topotypical approach is required to avoid further confusion in the complicated taxonomy of this group.

***Fukomys* (Kock, Ingram, Frabotta, Honeycutt and Burda 2006)**

Honeycutt et al. (1991) recognised six species: *C. foxi* (Thomas 1911) *C. zechi* (Matschie 1900) and *C. ochraceocinereus* (Heuglin 1864), which are distributed in the Sudanian savannahs and *Cryptomys bocagei* (De Winton 1897), *C. damarensis* (Ogilby 1838) and *C. mechowii* (Peters 1881) from the Zambezian region. Allozyme and karyological studies suggested the specific status of *C. darlingi* (Aguilar 1993), *C. amatus* (Macholán et al. 1998), *C. kafuensis* (Filippucci et al. 1994; Burda et al. 1999), *C. anselli* (Filippucci et al. 1994; Burda et al. 1999) and *C. whytei* (Chitaukali et al. 2001). Subsequently 11 new karyotypes were described from Zambia (Faulkes et al. 1997; Kawalika et al. 2001; Van Daele et al. 2004). Various other populations from the Zambezian region are known to be karyotypically different (PAAG Van Daele, unpublished data). Several of these may represent biological species. This is corroborated by studies showing considerable levels of allozyme divergence (Filippucci et al. 1994, 1997) and sequence diversity (Faulkes et al. 2004; Ingram et al. 2004; Van Daele et al., accepted).

Molecular sequence data show that a Sudanian clade is sister to six Zambezian clades (op. cit.). For the Sudanian clade, chromosomal ($2n = 66/70$; Williams et al. 1983) and sequence data (Ingram et al. 2004) are scarce but suggest that at least the three named forms deserve specific status. *Fukomys* distributed in the Zambezian phytochorion show extreme chromosomal variability ($2n = 40–78$), while levels of mitochondrial sequence divergence within and between subclades may vary considerably as detailed in Van Daele et al., accepted (Fig. 27.4). Remarkably the South Zambian phylogroup (micklemi clade: Fig. 27.4) containing *F. micklemi*, *F. anselli*, *F. kafuensis* and five other taxa with different karyotypes only show an average of 1.4% cyt b sequence divergence, similar to the widely distributed Southwestern Zambezian damarensis clade (five known karyotypes; 1% cyt b sequence difference). As opposed to these low levels of sequence divergence in the more arid parts of the Zambezian phytochorion, there appear to be high levels of genetic divergence between phylogroups from mesic regions (e.g. in the whytei clade).

Van Daele et al. (2004) noted that the distribution of the different karyotypical races correlates well with the extant river system configuration. They hypothesized that geomorphological changes and in particular river

Fig. 27.4. Phylogenetic relationships of *Fukomys* haplotypes: Cyt b maximum-likelihood phylogeny under TIM+I+Γ model of evolution. Phylogenetic relationships of *Fukomys* haplotypes reveal six distinct subclades. The *numbers in brackets* refer to the number of known chromosomal races. The *scale bar* indicates the number of substitutions per site (data adapted from Van Daele et al., accepted)

system dynamics in recent geological times (from the Neogene on) have played an important role in the chromosomal diversification and would have provided ample opportunities for speciation (including vicariance and allopatric processes) to occur. The phylogeographic history of Zambezian *Fukomys* would then have been driven by the joining and fragmentation of precursors of extant rivers driven by crustal flexion (both down warping and uplifting) and associated climatic shifts. This is corroborated to a large extend by estimates of divergence dates (Van Daele et al., accepted; Fig. 27.5).

27.6 Prospects (for Phylogenetic Studies and Beyond)

In addition to further molecular phylogenetic sequence analysis, morphometrical, ethological and detailed chromosomal studies using differential staining techniques are required to create a robust phylogenetic framework for examining intrageneric relationships and the identification of cryptic species. Differential patterns of chromosomal and molecular genetic evolution and speciation should be further explored using a multidisciplinary approach, starting with a topotypical sampling strategy. In an interesting overview of future directions of research on subterranean rodents, Lessa (2000) stresses that a phylogenetic framework is not only an important prerequisite for understanding macroevolutionary processes but also provides the necessary context to discuss the structure and function and the relative importance of adaptation. We believe that in particular *Fukomys* provides an evolutionary model that could yield fruitful results in the future, ultimately enriching evolutionary theory. Further DNA sequencing implies a better knowledge of gene structure, but it is an understanding of gene function that will enhance molecular genetic insight into evolutionary processes of speciation and adaptation. Using, e.g. microarray technology, *Fukomys* will undoubtedly be instrumental in the exploration of differential ageing patterns (Dammann and Burda 2006). Exemplatory at the organismal level are studies which have explored how morphometrical variation maps on phylogenetic trees. Preliminary data obtained from a geometric morphometrical analysis of the ventral side of the cranium of a number of closely related chromosomal races belonging to the *F. micklemi* phylogroup suggest that during raciation small localized shape changes (shape changes of so-called non-uniform origin; Zelditch et al. 2004) could modify the integration of structures with shifts in functionality. As expected, these shape changes do not seem to follow the cladogenetic sequence of events as depicted by a cyt b phylogeny (Van Daele et al. 2006). This fact could be further exploited to understand how either plastic responses to immediate ecological conditions (cf. Patton and Brylski 1987) and/or changes in shape associated with digging and chewing modes (reviews by Lessa 1990; Stein 2000) appear time and again in the evolution of subterranean lineages. Transposed on a phylogenetic framework morphometrical changes per se and other morphological characters (that may map better on phylogenetic trees) could help to elucidate the sequence and nature of morphological adaptation within the constraints set by the subterranean environment.

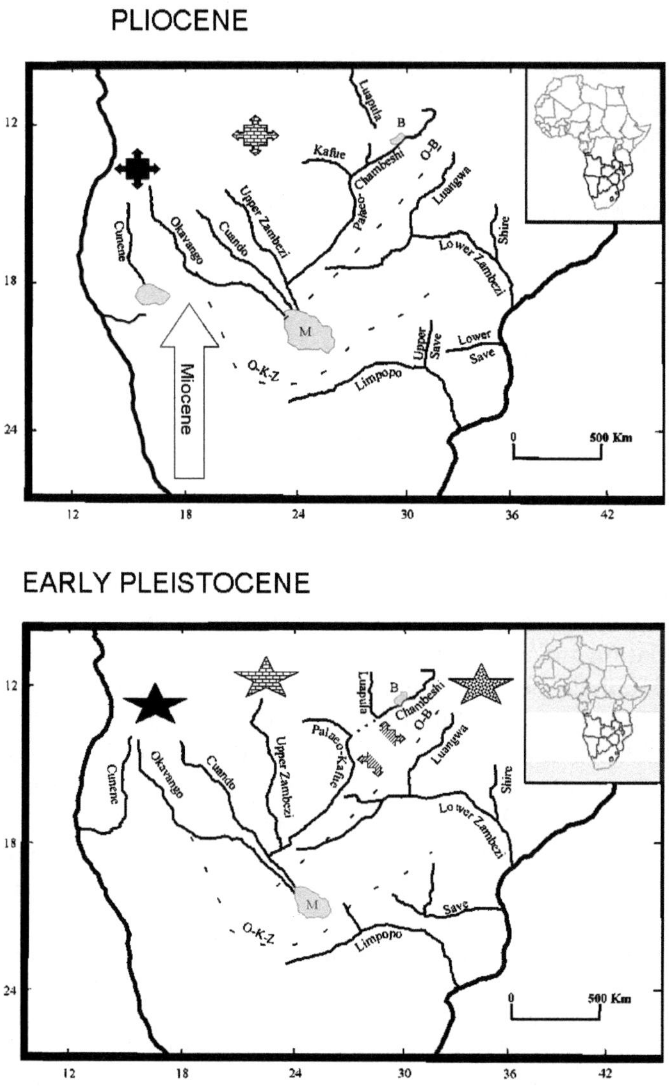

Fig. 27.5. Phylogeographic trends in Zambezian *Fukomys* estimated from Cyt b sequence differences using nonparametric rate smoothing and penalized likelihood methods as implemented in r8s (Sanderson 1997, 2002) (data adapted from Van Daele et al., accepted). The general distribution of the extant *Fukomys* clades would have been determined in the Holocene, when the riversystem configuration became similar to today's situation. (Drainage system evolution in South-Central Africa. Figure modified after Du Toit (1933), Moore and Larkin (2001) and Cotterill (2003). *M* Paleo-lake Makgadikgadi; *B* Lake Bangweulu; *OKZ* Ovamboland – Kalahari – Zimbabwe axis of crustal flexuring and *OB* Okavango – Bangweulu axis of crustal flexuring)

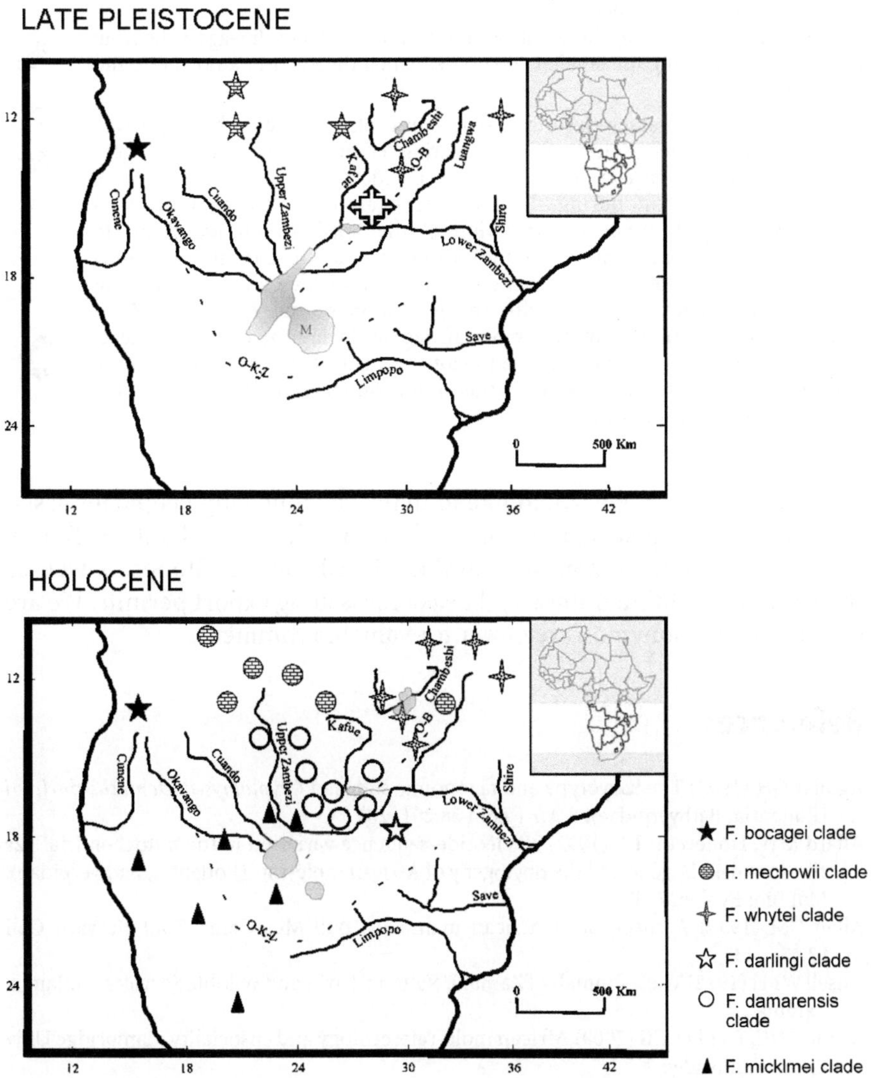

Fig. 27.5. Legend.

Divergence time estimates suggest a Pliocene origin of Zambezian *Fukomys*. The common ancestor of *Fukomys* would have colonized the lower latitudes dispersing along shifting vegetation belts during the late Miocene. As the Sudanian lineage (North of the tropical rainforest belt) became gradually isolated from the Zambezian lineage, the latter radiated in the Zambezian savanna belt. 

A first divergence occurred in the Early Pliocene, eventually giving rise to the basal Zambezian clade including *F. bocagei*, which dispersed throughout Angola. 

The common ancestor of the other Zambezian clades would have had ample opportunity during the Pliocene to disperse along the watershed between the precursors of the headwaters of the Zambezi and Zaire Rivers. 

Fig. 27.5. Legend. (continued)
During the Late Pliocene the giant mole-rat or *F. mechowii* lineage split from the lineage containing the smaller mole-rats, which diverged in a northern and southern lineage.

The Pleistocene is characterized by several consecutive radiation bursts, driven by the dynamic interaction of a fluctuating climate with associated geomorphological and vegetational changes. Coincident with reconfigurations of the major riversystem during the Pleistocene all major lineages of small mole-rats emerged. While the radiation of the northern lineage seems to be coupled to reconfigurations of the headwaters of the Luapula river during the Early Pleistocene, the divergence time estimates of the radiation of the southern clades correlates with the formation of the Kafue and Zambezi rivers in the South-Central Zambezian region during the Late Pleistocene. *F. damarensis*, being the most derived taxon, seems to have diverged from the *F. micklemi* clade after the final formation of the Zambezi, radiating around the shores of Paleo-lake Makgadikgadi.

Acknowledgements. We would like to thank the following people for assistance in collecting specimens or sending samples – we thank H. Burda, C. Ingram, V. Katanekwa, M. Kawalika, E. Sikaboa, S. Stoops as well as the Zambian Wildlife Authority, Lusaka for issuing export permits. We are grateful to an anonymous reviewer for valuable comments.

References

Aguilar GH (1993) The karyotype and taxonomic status of *Cryptomys-hottentotus-darlingi* (Rodentia, Bathyergidae). S Afr J Zool 28:201–204

Allard MW, Honeycutt RL (1992) Nucleotide-sequence variation in the mitochondrial 12s ribosomal-RNA gene and the phylogeny of African mole-rats (Rodentia, Bathyergidae). Mol Biol Evol 9:27–40

Allen GM (1939) A checklist of African mammals. Bull Mus Comp Zool Harvard Coll 83:425–433

Ansell WFH (1978) The mammals of Zambia. National Parks and Wildlife Services, Chilanga, Zambia

Bennett NC, Faulkes CG (2000) African mole-rats: ecology and eusociality. Cambridge Univ Press, Cambridge

Burda H, Zima J, Scharff A, Macholán M, Kawalika M (1999) The karyotypes of *Cryptomys anselli* sp nova and *Cryptomys kafuensis* sp nova: New species of the common mole-rat from Zambia (Rodentia, Bathyergidae). Int J Mamm Biol 64:36–50

Burda H, Honeycutt RL, Begall S, Locker-Grütjen O, Scharff A (2000) Are naked and common mole-rats eusocial and if so, why? Behav Ecol Sociobiol 47:293–303

Capanna E, Merani MS (1980) Karyotypes of Somalian rodent populations .1. *Heterocephalus glaber* Ruppell, 1842 (Mammalia, Rodentia). Mon Zool Ital Suppl 13:45–51

Chitaukali WN, Burda H, Kock D (2001) On small mammals of the Nyika Plateau, Malawi. IRD Editions, Collection colloques et séminaires, Paris, pp 415–426

Cotterill FPD (2003) Geomorphological influences on vicariant evolution in some African mammals in the Zambezi Basin: some lessons for conservation. Proceedings of the

Ecology and Conservation of Mini-antelope: An International Symposium on Duiker and Dwarf Antelope in Africa. Filander Verlag, Fürth, pp 11–58

Dammann P, Burda H (2006) Sexual activity and reproduction delay ageing in a mammal. Curr Biol 16:R117–R118

De Graaff G (1981) The rodents of Southern Africa. Butterworths, Durban

Du Toit AL (1933) Crustal movement as a factor in the geographical evolution of South Africa. S Afr Geogr J 6:14–20

Ellerman JR, Hayman RW, Holt GWC (1940) The families and genera of living rodents. Trustees of the British Museum (Natural History), London

Ellerman JR, Morrison-Scott TCS, Hayman RW (1953) Southern African mammals, 1758 to 1951: a reclassification. Trustees of the British Museum (Natural History), London

Faulkes CG, Bennett NC, Bruford MW, O'Brien HP, Aguilar GH, Jarvis JUM (1997) Ecological constraints drive social evolution in the African mole-rats. Proc R Soc Lond B 264:1619–1627

Faulkes CG, Verheyen E, Verheyen W, Jarvis JUM, Bennett NC (2004) Phylogeographical patterns of genetic divergence and speciation in African mole-rats (Family : Bathyergidae). Mol Ecol 13:613–629

Filippucci MG, Burda H, Nevo E, Kocka J (1994) Allozyme divergence and systematics of common mole-rats (*Cryptomys*, Bathyergidae, Rodentia) from Zambia. Int J Mamm Biol 59:42–51

Filippucci MG, Kawalika M, Macholán M, Scharff A, Burda H (1997) Allozyme differentiation and systematic relationship of Zambian giant mole-rats, *Cryptomys mechowi* (Bathyergidae, Rodentia). Int J Mamm Biol 62:172–178

George W (1979) Conservatism in the karyotypes of two African mole-rats (Rodentia, Bathyergidae). Int J Mamm Biol 44:278–285

Golding GB, Strobeck C (1983) Increased number of alleles found in hybrid populations due to intragenic recombination. Evol 37:17–29

Grant PR, Grant BR (1997) Genetics and the origin of bird species. Proc Natl Acad Sci USA 94:7768–7775

Harrison RG (1993) Hybrids and hybrid zones. In: Harrison RG (ed) Hybrid zones and the evolutionary process. Oxford Univ Press, New York, pp 3–12

Honeycutt RL, Edwards SV, Nelson K, Nevo E (1987) Mitochondrial-DNA variation and the phylogeny of African mole-rats (Rodentia, Bathyergidae). Syst Zool 36:280–292

Honeycutt RL, Nelson K, Schlitter DA, Sherman PW (1991) Genetic variation within and among populations of the naked mole-rat: evidence from nuclear and mitochondrial genomes. In: Sherman PW, Jarvis JUM, Alexander RD (eds) The biology of the naked mole-rat. Princeton Univ Press, Princeton, NJ, pp 195–208

Huchon D, Douzery EJP (2001) From the old world to the new world: a molecular chronicle of the phylogeny and biogeography of Hystricognath rodents. Mol Phyl Evol 20: 238–251

Ingram CM, Burda H, Honeycutt RL (2004) Molecular phylogenetics and taxonomy of the African mole-rats, genus *Cryptomys* and the new genus *Coetomys* Gray, 1864. Mol Phyl Evol 31:997–1014

Janecek LL, Honeycutt RL, Rautenbach IL, Erasmus BH, Reig S, Schlitter DA (1992) Allozyme variation and systematics of African mole-rats (Rodentia, Bathyergidae). Biochem Syst Ecol 20:401–416

Janse Van Rensburg L, Chimimba CT, Van Der Merwe M, Schoeman AS, Bennett NC (2004) Relative age and reproductive status in *Cryptomys hottentotus pretoriae* (Rodentia: Bathyergidae) from South Africa. J Mammal 85:1225–1232

Jarvis JUM (1981) Eusociality in a mammal – cooperative breeding in naked mole-rat *Heterocephalus glaber* colonies. Science 212:571–573

Jarvis JUM, Bennett NC (1990) The evolutionary history, population biology and social structure of African mole-rats: Family Bathyergidae. In: Nevo E, Reig OA (eds) Evolution of subterranean mammals at the organismal and molecular levels. Wiley-Liss, New York, pp 97–128

Jarvis JUM, Bennett NC (1993) Eusociality has evolved independently in two genera of bathyergid mole-rats – but occurs in no other subterranean mammal. Behav Ecol Sociobiol 33:253–260

Kawalika M, Burda H, Brüggert D (2001) Was Zambia a cradle of the genus *Cryptomys* (Bathyergidae, Rodentia)? African small mammals. IRD Editions, Collection colloques et séminaires, Paris, pp 253–261

Kock D, Ingram CM, Frabotta LJ, Honeycutt RL, Burda H (2006) On the nomenclature of Bathyergidae and *Fukomys* n. gen. (Mammalia : Rodentia). Zootaxa 51–55

Lavocat R (1978) Rodentia and Lagomorpha. Evolution of African mammals. Harvard Univ Press, Cambridge, Mass., pp 69–89

Lessa EP (1990) Morphological evolution of subterranean mammals: integrating structural, functional, and ecological perspectives. In: Nevo E, Reig OA (eds) Evolution of subterranean mammals at the organismal and molecular levels. Wiley-Liss, New York

Lessa EP (2000) The evolution of subterranean rodents: a synthesis. In: Lacey EA, Patton JL, Cameron G (eds) Life underground: the biology of subterranean rodents. Univ of Chicago Press, Chicago, pp 389–420

Macholan M, Scharff A, Burda H, Zima J, Grütjen O (1998) The karyotype and taxonomic status of *Cryptomys amatus* (Wroughton, 1907) from Zambia (Rodentia, Bathyergidae). Int J Mamm Biol 63:186–190

Moore AE, Larkin PA (2001) Drainage evolution in south-central Africa since the break-up of Gondwana. S Afr J Geol 103:47–68

Nedbal MA, Allard MW, Honeycutt RL (1994) Molecular systematics of hystricognath rodents: evidence from the mitochondrial 12S rRNA gene. Mol Phyl Evol 3:206–220

Nevo E, Capanna E, Corti M, Jarvis JUM, Hickman GC (1986) Karyotype differentiation in the endemic subterranean mole-rats of South-Africa (Rodentia, Bathyergidae). Int J Mamm Biol 51:36–49

Nevo E, Ben-Shlomo R, Beiles A, Jarvis JUM, Hickman GC (1987) Allozyme differentiation and systematics of the endemic subterranean mole-rats of South-Africa. Biochem Syst Ecol 15:489–502

Nowak R (1991) Walker's mammals of the world. Johns Hopkins Univ Press, Baltimore

Patton JL, Brylski PV (1987) Pocket gophers in alfalfa fields – causes and consequences of habitat-related body size variation. Am Nat 130:493–506

Roberts A (1951) The mammals of South Africa. Trustees of 'the mammals of South Africa' book fund, Johannesburg

Rosevear DR (1969) Rodents of West Africa. Trustees of the British Museum (Natural History), London

Ruedi M, Smith MF, Patton JL (1997) Phylogenetic evidence of mitochondrial DNA introgression among pocket gophers in New Mexico (family Geomyidae). Mol Ecol 6: 453–462

Sanderson MJ (1997) A nonparamateric approach to estimating divergence times in the absence of rate constancy. Mol Biol Evol 1218–1231

Sanderson MJ (2002) Estimating absolute rates of molecular evolution and divergence times: A penalized likelihood approach. Mol Biol Evol 19:101–109

Scharff A, Macholan M, Zima J, Burda H (2001) A new karyotype of *Heliophobius argenteocinereus* (Bathyergidae, Rodentia) from Zambia with field notes on the species. Int J Mamm Biol 66:376–378

Stein BR (2000) Morphology of subterranean rodents. In: Lacey EA, Patton JL, Cameron G (eds) Life underground: the biology of subterranean rodents. Univ Chicago Press, Chicago, pp 19–62

Steinberg EK, Patton JL (2000) Genetic structure and the geography of speciation in subterranean rodents: opportunities and constraints for evolutionary diversification. In: Lacey EA, Patton JL, Cameron G (eds) Life underground: the biology of subterranean rodents. Univ Chicago Press, Chicago, pp 301–331

Van Couvering JAH, Van Couvering JA (1976) Community evolution in East Africa during the late Cenozoic. In: Isaac G, McCown ER (eds) Human origins: Louis Leakey and the East African evidence. WA Benjamin, Menlo Park, California, pp 155–207

Van Daele PAAG, Dammann P, Meier JL, Kawalika M, Van De Woestijne C, Burda H (2004) Chromosomal diversity in mole-rats of the genus *Cryptomys* (Rodentia : Bathyergidae) from the Zambezian region: with descriptions of new karyotypes. J Zool Lond 264:317–326

Van Daele PAAG, Verheyen E, Corkery I, Adriaens D (2006) Trends in skull morphology in relation to differential molecular evolution in African mole-rats of the chromosomally hyperdivers genus *Fukomys* (Bathyergidae, Rodentia) from the Zambezian region. Ital J Mammal (n.s.) suppl.:143

Van Daele PAAG, Verheyen E, Adriens D, Brunain M (accepted) Cytochrome b sequence analysis reveals differential molecular evolution in African mole-rats of the chromosomally genus *Fukomys* (Bathyergidae, Rodentia) from the Zambezian region. Mol Phyl Evol

Van Zinderen Bakker EM (1967) The 'arid corridor' between southwest Africa and the Horn of Africa. In: Van Zinderen Bakker EM (ed) Paleoecology of Africa, vol 2. AA Balkema, Amsterdam pp 76–69

Walton AH, Nedbal MA, Honeycutt RL (2000) Evidence from intron 1 of the nuclear transthyretin (Prealbumin) gene for the phylogeny of African mole-rats (Bathyergidae). Mol Phyl Evol 16:467–474

Williams SL, Schlitter DA, Robbins LW (1983) Morphological variation in a natural population of *Cryptomys* (Rodentia: Bathyergidae) from Cameroon. Ann Mus R Afr Centr Sc Zool 237:159–172

Wilson DE, Reeder DM (2005) Mammal species of the world. Johns Hopkins Univ Press, Baltimore

Wood AE (1985) The relationships, origin, and dispersal of the hystricognathous rodents. In: Evolutionary relationships among rodents: a multidisciplinary analysis. NATO ASI Series, New York, pp 475–513

Zelditch ML, Swiderski DL, Sheets HD, Fink WL (2004) Geometric morphometrics for biologists: a primer. Elsevier Academic Press, London

28 Mosaic Evolution of Subterranean Mammals: Tinkering, Regression, Progression, and Global Convergence

Eviatar Nevo

28.1 Overview

The extensive convergent evolution of subterranean mammals across the planet (Nevo 1999) began during the global climatic transition from the middle Eocene to the early Oligocene, 45–35 million years ago. It involved seasonal stresses proceeding progressively throughout the Cenozoic. The ecological theater of open country biota set the stage for a rapid evolutionary play of recurrent adaptive radiations of unrelated mammals on all continents to the subterranean ecotope. The biota comprised marsupial and placental mammals (insectivores and herbivores) that involve several hundreds of small mammalian species belonging, at present, to fifty genera, eleven families, and three mammalian orders.

The stressful underground ecotope (solid soil, darkness, low productivity, hypercapnia, hypoxia, and high infectivity) provided subterranean mammals with shelter from extreme climatic (temperature and humidity) fluctuations and predators. All subterranean mammals share convergent molecular and organismal adaptations with a shared unique ecotope. By contrast, they display divergent adaptations to their separated herbivorous and insectivorous feeding niches and structures derived from their different phylogenies. The subterranean ecotope is simple, relatively stable, specialized, low in productivity, and discontinuous. Its major determinants are specialization, competition, and isolation. These characteristics led to convergent and divergent, regressive and progressive evolutionary patterns and processes, locally optimizing adaptations at all organizational levels for mammalian life underground.

Adaptive convergence comprises structural and functional reductions and expansions (hypertrophies) through molecular and organismal evolutionary tinkering. The reductions and expansions involve burrowing, energetics, and respiratory adaptations affecting molecular, morphological, physiological, and behavioural changes as compared with small aboveground mammals (Nevo 1999). Likewise, life in underground darkness led

Eviatar Nevo: University of Haifa, Institute of Evolution, Mount Carmel, Haifa 31905, Israel,
E-mail: nevo@research.haifa.ac.il

Subterranean Rodents: News from Underground
S. Begall, H. Burda, C.E. Schleich (Eds.)
© Springer-Verlag Berlin Heidelberg 2007

to the remarkable evolution of photoperiodic perception, affecting circadian rhythmicity through complex mosaic eye and brain evolution. Life underground in isolated territorial domains also triggered the evolution of aggression and seismic communication affecting the reorganization of the brain. The latter involves the expansion of the somatosensory into the visual cortex, at least in *Spalax*, but also possibly involves different degrees in other subterranean mammals. Life underground also selected for a suite of progressive adaptive traits to the subterranean ecotope compensating for the regressive loss of sight. The former involve vibratory or seismic, magnetic, auditory, chemical, nasal, and vomeronasal communication systems.

Adaptive convergence also implicates intraspecific territoriality and aggressive competition, circadian rhythmicity, food generalism, equilibrium populations and K-strategy, low genetic diversity and homoselection, interspecific competitive exclusion, and largely parapatric distributions between ecologically and genetically similar species. Adaptive divergence includes alternative patterns in herbivores and insectivores, respectively: large vs small body size; low vs high metabolic rates; high vs low thermal conductance; small vs large territories; small vs high gene flow; high vs low chromosomal speciation, and taxonomic diversity. Major phylogenetic differences relate to tooth and intestine structure and dietary specialization. Speciation involves all rates from gradual to rapid; neither punctuation nor stasis is the hallmark of speciation in subterranean mammals.

Subterranean mammals provide uniquely rich evolutionary global tests of speciation and adaptation, convergence, regression, and progression channeled by tinkering evolutionary processes due to subterranean multiple stress syndromes. Many unresolved evolutionary problems could be elucidated by studying mammalian evolution underground. The main challenge now is to explore the genetic basis of adaptation and speciation at the molecular levels of genomes and proteomes underlying convergence, regression and progression and relate it to organismal convergence. The Cenozoic global experiment of stressful mammalian life underground is a uniquely fruitful field for exploring predictions of evolutionary theory and, primarily, the effects of environmental stresses on the evolutionary process.

The revolution in molecular biology opens the door for relating genomics, proteomics and phenomics, and linking structure and function across all organization levels. These future studies, combining molecular and organismal facets, could highlight the global adaptive radiations of small mammals to life underground during most of the Cenozoic era to cope with an increasingly cold, dry, and seasonal planet. This is an ongoing drama of organism-environment evolution at its best.

Here, I demonstrate evolutionary organismal and molecular tinkering involving regression, progression, and convergence of eye evolution primarily in *Spalax*. Similar regression, progression, and convergent patterns could be demonstrated in all other systems, for example, genetic, morphologic, physiologic and behavioral (see Nevo 1999), but space limitations are restrictive. All citations until 2001 can be found in Nevo (1999) and Nevo et al. (2001). Later citations are fully presented in the references.

28.2 Visual Adaptations in Subterranean Mammals

The senses of subterranean mammals display impressive, variable, structural, and functional *convergent* adaptations required for life underground (Burda et al. 1990a, b). The sensory environment of subterranean mammals is deprived of the numerous aboveground complex stimuli (Nevo 1979). Diverse sensory communication modes are common to aboveground organisms: vision, photoperiodic perception, audition, smell, taste, touch, pressure, vibration, and electric and magnetic senses (Seboek 1977). Clearly, in extreme confinement to the underground dark environment, vision is largely excluded in many subterranean mammals, and it is replaced by other communication modalities. Here, I will focus on the visual apparatus, primarily on the extreme case of *Spalax*.

Lightless subterranean ecotopes lead commonly to differential *regression evolution* or to the disappearance of the vertebrate (Walls 1967) and invertebrate eye. The complete disappearance of the eyes has not occurred in the evolution of any natural population of mammals (Duke-Elder 1958). Some laboratory mice strains can be completely eyeless (Chase 1945), indicating that mutations in one or several genes can prevent eye development. Eyes of subterranean mammals display variable regressive stages in size, depending on whether the animal is strictly subterranean or occasionally emerges aboveground for foraging (Fig. 15i–n in Nevo 1999). Variation in eye size ranges from normal-medium (*Spalacopus*, Fig. 7.6aB in Nevo 1999) to medium-small (ctenomyids, geomyids, Figs.7.6aA and 7.2a–d in Nevo 1999) to small-minute (rhizomyids, subterranean cricetids, bathyergids, Figs. 14a and 17a–d in Nevo 1999) to minute (most talpids, Fig. 6.1ab in Nevo 1999) to vestigial and completely covered by skin (*Notoryctes*, *Talpa caeca*, chrysochlorids, and spalacids) (Figs. 5.5, 6.3, 7.4b and 15.3j in Nevo 1999).

Only some families possess completely subcutaneous eyes (spalacids, *Notoryctes*, chrysochlorids, and talpids) (Fig. 15.3b in Nevo 1999). Minute microphthalmic eyes (less than 2 mm) are also found in the echolocating microchiropteran bats, species of insectivores and aquatic shrews, in *Fukomys anselli*, and one cetacean, the blind river dolphin (*Platanista*) (Cooper et al. 1995). Reduction of the eyes is even more widespread in subterranean and cave-dwelling non-mammalian vertebrates including cyclostomes, teleosts, amphibians, and reptiles (Rochon-Duvigneaud 1943). Birds are the only vertebrates in which small eyes are absent. *Regression* of the eye can thus be viewed as resulting from diverse parallel and *convergent* evolutionary causes, potentially leading to a mosaic of consequences affecting the peripheral and central visual system.

Variable degrees of *reductions* were described in eye size, ocular muscles, fixed eyelids, cornea thickening, amorphous lens, fibrous vitreous body, minimized optic nerve, *regression* in size of brain visual centers, and retinal changes in size and structure. Generally, the structure of the visual apparatus *regresses* differentially and structurally in all subterranean mammals presumably correlated with confinement to life underground (see Burda et al. 1990a; see results across subterranean mammals in Fig. 15.3i–n in Nevo 1999). The following is a review of the detailed study in the *Spalax* eye.

28.2.1 The Mosaic Evolution of the *Spalax* Eye: Morphology, Physiology, and Molecular Biology

The entirely subcutaneous minute eye of *Spalax* has been extensively studied (see Nevo 1999, page 82). Concealed beneath the skin, the minute regressed eye is further embedded within a hypertrophied Harderian gland (Pevet et al. 1984). Thus, light can only penetrate to the retina with difficulty and the thick covering of skin precludes proper image formation. The size of the eye is extremely reduced (700 μm diameter) although the neuroretina appears to be structured normally with thin inner and outer plexiform and nuclear cell layers. Ganglion cells are reported to be sparse (Cei 1946; Sanyal et al. 1990; Cooper et al. 1993a, b, 1995), and the optic nerve contains less than 1000 unmyelinated fibers (Herbin et al. 1993, 1994). Development of the eye is initiated normally but at an early stage displays a number of *degenerative* features (Fig. 28.1). The optic fissure fails to close, the anterior chamber collapses, and the iris-ciliary body becomes hypertrophied and heavily pigmented. The lens remains vascularized and regressed (Cei 1946; Sanyal et al. 1990). In contrast, neurogenesis follows a normal course of development, but the ganglion cell and photoreceptor layers remain relatively thin. Photoreceptors have been shown to form normal synapses in the outer plexiform layer. The visual system of *Spalax* presents a mosaic of both regressive and progressive features (Cooper et al. 1993a, b). The retina projects bilaterally to all visual structures, described as receiving retinal afferents in non-fossorial rodents. Structures involved in form analysis and visually guided behaviors are reduced in size by more than 90%, receive a sparse retinal

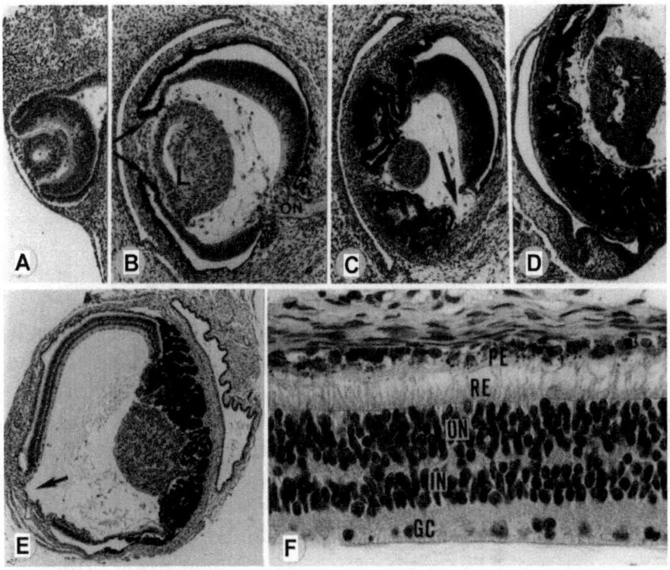

Fig. 28.1A–F. Light micrographs showing cross-sections of the developing eye of the mole-rat, *Spalax ehrenbergi*: **A** optic cup and lens vesicle initially develop normally (×100); **B** eye at a later embryonic stage. Note appearance of iris -ciliary body rudiment (arrows), and development of the lens nucleus (L). ON, optic nerve (×100); **C** eye at a still later fetal stage. Note massive growth of the iris-ciliary body complex colobomatous opening (*arrow*) (×100); **D** early postnatal stage. The iris-ciliary body complex completely fills the chamber. The lens is vascularized and vacuolated (×100); **E** adult eye. Eyelids are completely closed, and a pupil is absent. Note atrophic appearance of the optic disc region (*arrow*) (×65); **F** higher magnification of the adult retina. The different retinal layers are retained: PE, pigment epithelium; RE, receptor layer; ON outer nuclear layer; IN, inner nuclear layer; GC, ganglion cell layer (×500) (from Sanyal et al. 1990)

innervation, and are cyto-architecturally, poorly differentiated. The main regressive feature is the drastic relative reduction of retinal input to the superior colliculus (Fig. 28.2A–C). The volume of visual structures in *Spalax*, hamster, *Spalacopus*, rat, and microphthalmic rat appear in Fig. 28.2C.

In contrast to the foregoing, structures of the "non-image forming" visual pathway involved in photoperiodic perception are well developed in *Spalax*. The suprachiasmatic nucleus receives a bilateral projection from the retina and the absolute size, cyto-architecture, density, and distribution of retinal afferents in *Spalax* are comparable with those of other rodents (Fig. 28.2A–C). A relatively hypertrophied retinal projection is observed in the bed nucleus of the stria terminalis. The percentage of relative retinal input to the size of primary visual structures is presented in Fig. 28.2A. The percentage of relative difference of retinal input to the size of visual structure of *Spalax* contrasted with other rodents is given in Fig. 28.2B.

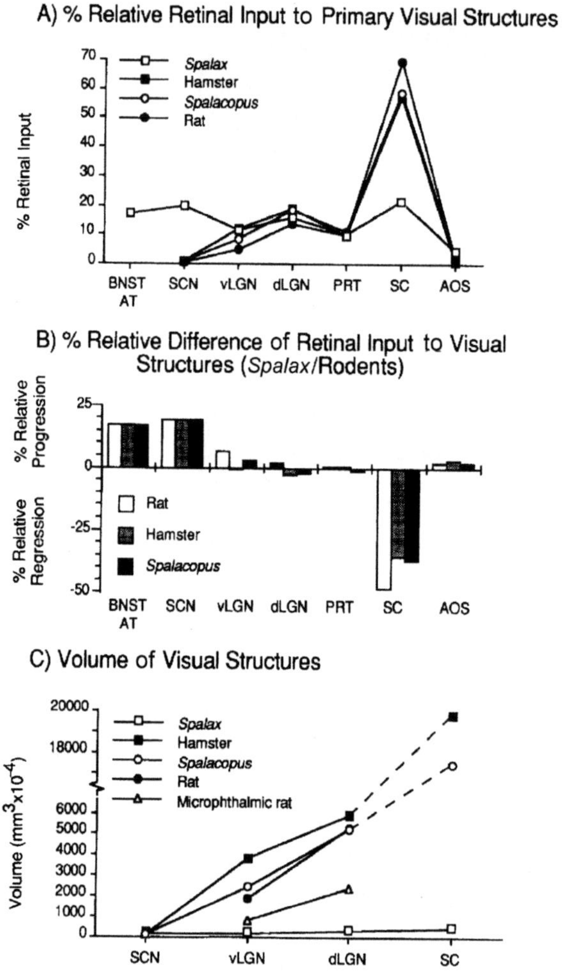

Fig. 28.2A–C. A) Relative degree of retinal input to primary visual structures in *Spalax*, hamster, rat, and *Spalacopus cyanus*. **B**) Relative degree of change in the proportions of retinal input to different primary visual structures in *Spalax* compared to measures obtained in other rodents. Relative progressive development in *Spalax* is seen in structures involved in photoperiodic and neuroendocrine functions. The main regressive feature is the drastic relative reduction of retinal input to the superior colliculus. **C**) Comparison of the absolute size (volume, $mm^3 \times 10^{-4}$) of several visual structures in *Spalax* and other rodents. The size of the SCN is equivalent in all species. The vLGN and dLGN are reduced by roughly 90% and 94%, respectively. The retino recipient layers of the superior colliculus are reduced by 98% (from Cooper et al. 1993a,b)

These results indicate that the apparently global morphological regression of the visual system conceals a selective progression of structures related to functions of photoperiodic perception and photo-neuroendocrine regulation. The evolution of an atrophied eye and a reduced visual system is an adaptively advantageous response to the unique subterranean environment. Factors favoring regression include mechanical aspects, metabolic constraints, and competition between sensory systems. The primary advantage of sensory atrophy is the metabolic economy gained by the reduction of visual structures that do not contribute significantly to the animal's fitness (Cooper et al. 1993a, b, 1995). The following is a description of the drastically changed lens α-crystallin of *Spalax* at the molecular level of existence and expression.

28.2.2 α-Crystallin: Structure, Expression, and Evolution

α-Crystallin apparently originated from the multigene small heat shock proteins (HSPs). The HSPs are induced upon stress and evolved in all eukaryotes by gene duplication and the subsequent sequence divergence that allows adaptation to novel functions, such as lens constituent. Because αA-crystallin is lens specific, we focused on its expression and structure at the DNA level (de Jong 1985; de Jong et al. 1990).

We determined the complete structure of the single-copy αA-crystallin gene of the blind mole-rat superspecies, *Spalax ehrenbergi*, in order to elucidate the evolutionary effects of the *loss* of vision on a lens-specific protein and its gene (Hendriks et al. 1987). The αA-crystallin gene of *Spalax* consists of four exons and three introns, as found in the mouse and hamster (King and Piatigorsky 1983; Van den Heuvel et al. 1985). The gene appears to have all the necessary transcriptional and translational signal sequences to be expressed in the rudimentary lens of the mole-rat and probably gives rise to two protein products by means of alternative splicing, as in rodents with normal vision (Figs. 1 and 2 in Hendriks et al. 1987). Comparisons of the blind mole-rat αA-crystallin sequence with αA-crystallin sequences from sighted rodents (involving representatives of 32 rodent families; see Fig. 3 in Hendriks et al. 1987) reveal a considerable *acceleration* of the substitution rate at nonsynonymous positions in the mole-rat lineage. These comparisons, however, demonstrate a selective constraint (see discussion in the Theory section in Nevo 1999).

28.2.3 Adaptive Evolution of Small Heat Shock Protein/αB-Crystallin Promoter Activity in *Spalax*

We have compared the tissue specificity of the small heat shock protein (shsp)/αB-crystallin promoter of the mole-rat superspecies, *Spalax ehrenbergi*, with that of the mouse (Figs. 1–6 in Hough et al. 2002). Earlier experiments showed that mouse shsp/αB-crystallin promoter/enhancer activity is high in the lens and moderate in the heart and skeletal muscle of transgenic mice. Using the firefly luciferase reporter gene in transgenic mouse experiments, we showed, despite relatively few changes in sequence (Fig. 1 in Hough et al. 2002), that the mole-rat shsp/αB-crystallin promoter/enhancer had selectively *lost lens activity* after 13.5 days of embryogenesis (E13.5) (Fig. 28.3). The ratios of mole-rat/mouse promoter activity were 0.01 for lens, 1.7 for heart, and 13.6 for skeletal muscle in eight-week-old transgenic mice. Our data indicate that the shsp/αB-crystallin

a

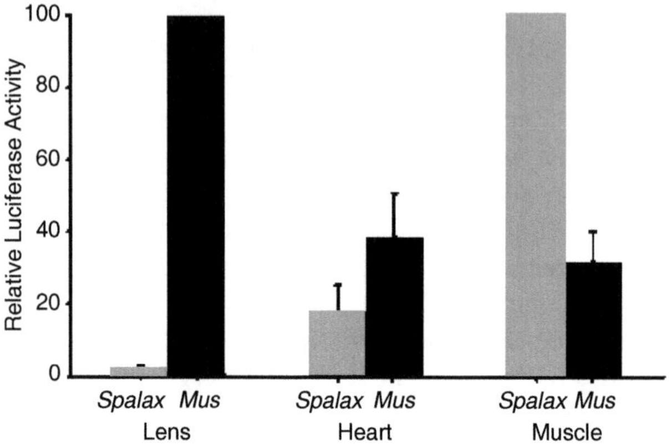

b

	Lens	Heart	Muscle
Spalax	29±26	782±180	8891±2778
Mus	2256 ±518	449 ±176	652 ± 296
Spalax/Mus	0.01	1.7	13.6

Fig. 28.3a–b. Mole-rat and mouse shsp/αB-crystallin promoter activity in eight-week-old transgenic mice: **a** (*filled gray bars*) average of four lines of transgenic mice with the mole-rat promoter, (*filled black bars*) average of three lines of transgenic mice with the mouse promoter. The highest activity using the mole-rat construct (muscle) and the highest activity using the mouse construct (lens) were given a value of 100. Setting the mole-rat muscle activity to 100 lowered the heart activity relative to the mouse even though the average luciferase activity of the mole-rat construct was actually higher than that of the mouse; **b** average relative luciferase units per μg of soluble protein, with standard deviations of the lens, heart, and skeletal muscle in the transgenic mice. The differences in activity between the mole-rat and the mouse promoters were statistically significant in the lens ($P < 0.001$) and in the muscle ($P < 0.02$), whereas in the heart the differences were not statistically significant (from Hough et al. 2002)

promoter/enhancer has undergone adaptive changes corresponding to the subterranean evolution of the blind mole-rat. We speculated that selective pressures on the metabolic economy may have contributed to these tissue-specific modifications of promoter/enhancer function during adaptation to life underground (Hough et al. 2002). Whereas the gene differentially loses its lens function (regression), it dramatically retains and even augments muscle function (progression) in transgenic mice (Fig. 28.3).

Intriguingly, it appears that minor sequence modifications (Fig. 2 in Hough et al. 2002), presumably associated with adaptation to life underground, can be associated with profound changes in promoter activity. These adaptive evolutionary changes may economize metabolic expenses (Cooper et al. 1993 a, b; Nevo 1999) by limiting expression of shsp/αB-crystallin in the degenerated lens, but augmenting expression in the muscle, thereby increasing fitness (Hough et al. 2002). Whereas the gene differentially loses its lens function (regression), it dramatically retains and even augments muscle function (progression) in transgenic mice (Fig. 28.3).

28.2.4 Retinal Photopigments in *Spalax*

Coneopsin

Remarkably, the blind eye of *Spalax* was retained as a circadian eye to assess photoperiodicity in total darkness underground. *Spalax* retina harbors circadian photopigments, coneopsin (Argamaso et al. 1995; Nevo 1999), rhodopsin (Janssen et al. 2000), and melanopsin (Hannibal et al. 2002) causing an action spectrum for circadian responses to light. The atrophied eyes of *Spalax* express a long wavelength sensitive (LWS) cone opsin. This photopigment is spectrally tuned to enhance photon capture in the red light environment of the eye (David-Gray et al. 1999). The LWS opsin of *Spalax* acts as a functional photopigment, and it is not a "residue" of the pre-subterranean visual system. As the eye of *Spalax* has only one known function, the entrainment of circadian rhythms to environmental light, the LWS photopigment is implicated in this task.

Rhodopsin

The rod visual pigment of *Spalax* shows > 90% homology with mammalian rod pigments (Janssen et al. 2000). This *Spalax* rod pigment is fully functional in vitro but also shows that all requirements for a functional pigment are present in vivo. During adaptation to a subterranean lifestyle, the visual system of *Spalax* has undergone mosaic reorganization, and the visual pigments have adapted to function in circadian photoreception (Janssen et al. 2000).

Melanopsin

We have shown that melanopsin, a recently identified opsin, is expressed in retinal ganglion cells that also co-store PACAP, a neurotransmitter of the retinohypothalamic tract (RHT) (Hannibal et al. 2002). The melanopsin-component of retinal ganglion cells in the *Spalax* retina is well conserved, resulting in a relatively higher density of melanopsin positive cells per area compared to the rat.

Adaptive loss of ultraviolet-sensitive/violet-sensitive (UVS/VS) coneopsin
Most mammals are dichromats possessing, in addition to a single rod photopigment, two classes of cone photopigment LWS and UVS/VS with differing spectral sensitivities, which mediate colour vision. We explored whether *Spalax* is a dichromat and has the potential to use colour discrimination for photoentrainment. Using immunocytochemistry and molecular approaches, we demonstrated that *Spalax* is a LWS monochromat (David-Gray et al. 2002). *Spalax* lacks a functional UVS/VS cone photopigment due to the accumulation of several deleterious mutational changes that have rendered the gene nonfunctional. Using phylogenetic analysis, we showed that the loss of this class of photoreceptor is likely to have arisen from the underground life of *Spalax* and is not an artifact of having an ancestor that lacked a functional UVS/VS cone photopigment. We concluded that colour discrimination is not a prerequisite for photoentrainment in *Spalax*.

Can genomic adaptive regulation be unraveled?

28.3 Evolutionary Adaptive Regulation of Wide Genome Gene Expression

The novel technique of wide genome gene expression through microarrays permits dramatic insights into gene activity across the genome. Can it highlight *Spalax* activity across climatic mesic-xeric gradients in Israel?

We applied genome-wide gene expression analysis to the evolutionary processes of adaptive speciation of the Israeli blind subterranean mole-rats of the *Spalax ehrenbergi* superspecies (Brodsky et al. 2005). The four Israeli allospecies climatically and adaptively radiated into the cooler-mesic northern domain (N) (*Spalax galili*, $2n = 52$, and *S. golani*, $2n = 54$) and the warmer-xeric southern domain(S) (*S. carmeli*, $2n = 58$, and *S. judaei*, $2n = 60$) (Nevo et al. 2001). The kidney and brain mRNAs of two N and two S animals were examined through cross-species hybridizations with two types of Affymetrix arrays (mouse and rat) and muscle mRNA of six N and six S animals with spotted cDNA mouse arrays. The initial microarray analysis was hypothesis-free, i.e., conducted without reference to the origin of animals.

Principal component analysis revealed that 20–30% of the expression signal variability could be explained by the differentiation of N-S species Fig. 28.4a,b. Similar N-S effects were obtained for all tissues and types of arrays: two Affymetrix microarrays using probe oligomer signals and the spotted array. Likewise, ANOVA and *t*-test statistics demonstrated significant N-S ecogeographic divergence and region-tissue specificity in gene expression. Analysis of differential gene expression between species corroborates previous results deduced by allozymes and DNA molecular polymorphisms (Fig. 28.4b). Functional categories show significant N-S ecological adaptive divergent up-regulation of genes highlighting a higher metabolism in N, and adaptive brain activity and kidney urine cycle pathways in S, as expected from environmental stresses (Brodsky et al. 2005; Nevo 1999). The results confirm ecologic-genomic separation of blind mole-rats into contrasting mesic (N) and xeric (S) ecological domains (Fig. 28.4a,b). Gene expression *regulation* appears to be central in the evolution of blind mole-rats.

Evolutionary change is primarily driven by ecological stress (Nevo 1999). Climatic stresses are instrumental in advancing the twin evolutionary processes of adaptation and speciation. This study highlights the fact that allelic diversity is not the only pattern that ecologically differentiates N from S *Spalax*. Most remarkably, differential genome-wide gene expression *regulation* in speciation and adaptation seems to be of paramount importance in *Spalax* evolution. This regulatory impact resides primarily in the noncoding genome (e.g., random amplified polymorphic DNAs (RAPD), amplified fragment-length polymorphisms (AFLP), variable number tandem repeats (VNTR), simple sequence repeats (SSR)), which has been repeatedly implicated in diverse environmental, climatic, respiratory (hypoxia – hypercapnia), energetic and infection stresses. Climatic selection seems to be instrumental in driving the noncoding and coding *Spalax* genome to adapt presumably to life underground and speciate in an N-S trend along a southward gradient of increasing aridity stress. Environmentally stressed regulation of genome-wide gene expression in *Spalax*, and probably more generally in other organisms across life, appears to be the hallmark of the ecological evolutionary process of speciation and adaptation.

We plan to explore deeply the nature of the divergent genes in *Spalax* and unravel their functional significance. Future studies also could elucidate the interaction of potential adaptive evolution of *Spalax* genomics, proteomics, and phenomics to local and regional stresses underground by focusing on SNP (single nucleotide polymorphism) sequencing and SNP of stress genes as well as their evolutionary environmental regulation (Brodsky et al. 2005). Differential regulation presumably underlies the processes of gene expression. Coupled with loss-and-gain-function mutations and homeotic genes, regulation underlies the dramatic global experiment of tinkering causing regression, progression, and convergence of subterranean mammals during the Cenozoic, displaying one of nature's dramatic evolutionary scenarios.

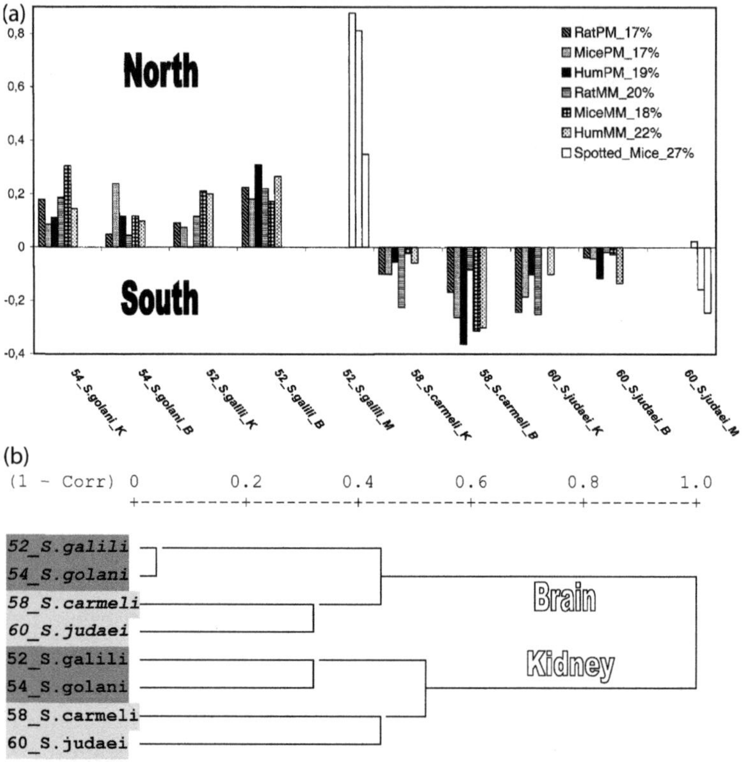

Fig. 28.4a–b. PCA and cluster analysis of probe oligomer (*Affymetrix arrays*) and gene signals (*spotted array*). Profiles of the second PC of PM and MM log signals for mouse and rat Affymetrix microarray experiments and mouse spotted-array experiment: **a** PC2 reflects the general signal difference between N and S animals for two types of Affymetrix microarrays [PM (*red bars*) and MM (*blue bars*) signals] and for the spotted array experiment (*green bars*). This PC2 component explains 20–30% of total signal variability; **b** the dendrogram of correlation distances between Affymetrix hybridizations (average linkage) after the cleaning of artifacts. All non-patterned probe oligomers with average signals of > 300 according to all eight hybridizations were taken. Clear N-S (*S. galili* plus *S. golani* vs *S. carmeli* plus *S. judaei*) separation is seen for both tissues as in all other evolutionary discrimination of *Spalax* species based on diverse molecular markers and morphological criteria (Figs. 16, 17, 19, 20–22, 23a,b, 31, and 33 in ref. 2 (from Brodsky et al. 2005))

References

Argamaso SM, Froelich AC, McCall MA, Nevo E, Provencio I, Foster RG (1995) Photopigments and circadian systems of vertebrates. Biophys Chem 56:3–11

Brodsky L, Jacob-Hirsch J, Avivi A, Trakhtenbrot L, Zeligson S, Amariglio N, Paz A, Korol AB, Rechavi M, Nevo E (2005) Evolutionary regulation of the blind subterranean mole rat, *Spalax*, revealed by genome-wide gene expression. Proc Natl Acad Sci USA 102: 17047–17052

Burda H, Nevo E, Bruns V (1990a) Adaptive differentiation of ear structures in subterranean mole rats of the *Spalax ehrenbergi* superspecies in Israel. Zool Jb Syst 117:369–382

Burda H, Bruns V, Müller M (1990b) Sensory adaptations in subterranean mammals. In: Nevo E, Reig OA (eds) Evolution of subterranean mammals at the organismal and molecular levels. Wiley-Liss, New York, pp 269–293

Cei G (1946) Ortogenesi parallela e degenerazione degli organi della vista negli Spalacidi. Monitori Zool Ital 55:69–88

Chase HB (1945) Studies in an anophthalmic strain of mice. J Comp Neurol 83:121–140

Cooper HM, Herbin M, Nevo E (1993a) Ocular regression conceals adaptive progression of the visual system in a blind subterranean mammal. Nature 361:156–159

Cooper HM, Herbin M, Nevo E (1993b) The visual system of a naturally microphthalmic mammal: the blind mole rat, *Spalax ehrenbergi*. J Comp Neurol 328:313–350

Cooper HM, Herbin M, Nevo E, Negroni J (1995) Neuroanatomical consequences of microphthalmia in mammals. Les seminaries Ophthalmologiques d'ISPEN. In: Christen Y, Doly M, Droy-Lefaix MT (eds) Vision et adaptation, vol 6. Elsevier, Paris, pp 127–139

David-Gray ZK, Cooper HM, Jannsen WH, Nevo E, Foster RG (1999) Spectral tuning of a circadian photopigment in subterranean "blind" mammals (*Spalax ehrenbergi*). FEBS Lett 461:343–347

David-Gray ZK, Bellingham J, Munoz M, Avivi A, Nevo E, Foster RG (2002) Adaptive loss of ultraviolet-sensitive/violet-sensitive (UVS/VS) cone opsin in the blind mole rat (*Spalax ehrenbergi*). Eur J Neurosci 16:1186–1194

De Jong WW (1985) Superordinal affinities of Rodentia studies by sequence analysis of eye lens protein. In: Luckett WP, Hartenberger J-L (eds) Evolutionary relationships among rodents. A multidisciplinary analysis. Plenum Press, New York, London, pp 211–227

De Jong WW, Hendriks W, Sanyal S, Nevo E (1990) The eye of the blind mole rat (*Spalax ehrenbergi*): regressive evolution at the molecular level. In: Nevo E, Reig OA (eds) Evolution of subterranean mammals at the organismal and molecular levels. Wiley-Liss, New York, pp 383–396

Duke-Elder S. (1958) System of ophthalmology. In: The eye in evolution. Vol 1. Henry Kimpton, London

Hannibal J, Hindersson P, Nevo E, Fahrenkrug J (2002) The circadian photopigment melanopsin is expressed in the blind subterranean mole rat, *Spalax*. NeuroReport 13:1411–1414

Hendriks W, Luenisson J, Bloemendal H, Nevo E, De Jong WW (1987) The lens protein αA crystalline in the blind mole rat, *Spalax ehrenbergi*: evolutionary change and functional constraints. Proc Natl Acad Sci USA 84:5320–5324

Herbin M, Rio JP, Reperant J, Cooper HM, Nevo E, Lemire M (1993) Ultrastructual de nerf optique chez un rongeur microphtalme, *Spalax leucodon*. CR Acad Sci (Paris) 316: 251–254

Herbin M, Rio JP, Reperant J, Cooper HM, Nevo E, Lemire M (1994) Ultrastructural study of the optic nerve in blind mole rats (Spalacidae, *Spalax*). Vis Neurosci 12:253–261

Hough RB, Avivi A, Joel A, Davis J, Nevo E, Piatigorsky J (2002) Adaptive evolution of small heat shock protein/αB-crystallin promoter activity of the blind subterranean mole rat, *Spalax ehrenbergi*. Proc Natl Acad Sci USA 99:8145–8150

Janssen JW, Bovee-Geurts PH, Peeters ZP, Bowmaker JK, Cooper HM, David-Gray JK, Nevo E, DeGrip WJ (2000) A fully functional rod visual pigment in a blind mammal: a case for adaptive functional reorganization? J Biol Chem 275:38674–38679

King CR, Piatigorsky J (1983) Alternative RNA splicing of the murine αA-crystallin gene: protein coding information within an intron. Cell 32:707–712

Nevo E (1979) Adaptive convergence and divergence of subterranean mammals. Ann Rev Evol Syst 10:269–308

Nevo E (1999) Mosaic evolution of subterranean mammals: regression, progression and global convergence. Oxford Univ Press, Oxford
Nevo E, Ivanitskaya E, Beiles A (2001) Adaptive radiation of blind subterranean mole rats: naming and revisiting the four sibling species of the *Spalax ehrenbergi* superspecies in Israel: *Spalax galili* ($2n = 52$), *S. golani* ($2n = 54$), *S. carmeli* ($2n = 58$), *S. judaei* ($2n = 60$). Backhuys Publishers, Leiden, The Netherlands
Pevet P, Heth G, Haim A, Nevo E (1984) Photoperiod perception in the blind mole rat (*Spalax ehrenbergi* Nehring): Involvement of the harderian gland, atrophied eyes and melatonin. J Exp Zool 232:41–50
Rochon-Duvigneaud A (1943) Les yeux et la vision des vertebres. Masson, Paris
Sanyal S, Jansen HG, De Grip WJ, Nevo E, De Jong WW (1990) The eye of the blind mole rat, *Spalax ehrenbergi*: rudiment with hidden function. Invest Ophthalmol Vis Sci 31: 1398–1404
Seboek TA (1977) How animals communicate. Indiana Univ Press, Bloomington, London
Van den Heuvel R, Hendriks W, Quax W, Bloemendal H (1985) Complete structure of the hamster αA crystalline gene. J Mol Biol 185:273–284
Walls GL (1967) The vertebrate eye and its adaptive radiation. Hafner Pub Co, New York

Subject Index

GPSR Compliance

The European Union's (EU) General Product Safety Regulation (GPSR) is a set of rules that requires consumer products to be safe and our obligations to ensure this.

If you have any concerns about our products, you can contact us on ProductSafety@springernature.com

In case Publisher is established outside the EU, the EU authorized representative is:

Springer Nature Customer Service Center GmbH
Europaplatz 3
69115 Heidelberg, Germany

Batch number: 10755034

Printed by Printforce, the Netherlands